Data-Driven Applications for Emerging Technologies

Data-Driven Applications for Emerging Technologies explores the practical use of data science in AI, healthcare, sustainability, and security. It covers key topics like predictive modelling, deep learning, and natural language processing, offering a mix of theory and hands-on applications. The book highlights how data-driven techniques can improve decision-making, optimize processes, and solve real-world problems.

Each chapter includes research contributions from academics and industry professionals, making the content both relevant and accessible. Readers will find practical insights into applying machine learning frameworks, data preprocessing techniques, and emerging technologies across different domains.

Designed for researchers, professionals, and students, this book provides a solid foundation in data-driven methods without being overly technical. Whether you're looking to enhance your understanding of AI and machine learning or apply data science in real-world scenarios, this book serves as a useful and practical resource.

Data-Driven Applications for Emerging Technologies

Edited by
Nazmul Siddique, Mohammad Shamsul Arefin,
Ahmed Wasif Reza, and Aminul Haque

CRC Press
Taylor & Francis Group
Boca Raton London New York

CRC Press is an imprint of the
Taylor & Francis Group, an **informa** business

Designed cover image: Shutterstock_2542410727

First edition published 2026
by CRC Press
2385 NW Executive Center Drive, Suite 320, Boca Raton FL 33431

and by CRC Press
4 Park Square, Milton Park, Abingdon, Oxon, OX14 4RN

CRC Press is an imprint of Taylor & Francis Group, LLC

ISBN: 978-1-032-99693-6 (hbk)
ISBN: 978-1-032-99694-3 (pbk)
ISBN: 978-1-003-60550-8 (ebk)

DOI: 10.1201/9781003605508

Typeset in Latin Modern font
by KnowledgeWorks Global Ltd.

Publisher's note: This book has been prepared from camera-ready copy provided by the authors.

This book is dedicated to all the researchers, professionals, and students who are continuously trying their best for the advancement of technologies for the betterment of humanity.

Contents

Preface

The exponential growth of data has revolutionized the way we approach challenges and opportunities in modern society. Emerging technologies, powered by data-driven methodologies, have enabled groundbreaking advancements in fields such as artificial intelligence, machine learning, healthcare, sustainability, and security. This book, *Data-Driven Applications for Emerging Technologies*, serves as a comprehensive exploration of innovative techniques, methodologies, and applications that harness the potential of data to address complex problems across diverse domains.

The content of this book is thoughtfully curated to provide readers with a rich blend of theoretical insights and practical applications. The chapters cover a range of topics, including advanced predictive modeling, data preprocessing techniques, machine learning frameworks, and the application of cutting-edge technologies such as deep learning and natural language processing. Readers will gain an in-depth understanding of how data can be leveraged to optimize processes, enhance decision-making, and create impactful solutions for real-world challenges.

Each chapter represents the collaborative effort of dedicated researchers and practitioners who have contributed their expertise to illuminate the potential of data-driven approaches. The case studies, experiments, and methodologies presented in this book are rooted in rigorous research and designed to offer actionable insights. These contributions highlight the diverse applications of data-driven techniques, from improving healthcare outcomes and optimizing transportation systems to developing eco-friendly practices and enhancing cybersecurity.

This book is intended to serve as a valuable resource for a wide audience, including researchers, industry professionals, and students. By showcasing the interplay between data science and emerging technologies, it aims to inspire innovation and foster a deeper understanding of the transformative impact of data-driven approaches.

We extend our heartfelt gratitude to all the authors and contributors whose dedication has made this publication possible. Their commitment to excellence and innovation has resulted in a work that reflects the forefront of data-driven research and development.

It is our hope that this book sparks curiosity, promotes interdisciplinary collaboration, and empowers readers to contribute meaningfully to the ever-evolving landscape of emerging technologies.

Contributors

Ajoy Deb Nath
Department of Computer Science and Engineering,
CUET, Bangladesh

Md. Mokammel Haque
Department of CSE,
CUET, Bangladesh

Silvia Sifath
Department of CSE,
University of Barishal, Bangladesh

Md Manjur Ahmed
Department of CSE,
University of Barishal, Bangladesh

Samir B. Belhaouari
College of Science and Engineering,
Division of Information and Computing Technology,
Hamad Bin Khalifa University, Doha, Qatar

K. M. Azharul Hasan
Department of CSE,
KUET, Bangladesh.

Kawshik Banerjee
Department of CSE,
RUET, Bangladesh

Azmain Yakin Srizon
Department of CSE,
RUET, Bangladesh

Md. Farukuzzaman Faruk
Department of CSE,
RUET, Bangladesh

S. M. Mahedy Hasan
Department of CSE,
RUET, Bangladesh

Md. Rakib Hossain
Department of ETE,
RUET, Bangladesh

Soumit Das
Department of CSE,
RUET, Bangladesh

Ashraf-Ul-Alam
Department of CSE,
RUET, Bangladesh

Sudipta Progga Islam
Department of CSE,
RUET, Bangladesh

Azmain Yakin Srizon
Department of CSE,
RUET, Bangladesh

Md. Nazmus Sakib
Institute of Information Technology,
Jahangirnagar University, Bangladesh

Fahima Tabassum
Institute of Information Technology,
Jahangirnagar University, Bangladesh

Md Sayem Ahamed
Deptartment of Computer Science,
United International University, Bangladesh

Md Shahzamal
Independent Researcher

Md. Golam Morshed Raju
Department of CSE,
KUET, Bangladesh

Pintu Chandra Shill
Department of CSE,
KUET, Bangladesh

S.M. Shadman Sakib
Department of CSE,
KUET, Bangladesh

Md. Jannatun Nayeem
Department of CSE,
KUET, Bangladesh

Abdul Aziz
Department of CSE,
KUET, Bangladesh

Abu Zafar Md Nuruzzaman Abir
Tata Consultancy Services Limited,
United States of America

Jakaria Rabbi
Department of CSE,
KUET, Bangladesh

Diksha Sharm
Computer Science Engineering,
Chandigarh University, Mohali, Punjab, India

Kulvinder Singh
Computer Science Engineering,
Chandigarh University, Mohali, Punjab, India

Arafat Bin Hossain
North Dakota State University, USA

Koushik Howlader
North Dakota State University, USA

Muhammad Zubair Malik
North Dakota State University, USA

Jafrin Iqbal Chowdhury
Department of CSE, IIUC,
Chattogram, Bangladesh

Shahanaz Ahmed Nishi
Department of CSE, IIUC,
Chattogram, Bangladesh

Umma Aymon
Department of CSE, IIUC,
Chattogram, Bangladesh

Tanveer Ahsan
Department of CSE, IIUC,
Chattogram, Bangladesh

Shayhan Ameen Chowdhury
Department of CSE, IIUC,
Chattogram, Bangladesh

Muhammed J. A. Patwary
Department of CSE, IIUC,
Chattogram, Bangladesh

Md. Asfakul Islam
Department of CSE,
East West University, Dhaka, Bangladesh

Nafiz Faiyaz
Department of CSE,
East West University, Dhaka, Bangladesh

Mehedi Hasan Adnan
Department of CSE,
East West University, Dhaka, Bangladesh

A. M. Feroz Ehses Shishir
Department of CSE,
East West University, Dhaka, Bangladesh

Faysal Ahmed
Department of CSE,
East West University, Dhaka, Bangladesh

Farhana Alam
Department of CSE,
East West University, Dhaka, Bangladesh

Ahmed Wasif Reza
Department of CSE,
East West University, Dhaka, Bangladesh

Izazul Hoq Imran
Department of CSE,
East West University, Dhaka, Bangladesh

Tropa Das
Department of CSE,
East West University, Dhaka, Bangladesh

Sharmin Akther Rima
Department of CSE,
East West University, Dhaka, Bangladesh

Meftahul Zannat
Department of CSE,
East West University, Dhaka, Bangladesh

Ahmed Wasif Reza
Department of CSE,
East West University, Dhaka, Bangladesh

Paresh Dhakan
Intelligent Systems Research Centre;
University of Ulster, Londonderry, UK

Md. Tahmid Zoayed
Department of CSE,
East West University, Dhaka, Bangladesh

Sayma Haque Arshe
Department of CSE,
East West University, Dhaka, Bangladesh

Taohid Hossain Ayon
Department of CSE,
East West University, Dhaka, Bangladesh

Ayesha Akter
Department of CSE,
East West University, Dhaka, Bangladesh

Ahmed Wasif Reza
Department of CSE,
East West University, Dhaka, Bangladesh

M. M. Akash*
visie.tech, Dhaka, Bangladesh

Rahul Deb Mohalder
Khulna University, Khulna, Bangladesh

Md. Al Mamun Khan
visie.tech, Dhaka, Bangladesh

Laboni Paul
Khulna University, Khulna, Bangladesh

Ferdous Bin Ali
visie.tech, Dhaka, Bangladesh

Arifur Rahman
Department of CSE,
KUET, Bangladesh

Anik Mahmud
Department of CSE,
KUET, Bangladesh

Abu Bakar Siddique
Department of Mechanical Engineering,
CUET, Bangladesh

Avizit Nandi
Department of ECE,
KUET, Bangladesh

Dipanjali Kundu
Department of CSE, National Institute of Textile Engineering and Research (NITER),
Constituent Institute of the University of Dhaka, Savar, Dhaka, 1350, Bangladesh

Diganta Das
Department of Statistics,
Jahangirnagar University, Dhaka, Bangladesh.

Tanoy Debnath
Department of CSE, Green University,
Dhaka, Bangladesh

Anichur Rahman
Department of CSE, National Institute of Textile Engineering and Research (NITER),
Constituent Institute of the University of Dhaka, Savar, Dhaka, 1350, Bangladesh

Solaiman Ahmed
Department of CSE, DUET,
Gazipur, Bangladesh

Md. Mostafizer Rahman
Department of CSE, DUET,
Gazipur, Bangladesh

Editors

Dr. Nazmul Siddique is affiliated with the School of Computing, Engineering and Intelligent Systems, Ulster University. He earned a Dipl.-Ing. degree in Cybernetics from the Dresden University of Technology, Germany, an MSc. in Computer Science from Bangladesh University of Engineering and Technology, and a Ph.D. in Intelligent Control from the Department of Automatic Control and Systems Engineering, University of Sheffield, England. His research interests include cybernetics, computational intelligence, nature-inspired computing, stochastic systems and vehicular communication. He has published over 170 research papers including five books published by John Wiley, Springer and Taylor & Francis. He guest-edited eight special issues of reputed journals on Cybernetic Intelligence, Computational Intelligence, Neural Networks and Robotics. He is on the editorial board of seven international journals including *Nature Scientific Research*. He is a Fellow of the Higher Education Academy, a senior member of IEEE and a member of different committees of the IEEE SMC Society and UK-RI Chapter. He was involved in organising many national and international conferences and co-edited seven conference proceedings.

Dr. Mohammad Shamsul Arefin is a Professor at the Department of Computer Science and Engineering at Chittagong University of Engineering and Technology (CUET), Bangladesh and currently serving as the Dean, the Faculty of Electrical and Computer Engineering at CUET. Earlier he was the Head of the Department of CSE, CUET. Prof. Arefin earned his Doctor of Engineering Degree in Information Engineering from Hiroshima University, Japan with the support of the scholarship of MEXT, Japan. As a part of his doctoral research, Dr. Arefin worked with IBM Yamato Software Laboratory, Japan. His research includes privacy-preserving data publishing and mining, distributed and cloud computing, big data management, machine learning, green computing and IT for agriculture and the environment. Dr. Arefin has more than 170 referred publications in international journals, book series and conference proceedings. He is a senior member of IEEE, Member of ACM, Fellow of IEB and BCS. Dr. Arefin is the general chair of COMPAS 2024, MIET 2024, BIM 2023; Organizing Chair of ECCE 2025, BIM 2021; TPC Chair, ECCE 2017; Organizing Co-Chair, ECCE 2019; and Organizing Chair, BDML 2020. Dr. Arefin visited Japan, Indonesia, Malaysia, Bhutan, Singapore, South Korea, Egypt, India, Saudi Arabia, Nepal and China for different professional and social activities.

Ahmed Wasif Reza (Ph.D., M.Eng.Sc., B.Sc. Engg. (Hons.), CEng (UK)) is a Professor at the Department of Computer Science and Engineering (CSE), East West University (EWU), Bangladesh. Moreover, he is currently the Additional Director,

Quality Assurance, Institutional Quality Assurance Cell (IQAC). Besides, he is serving as a member of the Evaluation Team for the Accreditation of different programs of various universities, appointed by the Board of Accreditation for Engineering and Technical Education (BAETE), The Institution of Engineers, Bangladesh. He is also serving as an Academic Auditor, appointed by the Bangladesh Accreditation Council. He has been working in the fields of Wireless communications, Bioimaging, Artificial Intelligence, Machine Learning, Deep Learning, Robotics and Internet of Things, Green Computing, Brain-Computer Interface, and Biomedical signal Processing, for almost 20 years, both in industrial exposure and academically research valued work. He has authored and co-authored several SCI/WoS and Scopus-indexed journals and conference papers (250+ papers).

Dr. Aminul Haque earned a B.Sc. degree from Shahjalal University of Science and Technology, Bangladesh, then attended MONASH University for his Ph.D. He is a professor and associate head at the Department of Computer Science and Engineering of Daffodil International University (DIU), Dhaka, Bangladesh. He has also presented his research work at multiple international peer-reviewed conferences and journals. Dr. Haque was instrumental in the development of a national skill-based curriculum for courses associated with big data and data science. His research interests include machine learning, the Internet of Things, and artificial intelligence. He is also a member of the Program Self-Assessment Committee (PSAC), which is an internal quality assurance committee for the CSE department. He also works as a departmental coordinator for the Center of Excellence of the Future DIU program. In addition, the department has 4IR initiatives where he supervises the IoTLab and the FabLab that foster innovation and technology.

CHAPTER 1

Toward Deep Autoencoder for Recommendation System Using Implicit Feedback

Ajoy Deb Nath

Chittagong University of Engineering & Technology, Dept. of Computer Science and Engineering, Chittagong 4349, Bangladesh.

Md. Mokammel Haque

Chittagong University of Engineering & Technology, Dept. of Computer Science and Engineering, Chittagong 4349, Bangladesh.

CONTENTS

RECENT years have seen an exponential increase in the use of recommendation systems. It is necessary to filter, prioritize, and effectively convey relevant information to consumers in order to suit their tastes and preferences on the online platform, where the sheer volume of options is available. In this chapter, we present a recommendation model that yields top-N recommendations based on collaborative filtering. We strengthen the `Variational Autoencoder` (VAE) by incorporating residual connections in encoder architecture, improvising mixture prior distribution using `hierarchical priors`, and introducing β divergence as the objective function.

DOI: 10.1201/9781003605508-1

Empirically, we analyze the impact of β on the objective function. Through a number of experiments, we also emphasise the significance of latent space in the model. By considering implicit feedback, we construct a user-item interaction matrix and feed the matrix to our autoencoder model. In our investigation, we use two datasets, and the results show that the MovieLens-20M dataset produces the best outcome, while the Netflix Prize dataset provides competing results. On the ML-20 evaluation dataset, our proposed model receives a Normalized Discounted Cumulative Gain ($nDCG$) score of 0.49, indicating the effectiveness of our method in capturing relevant recommendations within the top 100 ranked items.

1.1 INTRODUCTION

Given the massive expansion of available digital information and the growing number of Internet users, there has been a probability of information overload, which impedes quick and easy access to things of interest on the Internet. A recommendation system is an information filtering system that anticipates user preferences and offers suggestions for items they might like.

According to 54% of merchants, product suggestions are the primary driver of AOV (average order value) in consumer purchases [1]. 75% of what people are watching on Netflix comes from recommendations, according to McKinsey [2]. Several forms of recommendation systems based on textual assessment, correlative opinion, user ratings, purchasing patterns, user profiles, and the like have been developed in recent years. There are numerous types of recommendation approaches [4], such as content-based, collaborative filtering, demographic, knowledge-based, community-based, and hybrid recommendation systems. Some recommender systems rely on text-based similarity [9] [18]. The collaborative filtering technique is very widely known among all methods and has been utilized in most cases with content-based probabilistic topic modeling [10], deep neural networks [19], and so on. However, due to the limits of such models in capturing individuals' preferences and item attributes, the improvement in recommendation effectiveness is still not noteworthy. Autoencoder (AE) has evolved into one of the most powerful techniques for capturing key features of data. To better accommodate consumers, AE has enhanced the recommendation architecture, introducing us to more possibilities to reassess user experiences.

In this work, we propose an improved VAE model for top-K recommendations using implicit feedback. Our model incorporates residual connection in encoder architecture and composite prior with hierarchical distribution to encourage the encoder network to generate compact representation. The skip connections help the network retain information about low-level features of the input data that may lose in deeper layers. Depending on the compact representation, our model learns the complex user behavior more precisely. We also introduce β divergence as the objective function. Our study has found a significant impact of hyperparameter β in the outcome. Finally, we have compared our model with autoencoder-based models on the benchmark datasets including *MovieLens-20M* (ML-20m) and *Netflix Prize Dataset*. As per the result, our model outperforms several autoencoder-based models on ML-20m dataset and shows a competing result on the Netflix Prize dataset.

1.2 RELATED WORK

Deep neural networks have achieved incredible success in speech recognition, computer vision, and natural language processing in recent years. `Neural network-based Collaborative Filtering` (NCF) [11] was proposed as a general framework for the recommender system. The paper suggested a multi-layer perceptron to learn the user-item interaction function to boost up NCF modeling with non-linearities. According to their study, deeper layers of neural networks provide greater recommendation performance. In the paper [12], the matrix factorization method was combined with a deep neural network, and both implicit and explicit interactions were taken into account. The paper describes a deep learning architecture for learning a uniform low-dimensional space for user and item representations. They also considered developing a new loss function for the training process. The experiment results show that a deeper layer in the architecture lowers performance. In another work, deep autoencoders with iterative output refeeding techniques were proposed [20]. The model used Scaled Exponential Linear Units (SELUs) and the dropout rate was high. The network was much deeper (6 layers). The Root Mean Squared Error was not very much competitive than the recent deep learning approach.

A deep learning-based collaborative filtering approach was proposed where they used a deep learning approach instead of matrix factorization [5]. The model employs the forward propagation approach to predict rating scores after learning the latent variables of the user and the item. The model performed better than other traditional approaches. `Collaborative Dual Attentive Autoencoder` (CATA++) is another method proposed in paper [21]. Two parallel autoencoders are used to learn the latent space of an item using its content. In order to represent the things in latent space, an attention mechanism was applied in the middle of the two autoencoders to obtain better conceptual information. The model only functioned with contextual data and was only evaluated for the CiteULike dataset, which does not contain any user data.

`Variational autoencoders` (VAEs) [13] are extended to collaborative filtering for implicit feedback. The authors proposed a nonlinear probabilistic model which uses multinomial likelihood alongside with Bayesian inference. Moreover, they used annealing technique to estimate the hyper-parameters. In paper [14], the authors incorporated a ranking loss that optimizes the model for top-N recommendation tasks. The model presents various novel techniques to improve the VAE model. In another work [15], the author proposed a novel top-N recommendation strategy based on a deep variational autoencoder (VAE) architecture with a shallow parallel path. To capture both high-level and low-level aspects of user-item interactions, the suggested `VASP` model paired a deep VAE with a shallow parallel path. By modeling the interactions between users and items, the deep VAE was able to learn high-level representations of them, but the shallow parallel route was able to capture low-level aspects by modeling the user and item features independently. They achieved the best results among all other recent models. A comprehensive explanation of our proposed method for the recommendation system will be found in the forthcoming section.

1.3 METHODS

The present section presents the proposed methodology of our work, improvising the Variational Autoencoder (VAE) [13] model with the help of hierarchical prior distribution and β divergence as the objective function. Various traditional autoencoder model has been proposed in recent years [6]. In our proposed method, we focus on regularizing the autoencoder model using hierarchical prior distribution and incorporated β divergence with KL divergence as the objective function to provide an effective recommendations. We also depict our way of residual connections in the encoder network.

1.3.1 Data Pre-processing

In this chapter, We take the recommendation system as a collaborative filtering problem. We consider user-item interaction matrix $X_{U,I}$ as our input to the model where U and I denote sets of users and items respectively. We also consider the rating as our implicit feedback and binarize the the data.

1.3.2 Proposed Approach

Our proposed framework is illustrated in Fig. 1.1. The following key steps make up the framework: (1) Pre-processing the dataset, (2) Data splitting (train-validation- test), (3) Training the autoencoder model, and (4) Recommend top-K items. For the work, we used several real world datasets. Our pre-processing steps include data filtering and creating a user-item interaction matrix taking account of implicit feedback. Implicit feedback can be views, clicks, times spent on a page, etc. Then we feed the matrix data into our Autoencoder model and we get the same dimensional matrix which contains the likelihood probability of the items for each user. Our proposed model includes Variational Autoencoder with composite prior using hierarchical distribution to regularize the encoder network and β divergence as reconstruction loss.

1.3.2.1 Variational Autoencoder (VAE)

Variational Autoencoder (VAE) is a neural network design that can be used for collaborative filtering. The input data in a VAE for collaborative filtering contains user-item interactions, with each user having a collection of objects with which they

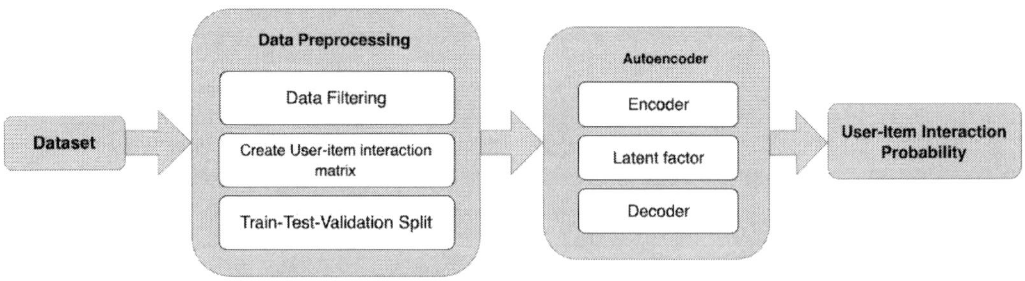

Figure 1.1: Block diagram of proposed framework

engaged. An encoder network maps user-item interactions to a latent representation, while a decoder network maps the latent representation back into the original user-item space. The encoder network generates two vectors, mean and variance vector, which are used to construct a sample in the latent space from a Gaussian distribution. Our autoencoder architecture is deep in nature. We use the residual connection to capture all relevant information about the input data. We use the residual connection in a way that connects the output of the encoder to intermediate layers within the encoder network. This helps preserve information about the input data as it is processed by the network and improves the performance of the network. The posterior distribution is computed as:

$$h_k = f(W_k x + b_k) \tag{1.1}$$

$$h_{k+1} = f(W_{k+1} h_k + b_{k+1}) \tag{1.2}$$

$$\mu = W_\mu h_n + b_\mu \tag{1.3}$$

$$\log(\sigma^2) = W_\sigma h_n + b_\sigma \tag{1.4}$$

$$z = \mu + \sigma \odot \epsilon \tag{1.5}$$

where x is the input data, W and b are the weight matrices and biases, respectively, of the fully connected layers, f is the activation function (Swish), z is the latent variable, and ϵ is a sample from the standard normal distribution. The mean μ and log-variance $\log(\sigma^2)$ of the approximate posterior distribution $q(z|x)$ are also calculated using the output of the last hidden layer of the encoder presented in the equation (14.3) and (14.4).

$$\hat{h}_k = f(U_k z + c_k + h_k) \tag{1.6}$$

The residual connections between the latent variable z and the earlier layers of the encoder network are implemented using learnable parameters U and biases c. The decoder network is a mirror image of the encoder network, and it takes the latent variable z as input and produces a reconstruction of the input data $\hat{x}$. The decoder network uses the sampled latent vector to predict the user-item interactions. The VAE is given training by minimizing the reconstruction loss, which calculates the difference between expected and actual user-item interactions. The VAE incorporates, in addition to the reconstruction loss, a regularization element known as the KL divergence, which promotes the latent space to follow a prior distribution, such as a standard Gaussian distribution.

1.3.2.2 Hierarchical Modeling with Composite Prior

Overfitting is quite a big issue in case of training a complex model. In our case, the model may learn more accurate representation for some users, while learning complex latent representation. When a new user arrives to the system, the model may face difficulties adjusting the distribution value. As a result, the model can't generate accurate recommendation. Mult-VAE [13] employs a Gaussian prior as well as a separate KL regularizer, which adjusts posterior variational features to their observed value. In paper [15], author introduced a novel composite prior distribution for the latent variables to improve VAE model. Our proposed technique incorporates a composite prior distribution which includes the standard prior, posterior prior, uniform prior and empirical prior [7] distribution. Formula (14.1) presents the general equation for calculating prior distribution. Then, we modified the composite prior distribution to incorporate the hierarchical prior.

$$\log(\eta(x|\mu, \sigma)) = 0.5 * [\log(\sigma) + \log(2\pi) + (x - \mu)^2/\sigma] \tag{1.7}$$

In the encoder network, we map the observed data to the mean and the log variance of latent variables to sample the latent variables from the composite prior distribution at each level of the hierarchy. The latent variables then pass up to the next level of the hierarchy, where they act as inputs to the hierarchical prior distribution. The hierarchical prior allows information to be shared across different levels of the model. It handles the estimation of parameters at each level while taking into account the dependence and linkages between the parameters at each level.

1.3.2.3 Objective Function

According to the [3], β divergence plays a vital role at creating effective latent representation with more diverse factors autoencoder. We incorporated β divergence alongside KL divergence as our objective function. From our study, it generates better results for our recommendation system while applying the value of β>1. The impact of β is presented in the later section. The β divergence loss for a single user-item rating pair is defined as:

$$\mathcal{L}_\beta(r \mid\mid \hat{r}) = \frac{1}{\beta(\beta - 1)}(\hat{r}^\beta + (\beta - 1)r^\beta - \beta \times \hat{r} \times r^{\beta-1}) \tag{1.8}$$

where the predicted rating($\hat{r}$) is the rating predicted by the VAE for the user-item pair and the actual rating(r) is the true rating. In paper [15], authors also introduced the β hyper-parameter. The β hyper-parameter controls the weighting of the KL divergence term in the objective function, which is used as a regularization term to encourage the learned latent space to follow a Gaussian distribution. In our model, the β divergence term regularizes the reconstruction error of the model. It allows to control the degree of emphasis placed on the distance measure (squared Euclidean distance) in the divergence calculation through the choice of beta hyper-parameter. To use the beta divergence as the reconstruction loss, we modify the reconstruction loss term and the KL divergence as the regularizer in the objective function which can be written as:

$$\mathcal{L}(\theta, \phi) = \frac{1}{N_u}\sum_{u=1}^{N_u} \mathcal{L}_\beta(r_u \mid\mid \hat{r_u}) + \beta\mathcal{L}_{KL} \tag{1.9}$$

where θ and ϕ are the parameters of the decoder and encoder networks, respectively. There are two instances of the hyper-parameter β in the objective function: one in the beta divergence term, which controls the degree of distortion between the true ratings and the reconstructed ratings, and another with the KL divergence term,

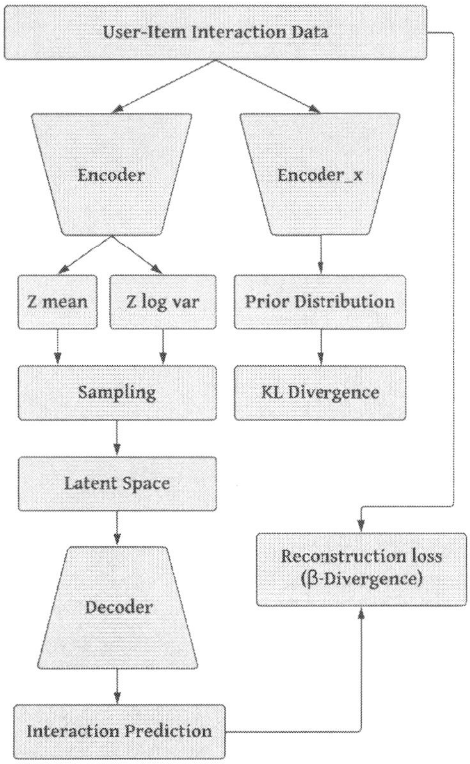

Figure 1.2: Proposed Model Architecture

which controls the trade-off between the reconstruction loss and the regularization term. In Fig. 11.3, the architecture of our proposed model is illustrated.

1.4 EXPERIMENTAL ANALYSIS

In this section, we demonstrate our experiments and result analysis about our proposed model. We also compare our work with other baseline models.

1.4.1 Evaluation Metrics

According to [24], before being officially deployed, a new recommender system typically has to complete the evaluation process of three stages: offline analytics, user analysis, and online experiment. Offline analytics just needs datasets to produce the associated evaluation metrics, including forecast accuracy and coverage. Offline analytics does not require any user participation. Recommendation problems can be classified as rating prediction and ranking prediction problems. The quality of a recommendation system may be analyzed using many sorts of measurement techniques, such as accuracy or coverage. The metrics considered are determined by the type of problem. For our top-N ranking prediction model, we have used Normalized Discounted Cumulative Gain (nDCG) [8] and Recall. In general, nDCG evaluates a list of items' quality by contrasting it to a perfect rating. It considers both the significance of the ranking's items and where they fall on the list. On the other side, Recall@K is defined more precisely as the proportion of recommended items that are relevant to all relevant items in the dataset.

1.4.2 Datasets

To evaluate our model, We have used two real world datasets, including MovieLens-20M[1] [22] and Netflix Prize dataset[2] [23] for experiments. These datasets contain interaction history between users and movies. The ML-20m dataset contains 20000263 ratings and 465564 tag applications across 27278 movies rated by 138493 users. On the other hand, around 100,000,000 ratings for 17,770 films, provided by 480,189 users, constitute the Netflix Prize Dataset.

1.4.3 Experimental Setup

For implementing our proposed work, we have used Pytorch[3] and Kaggle[4] which is an open-source platform to build machine learning & deep learning model. For training our model, the Adam optimizer with a learning rate of 0.0005 is used as a parameter. The rate is similar for both the encoder and decoder network training. The Encoder Dropout value is 0.5 and the batch size is 500. To prevent overfitting or underfitting, we have used earlystopping which considers the loss of validation in the training phase of the model. Choosing all the best possible parameters, we compiled the model with 30 epochs for the ML-20M dataset and 35 epochs for the Netflix Prize dataset. We performed Hyper-parameter tuning on the model for the ML-20M dataset.

1.4.4 Evaluation of Performance

The experimental results on the basis of our previously explained evaluation criterion are shown in Table 1.1. The result is shown for all three datasets that we used to

Table 1.1: Experiment Result of Our Proposed Model

Dataset	nDCG@100	Recall@20	Recall@50
ML-20M	0.489	0.461	0.552
Netflix Prize	0.401	0.379	0.463

train and evaluate our model. We can see that our model performs individually on each dataset as each of them is different in size and context. Our model outperforms previous models on the ML-20M dataset and shows a competitive result on Netflix Dataset.

[1]https://grouplens.org/datasets/movielens/20m
[2]https://www.netflixprize.com
[3]https://pytorch.org
[4]https://www.kaggle.com

1.4.5 Analysis

In this section, we discuss the following two questions regarding our model performance and dependency on some key parameters:

1. What is the importance of latent factors in our proposed model?
2. How does the β parameter impact the model performance?

1.4.5.1 Importance of Latent Space

A latent space is characterized in formal terms as an abstract multi-dimensional space that stores a meaningful internal representation of externally observable events. In the latent space, samples that are similar in the external world are positioned close to each other. Our model success very much depends on latent representation learning. To highlight the significance of the latent factors in our work, we try to visualize the latent space on varying dimensions which shows the ability to capture behavioral patterns between different users. The experiment is carried out with 10000 heldout users where the higher dimensional latent representation of the users is reduced to a lower dimension for the benefit of visualization.

From Fig. 1.3, we can see that our model captures different pattern of users behavior. By changing the dimension of latent space, we observe the patterns between users for various sizes. User data points of the same behavioral pattern, stay close to each other in the latent space. The more relevant latent factors, the more accurate the output decoder can produce. Moreover, we also depict the importance of latent factors of various dimension based on `F1-score` and `mean average precision` (mAP). In Table 1.2, we highlight the impact of latent factors on our model performance. As we can see, our model shows better `F1-score` and `mAP` for 100-dimensional latent space.

1.4.5.2 Impact of β

In the following Fig. 1.4, we have highlighted the variation of β hyper-parameter for training our model. From the visualization, we can observe the impact of β divergence and prove the effectiveness of our proposed technique to improve the VAE framework. We can see that having $1 < \beta < 2$ improves the result significantly.

1.4.6 Comparison with other baselines

`Multinomial Denoising Autoencoder` (Mult-DAE) and `Multinomial Variational Autoencoder` (Mult-VAE PR) [13] are based on autoencoder architecture and popular methods for collaborative filtering. Mult-DAE uses a denoising autoencoder to model the interactions as binary ratings. Mult-VAE approximates the true posterior distribution over the latent representations. `H+Vamp Gated` [16] model had experimented with the `VampPrior` [17], originally proposed for image generation, to examine the effect of flexible priors in Collaborative filtering. `VASP` [14] proves to be one of the state-of-the-art models showing a tremendous result on ML-20M dataset.

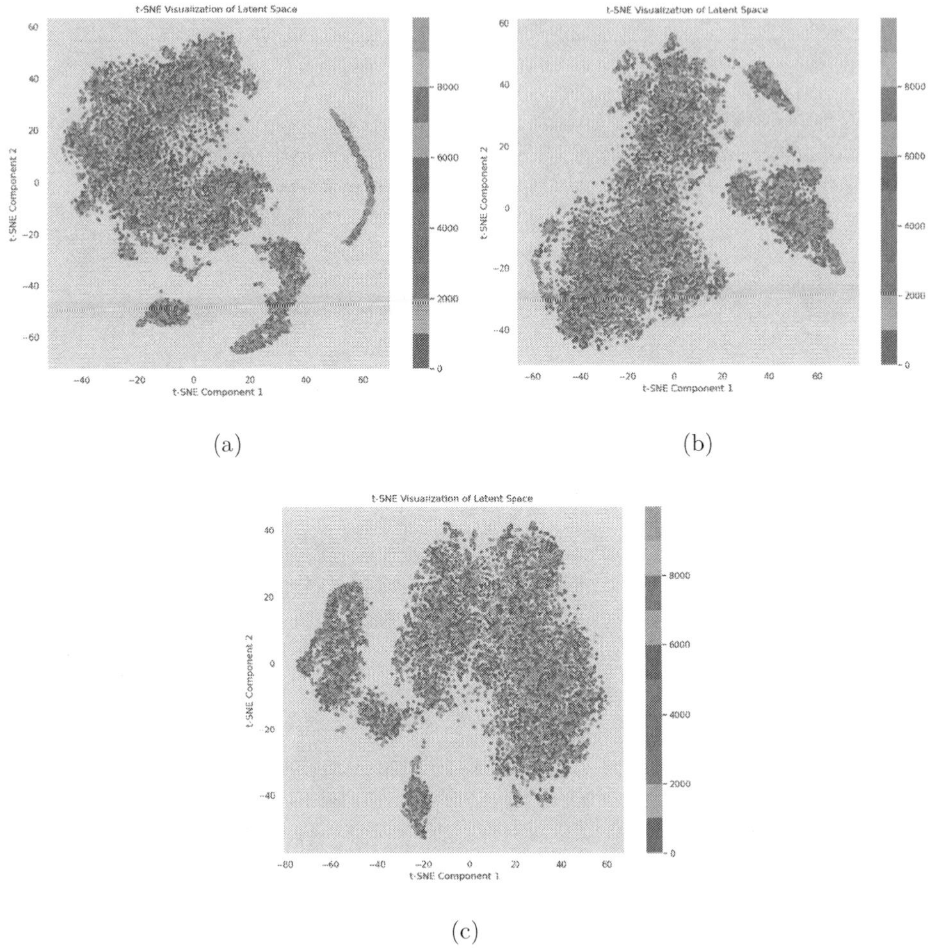

Figure 1.3: Latent space visualization of various dimensions (a) 50, (b) 100, and (c) 300 size

All of the above models use implicit feedback as the input of their model. Data pre-processing is also similar as all the autoencoder model works only with interaction data. Overall comparison of different performance metric of the existing method and our proposed method for both the datasets are illustrated in the Tables 17.2 and 17.3. Graphical illustrations of these comparisons are presented in Figs. 1.5 and 1.6.

Table 1.2: F1-score and mAP based on various sizes of latent space

Latent space (size)	F1-score@10	mAP@10
50	0.228	0.132
100	0.230	0.135
300	0.224	0.129

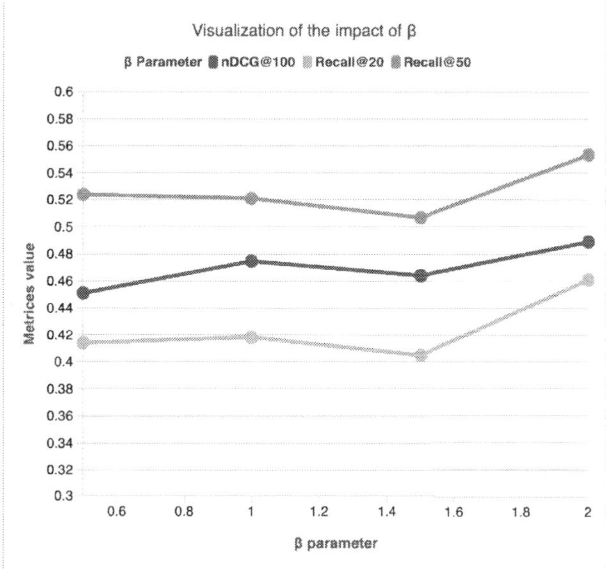

Figure 1.4: β hyper-parameter tuning

Table 1.3: Overall performance analysis of the existing method & proposed method for ML20M. The best results are highlighted in bold.

Performance metrics	nDCG@100	Recall@20	Recall@50
Mult-DAE	0.419	0.387	0.524
Mult-VAE PR	0.426	0.395	0.537
H+Vamp Gated	0.445	0.413	0.551
VASP	0.448	0.414	**0.552**
Proposed method	**0.489**	**0.461**	**0.552**

nDCG@100 and recall@20 of our model are 0.489 and 0.461 respectively which is far better than all other mentioned state-of-art models in the case of ML-20M dataset. On the Netflix Prize dataset, our model still competes with the state of art models. Better parameter tuning can help us to gain better results. So, we need to optimize our model based on the datasets.

1.5 CONCLUSION

In today's internet services, recommendation systems are taking on a bigger and bigger significance. They represent an effective way of allowing consumers to browse enormous information and product domains. In this work, we have attempted to develop a recommendation system by improvising the Variational Autoencoder (VAE). We used two benchmark datasets, such as MovieLens-20M and Netflix Prize. Our

Table 1.4: Performance analysis of the existing method & proposed method for Netflix Prize. The best results are highlighted in bold.

Performance metrics	nDCG@100	Recall@20	Recall@50
Mult-DAE	0.380	0.344	0.438
Mult-VAE PR	0.386	0.351	0.444
H+Vamp Gated	**0.409**	0.376	**0.463**
VASP	0.406	0.372	0.457
Proposed method	0.401	**0.379**	0.449

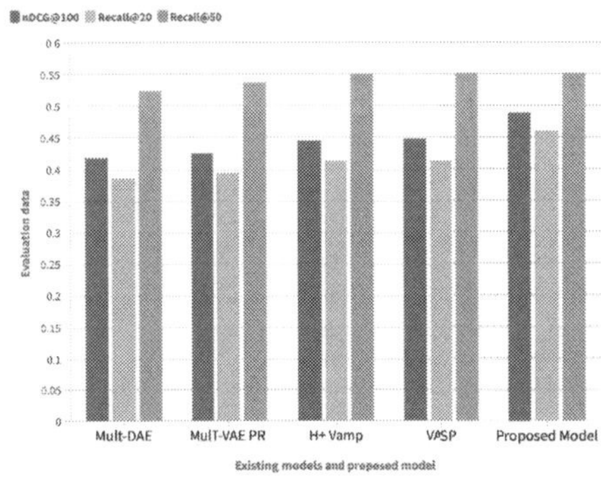

Figure 1.5: Graphical illustration of performance for ML-20M (existing models vs proposed model)

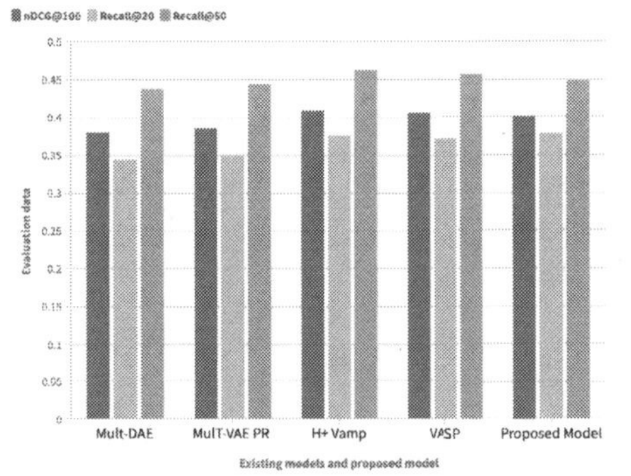

Figure 1.6: Graphical illustration of performance for Netflix Prize (existing models vs proposed model)

goal was to comprehend prospective user preferences and make appropriate product or item recommendations to the user. We present several improvements over VAE models by introducing hierarchical modeling with composite prior, residual connection inside the architecture, new objective function, and proper tuning of hyperparameter β and achieved an optimal result from our proposed methods. We achieve nDCG@100 of 0.489 for ML-20 Dataset and 0.401 for the Netflix Prize. We also highlight the significance of latent space in our model. Finally, we compare the results of our experiments with baseline models to prove the effectiveness of the proposed methods.

The recommendation system lacks an explanation of recommendations which is essential to gain the trust of the users of the system. In the future, the introduction of explainability can be a research direction for autoencoder-based models.

Bibliography

[1] The power of product recommendations: 30 must-know statistics for 2022, `https://blog.clerk.io/product-recommendations-statistics-2022`.

[2] How retailers can keep up with consumers, https://www.mckinsey.com/industries/retail/our-insights/how-retailers-can-keep-up-with-consumers.

[3] Weng, Lilian. "From autoencoder to beta-vae." lilianweng.github.io/lil-log (2018).

[4] Ricci, Francesco, Lior Rokach, and Bracha Shapira. "Introduction to recommender systems handbook." Recommender systems handbook. Boston, MA: springer US, 2010. 1–35.

[5] Aljunid, Mohammed Fadhel, and Manjaiah Dh. "An efficient deep learning approach for collaborative filtering recommender system." Procedia Computer Science 171 (2020): 829–836.

[6] Zhang, Guijuan, Yang Liu, and Xiaoning Jin. "A survey of autoencoder-based recommender systems." Frontiers of Computer Science 14 (2020): 430–450.

[7] Friston, K., and W. Penny. "CHAPTER 23–Posterior Probability Maps. Statistical Parametric Mapping." London, Academic Press 295 (2007): 302.

[8] Järvelin, Kalervo, and Jaana Kekäläinen. "IR evaluation methods for retrieving highly relevant documents." ACM SIGIR Forum. Vol. 51. No. 2. New York, NY, USA: ACM, 2017.

[9] Tavakoli, Mohammadreza, et al. "A recommender system for open educational videos based on skill requirements." 2020 IEEE 20th International Conference on Advanced Learning Technologies (ICALT). IEEE, 2020.

[10] Wang, Chong, and David M. Blei. "Collaborative topic modeling for recommending scientific articles." Proceedings of the 17th ACM SIGKDD international conference on Knowledge discovery and data mining. 2011.

[11] He, Xiangnan, et al. "Neural collaborative filtering." Proceedings of the 26th International World Wide Web Conference, Perth Australia, April 3 - 7, 2017, pp. 173–182.

[12] Xue, Hong-Jian, et al. "Deep matrix factorization models for recommender systems." IJCAI. Vol. 17. 2017.

[13] Liang, Dawen, et al. "Variational autoencoders for collaborative filtering." Proceedings of the 2018 Web Conference, April 23–27, 2018, Lyon, France, pp. 689–698.

[14] Shenbin, Ilya, et al. "Recvae: A new variational autoencoder for top-n recommendations with implicit feedback." Proceedings of the 13th international conference on web search and data mining. 2020.

[15] Vančura, Vojtěch, and Pavel Kordík. "Deep variational autoencoder with shallow parallel path for top-N recommendation (VASP)." Artificial Neural Networks and Machine Learning–ICANN 2021: 30th International Conference on Artificial Neural Networks, Bratislava, Slovakia, September 14–17, 2021, Proceedings, Part V 30. Springer International Publishing, 2021.

[16] Kim, Daeryong, and Bongwon Suh. "Enhancing VAEs for collaborative filtering: flexible priors & gating mechanisms." Proceedings of the 13th ACM Conference on Recommender Systems. 2019.

[17] Tomczak, Jakub, and Max Welling. "VAE with a VampPrior." International Conference on Artificial Intelligence and Statistics. PMLR, 2018.

[18] Philip, Simon, Peter Shola, and A. Ovye. "Application of content-based approach in research paper recommendation system for a digital library." International Journal of Advanced Computer Science and Applications 5(10) (2014).

[19] Magron, Paul, and Cédric Févotte. "Neural content-aware collaborative filtering for cold-start music recommendation." Data Mining and Knowledge Discovery 36.5 (2022): 1971–2005.

[20] Kuchaiev, Oleksii, and Boris Ginsburg. "Training deep autoencoders for collaborative filtering." arXiv preprint arXiv:1708.01715 (2017).

[21] Alfarhood, Meshal, and Jianlin Cheng. "CATA++: a collaborative dual attentive autoencoder method for recommending scientific articles." IEEE Access 8 (2020): 183633–183648.

[22] Harper, F. Maxwell, and Joseph A. Konstan. "The movielens datasets: History and context." Acm transactions on interactive intelligent systems (tiis) 5.4 (2015): 1–19.

[23] Bennett, James, and Stan Lanning. "The netflix prize." Proceedings of KDD cup and workshop. Vol. 2007. 2007.

[24] Chen, Mingang, and Pan Liu. "Performance evaluation of recommender systems." International Journal of Performability Engineering 13.8 (2017): 1246.

CHAPTER 2

Intelligent Learning Behavior Analysis for Student Based on Fuzzy Agent Model

Silvia Sifath

University of Barishal, Dept. of CSE, Barishal, Bangladesh. Email: ssifath17.cse@bu.ac.bd

***Md Manjur Ahmed**

University of Barishal, Dept. of CSE, Barishal, Bangladesh. Email: mmahmed@bu.ac.bd

***Samir B. Belhaouari**

Hamad Bin Khalifa University, College of Science and Engineering, Division of Information and Computing Technology, Doha, Qatar. Email: sbelhaouari@hbku.edu.qa

K. M. Azharul Hasan

Khulna University of Engineering and Technology, Dept. of CSE, Khulna, Bangladesh. Email: az@cse.kuet.ac.bd

CONTENTS

Learning style plays a significant role in improving a student's educational performance and finding strategies to adapt to new concepts. It is the process of evaluating the way of learning of individuals. Several factors, such as interaction with fellow students and teachers, emotional and social influences, interaction with the environment, teaching style, and personal characteristics, play vital roles in determining

DOI: 10.1201/9781003605508-2

a student's learning behavior. This study introduces a methodology to determine student's learning styles automatically using a fuzzy agent model. Fuzzy logic is a good solution for handling uncertainty and vagueness and is therefore suitable for use in an intelligent learning behavior model. However, in a model that contains human behavior, complexities arise in the interaction of the factors in the system. Agent-oriented modeling can handle system complexities that are not considered in fuzzy logic. In this paper, we analyze students' behavior, present the concept of fuzzy logic and agent-based modeling, and finally develop a fuzzy agent model to implement an intelligent model. This implementation will help students increase interaction with education and knowledge, expand awareness of the student about their suitable way to learn, and enlarge student performance. The main contribution of this paper is to develop a fuzzy agent model and determine an individual's learning style automatically and accurately, which supports the student in acquiring knowledge appropriately and the teacher in suitably teaching the students. For this purpose, a real dataset is extracted from final-year CSE students at the University of Barishal, Bangladesh, that contains 35 students' required information. The proposed model predicts learning styles with 91% accuracy, which is satisfactory for contributing to the study of learners.

2.1 INTRODUCTION

At present, the number of students is increasing, but the majority of them feel difficulty at different stages of their learning. They can't understand the topic or do not know if their strategy is right or wrong. Some students are extrovert types, and they do not express themselves. As a result, students drop out of school or perform poorly on assessments. To overcome this situation and lead to faster and more satisfactory improvement, it is highly needed to know the learning behavior of individuals, which is suggested by many researchers. For this purpose, we propose an intelligent system to evaluate an individual's learning style automatically that is developed using fuzzy logic and an agent-based model.

Research on learning behavior in students faces many gaps and shortcomings. Learning behavior is fraught with uncertainty, and in real life, there are complexities, such as human behavior. Without consideration of uncertainty and complexity, an accurate result cannot be determined in a learning style. Researchers have proposed many solutions to identify students' learning styles. Some researchers develop methods for e-learning systems using machine learning that don't cover the uncertainty and complexities appropriately [1]. Some researchers only focus on two or three dimensions of the Felder-Silverman learning style (FSLM) [2]; otherwise, they use different learning style models instead of FSLM [3]. As most of the researchers work with machine learning and e-learning systems, taking information from online activities, they can't directly focus on self-motivation, but these elements are very crucial for learning styles. Besides, the interaction among the input factors cannot be taken into consideration to determine the learning style, and many researchers didn't achieve higher accuracy.

To solve the gaps and limitations, we develop a fuzzy agent-based model to solve the challenge of uncertainty and complexity, which are inherent in learning behavior.

The Felder-Silverman learning style model is used, and all of the dimensions are taken into consideration in this study. In this experiment, the relationship between the input agents is considered to obtain an accurate result. The proposed method also considers self-motivation as an important factor in evaluating a learning style, which escalates the possibility of increasing the accuracy of the learning style in the proposed system.

A fuzzy agent-based model is necessary for detecting student learning styles automatically. A fuzzy agent-based model helps deal with complex and uncertain problems. In this paper, agent-based models and fuzzy logic are integrated to evaluate students' learning styles. Information about students is collected using questionnaires. Information is processed, and the proposed model takes input as crisp values and converts them into fuzzy values using fuzzy techniques. In this study, an agent interaction model is developed to define the rules for the proposed model. According to the rules, the impact of interaction among agents is calculated. By considering all the rules, we get a fuzzy value that is converted into a crisp value to detect individual learning styles.

In this paper, we propose an intelligent learning behavior profiling system for students based on a fuzzy agent-based model. This study aims to evaluate individuals' learning styles automatically and accurately to support teachers and students as well as educational institutions to help make productive and effective students. In this study, a fuzzy agent-based system with optimized rules is used due to the complex nature of the learning style. The contribution of this proposed method is as follows:

1. By adopting a fuzzy agent-based approach, an intelligent model is proposed for automatically detecting learning styles.

2. An interaction model is developed in the proposed model with the assistance of input agents, which aid in establishing optimized rules and ensuring accuracy.

3. Our newly proposed method performs well in terms of accuracy and solving uncertainties and complexities.

In the following sections, the concepts of learning styles, methodology, implementation, evaluation, and discussion are presented. The other section of the article is ordered as follows: In Section 2.2, we discuss learning style, fuzzy logic, and the agent based model. In Section 2.3, we discuss the proposed method. In Section 2.4, we discuss the implementation of the model, result validation, and finally the analysis of the results. Section 2.5 represents the conclusion and future research opportunities.

2.2 LITERATURE REVIEW

Before discussing the methods used to identify the learning style, a short discussion of the characteristics of learning styles, the Felder-Silverman learning style, agent-based model, and fuzzy logic is presented. Agent-based modeling (ABM) is an appropriate approach to analyzing and simulating a complex system by developing interaction with agents and the environment [4]. It is an approach to simulating the real world that resembles human behavior [5]. Individual agents have different internal characteristics and behaviors that allow them to be autonomous, accomplish to dynamically detect any conditions occurring in the model, and act on the appropriate behavior in response. ABM maintains the evolution strategy of a system with degrees of freedom in the structure and intensity of agent interactions, agent learning, and adaptation [6, 7]. Besides, ABM helps to simulate complex social science, economics, geography, ecology, cognitive science, and behavioral problems [6, 7].

Lotfi A. Zadeh introduced the concept of fuzzy logic, which is a multivalued logic concept that takes into account the uncertainty of any decision. This means it helps to define different degrees of the intermediate values between normal evaluations [8], and this concept makes the computer system more intelligent [9]. The basic section of fuzzy is the fuzzy set. For solving decision reasoning, a fuzzy set is very important as it can measure the intermediate situation of two crisp values. At present, fuzzy logic is integrated with various fields to solve problems, and by using it, an increasing number of works are completed successfully. Moreover, it helps to implement different kinds of frameworks (micro or macro), natural language [10], various intelligent systems for decision-making [11], and system control [11]. Also, fuzzy logic is widely used in identification, optimization, and pattern reorganization [12].

Learning style refers to the preferred way of acquiring knowledge. Learning styles are not preferences for a certain form of activity but rather describe a set of student preferences for how learning materials are distributed, how they process data, and how they distribute information [13]. There are many definitions of learning styles, such as "describing the attitudes and behaviors that determine an individual's preferred learning style" [14], "characteristic strengths and preferences in how they [learners] receive and process information" [15], and "a complex way and conditions in which learners perceive, process, store, and remember what they attempt to learn in a manner most effective and efficient" [16]. Individuals learn in different ways [17]. Some learners may learn best when they show something, while some students may learn best when they hear something. Student learning styles depend on some factors such as learner cognitive load, learner emotions, self-motivated capability, teaching style, etc. The primary contributions of learning styles involve increasing awareness among students about their ways of learning, facilitating suitable interactions between students and knowledge, enhancing student accomplishment, and helping students to know their strengths and weaknesses. In the next section, the investigation to determine the learning styles as well as the computational concept and the connection of learning style with other factors will be presented. According to the Felder-Silverman learning style model, the dimensions have some characteristics, and those characteristics help to establish rules and models.

Active/reflective dimension. The preference to process information is determined in this dimension. Active learners are searching for the main idea in the learning materials. They enjoy making groups as well as discussing and explaining questions. On the other hand, reflective learners like to work alone or with a familiar partner [18]. While active learners show enthusiasm for doing assignments and solving problems, reflective learners prefer to observe the content that is provided in the learning resources. They spend their maximum learning time reading and thinking about theory but exercise less frequently [19].

Sensing/intuitive dimension. The sensory-intuitive dimension represents the learner's preference for perceiving or assimilating information. Sensing learners want to think about real situations and data by using examples, while intuitive learners learn the basic and deep concepts to do challenging and creative things [18]. Sensitive learners are more conscious about their work [19].

Visual/verbal dimension. This dimension determines how learners like the information presented. Visual learners prefer to learn information that is visually presented. Pictures, diagrams, and flow charts are the ways of representing information visually that learners prefer [18]. Verbal learners prefer explanations, and it can be writing, reading, or any lecture; that's why they like to engage with different groups, forums, etc. Even they like to do their assignments and projects and grasp concepts by discussing, reading, and writing [20].

Sequential/global dimension. This dimension determines how learners prefer to organize and progress toward understanding information. Sequential learners prefer to think straightforwardly, and they learn concepts by dividing them into small modules to capture all the details [18]. Global learners prefer to think in terms of an entire idea. They take a systematic approach to observe, connect, and relate the focusing concept with other disciplines [20].

For the automatic detection of learning styles, numerous solutions have been proposed. A few of these solutions use fuzzy approaches to detect students' learning styles. Most of the works use machine learning algorithms to evaluate learning styles. According to Rasheed [1], an e-learning platform was used for experiments where questionnaires and student behavior in the system were both used. The author used feature extraction and removal of attribute techniques to get better accuracy. A machine learning classifier was used in this work, where the algorithms were decision trees, support vector machines, K-nearest neighbor, Nave Bayes, linear discriminant analysis, random forest, and logistic regression. Four hundred ninety-eight samples were taken for this experiment. Aissaoui [21] proposed an approach to detect learning styles using web usage mining techniques and machine learning algorithms. Using web mining techniques, the log file of the e-learning environment was extracted, and the learner's sequence was captured. These sequences were applied to the K-mode clustering and clustered into 16 FSLM learning style combinations. A Naive Bayes algorithm was used in this paper to predict the learning styles of students. Zhang [22] used a deep belief network for online education to identify student learning styles. This work utilized soliciting questionnaires to determine the learning style based on the ILS theory by Felder and Soloman [13]. Li et al. [23] used a tree-augmented naive Bayesian network to identify learning styles automatically. The author used 36 students' data to train the model. El Fazazi et al. [24] proposed a model using fuzzy c-means and an artificial neural network algorithm to determine the learning styles. Web log mining is the technique used in this paper to collect data from an e-learning platform.

2.3 PROPOSED METHOD FOR ANALYZING LEARNING BEHAVIOR

In this work, we have proposed an approach that uses fuzzy logic and an agent-based model to build a model (Figure 2.1) to detect learning styles. This model uses students' information as input to identify learning styles. The student's information is collected from questionnaires. After processing the data, these are converted into crisp values. These crisp values are used as the input of the proposed fuzzy model.

In this study, the Mamdani fuzzy system is used. Fuzzification is the first step in the fuzzy expert system that converts crisp values into fuzzy values. For this, the membership function needs to be established, and the function can be constructed using the existing formula.

According to the input and output, a set of fuzzy rules is established using the proposed interaction model. Defuzzification is described using the centroid method to develop the fuzzy model. Figure 2.1 shows the process of the fuzzy model for detecting learning behavior used in this study.

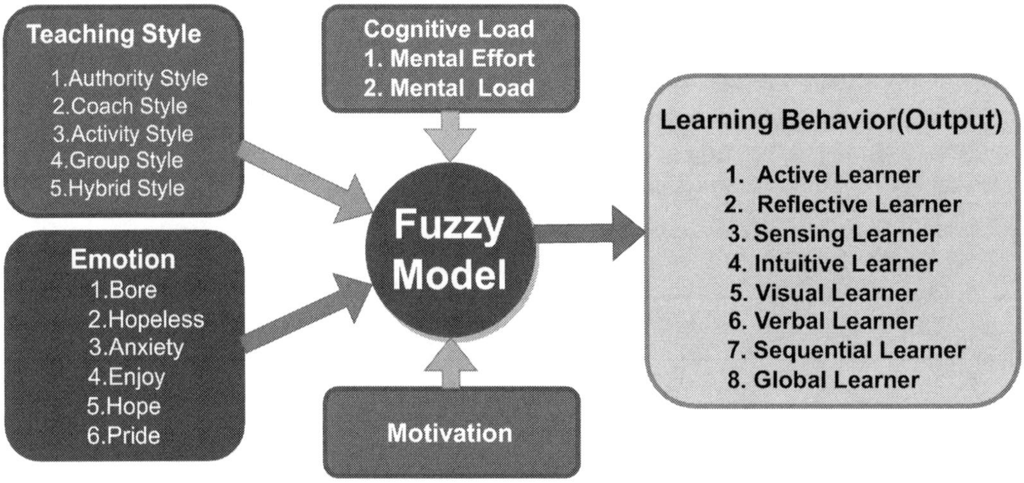

Figure 2.1: Proposed Model for evaluating learning behavior.

2.3.1 Fuzzy Input Agent for the Proposed Model

Vagueness and uncertainty are the nature of human behavior [25, 26]. For evaluating learning behavior, vagueness and uncertainty need to be considered, and these problems can be solved by fuzzy logic [27]. For this reason, fuzzy logic plays an important role in making systems more accurate. In this experiment, the Felder-Silverman learning style model is used, as this model represents the dimensions of the learning behavior and presents information about the learning characteristics of the specific learner that helps to implement the proposed model. Based on the literature, four inputs and one output were used in this study. They are described in brief below:

Cognitive load (mental load and mental effort): Cognitive load theory [28] is an instructional design theory. It builds on our rapidly expanding knowledge of human perception and is well-suited to providing guidance suggesting which educational technologies are likely to be effective and how they should be used. Cognitive payload refers to an individual's cognitive capacity used to perform a task, learn, or solve a problem [29].

Motivation: In this study, "motivation" refers to the power of an intrinsic state to do work or accomplish a task. The motivational (intrinsic) level varies from learner to learner. It is thought that everyone is born with the power of intrinsic motivation to learn. Motivation has a direct effect on learning behavior [30].

Teaching style: Teachers use strategies to teach learners. Different strategies are recommended by the researchers to apply in teaching. Teaching style refers to the way and strategies the lecturer uses to teach the student. Two major approaches to teaching style are used: the teacher-centered approach and the student-centered approach [31]. Authority, Coach, Activity, and Group styles are the teaching styles that individually indicate one of the teacher-centered approaches or the student-centered approaches. The hybrid teaching style is the combined approach of all teaching styles that are mentioned here.

Emotion: Emotion is a powerful instinctive or intuitive feeling that is distinguished according to surroundings, mood, and relationships with others. Emotions have a significant influence on human cognitive processes, including perception, attention, learning, memory, reasoning, and problem-solving [32].

Learning behavior (output): Learning behavior is the output of this study. For effective student performance, different ways need to be accomplished as individual students' learning styles are different. The way or preference with which a person concentrates, absorbs, and retains new or challenging knowledge or abilities is referred to as their learning style. It can be a positive or negative preference [17]. However, in this research, the Felder-Silverman [18] learning style is considered. According to Felder-Silverman, the learning styles are Active, Reflective, Sensing, Intuitive, Visual, Verbal, Sequence, and Global learning style.

2.3.2 Fuzzy Agent Interaction Model for Learning Behavior

Agents are those who have some particular characteristics. In this study, input and output elements worked as agents. They are correlated with others, and they can observe the environment. They are defined by particular behaviors and some rules. The correlation between the agents is shown in Figure 2.2. To make an intelligent system with a fuzzy agent-based model, it is essential to find out the interaction and connection between the input and output. In a fuzzy expert system, rule generation is very important. Rules control the whole system's behavior. In this study, rules are generated according to the behavior of agents and their influence on other agents. Many rules needed to be established according to the agents used in this study. The consideration of the relationship among the agents to establish rules makes a system more accurate. However, one agent can have an impact on the other. As an example, for emotions, negative emotions decrease cognitive load and positive emotions can increase cognitive load, but when it is the case of anxiety, the mental load can be decreased but the mental effort may be decreased or increased. Emotion also has effects on motivation. Positive emotions increase motivation to learn anything, and negative emotions decrease motivation [33]. For cognitive load, a positive correlation exists between mental load and mental effort. Teaching style can influence the emotions of the learners. Dunn shown that when the teaching mode is through lecture (auditory), visually preferring learners may have difficulty learning, whereas auditory-preferring learners find easy to learn [34].

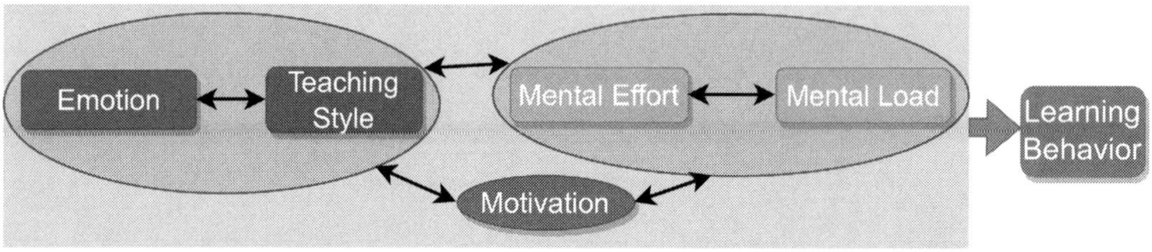

Figure 2.2: Agent interaction model according to the input-output.

The literature supports the establishment of Figure 2.2. The components of cognitive load are mental effort and mental load, and they are positively connected [28]. They are grouped in the interaction model as a result. The learner's emotions during the learning process can be managed by the teacher's style of instruction and vice versa. Emotions and teaching methods are closely associated [35]. Cognitive load, motivation, and learning behavior are all strongly correlated with emotion [32]. Cognitive load [36], learning behavior [30], and emotions [37] exhibit relationships with motivation.

2.3.3 Membership Functions and Fuzzy Rules for the Proposed Model

The main target of this study is to automatically and accurately evaluate an individual's learning behavior. For this purpose, several factors related to learning behavior are studied to examine those factors' impact on students' learning styles. The factors are mental load, mental effort, motivation, teaching style, and emotions, which are used as inputs in this study. Furthermore, learning behavior is used as an output. By analyzing factors related to learning styles, questionnaires are established to gather information from students. Considering the eight-learning styles of the Felder-Silverman learning, eight questions for each input are developed. The questionnaires are given in the appendix section. In the proposed model, membership functions and their ranges are established according to questionnaires and inputs. For mental load, mental effort, and motivation, three membership functions are used, and the range is [0, 5]. For teaching style, five membership functions are used, and the range is [0, 19]. The number of membership functions for emotion is five, and the range is [0, 6.5]. Eight membership functions are also used for output (learning behavior). According to the fuzzy linguistic variable for each input variable, the number of membership functions is determined. The membership functions of mental load and learning behavior are presented in Figure 2.3 and Figure 2.4 respectively.

Each question strongly represents a learning style. It is a very crucial step to find out the accurate membership function for each input agent. A fixed mark for each option of a question is approved according to the other option of that question. By calculating all eight question marks and adding them, we get the crisp value for the input. We check the value represented by the membership function and the intersection point of the membership function and find the fuzzy value for that input. A strong and optimized rule base is established to get accurate results. According to the combination of input-output factors and literature, 130 rules are established to operate the proposed fuzzy agent system. The rules are developed using the agent interaction model. Some rules are listed in Table 2.1.

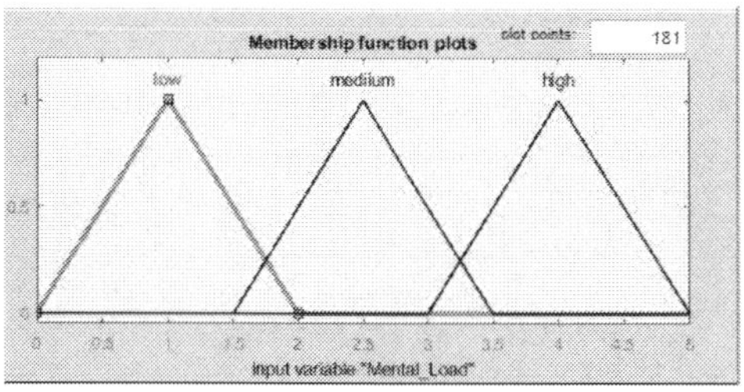

Figure 2.3: Membership function of "Mental Load".

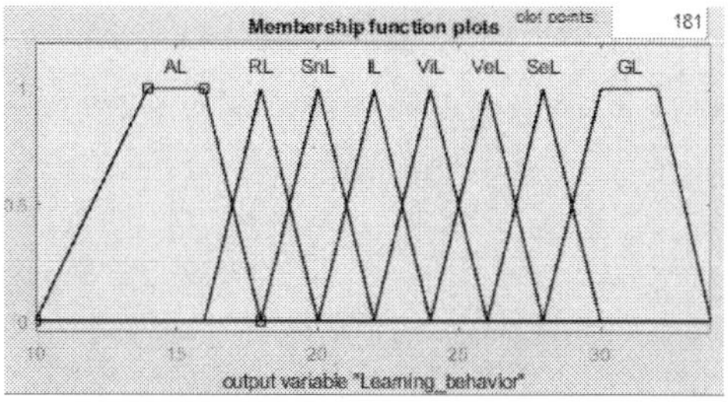

Figure 2.4: Membership function of "Learning Behavior.".

Table 2.1: Rules for evaluating learning styles (example).

NO.	ML	ME	MO	TS	Em	Learning Style
1	M	M	H	AC	E	IL
2	H	H	H	GS	P	GL
3	M	M	M	HS	E	VeL
4	H	H	H	HS	H	GL
5	H	H	H	HS	E	GL

In Table 2.1, ML = Mental load, ME = Mental effort, MO = Motivation, TS = Teaching style, Em = Emotion, LB = Learning behavior, GS = Group style, HS = Hybrid style, CS = Coach style, AC = Activity style, AS = Activity style, VeL = Verbal Learner, GL = Group Learner, SeL = Sequential Learner, ViL = Visual Learner, InL = Intuitive Learner, AL = Active Learner, SnL = Sensing Learner. For emotion, E = Enjoy, P = Pride, H = Hope, and An = Anxiety. For Mental load, Mental effort and Motivation, M = Medium, H = High and L = Low.

2.3.4 Flowchart and Algorithm of Proposed Methodology

The flowchart is a method of showing the sequence of actions. This part shows the flowchart of the proposed method, which is implemented in Matlab.

In this section, the algorithm of the proposed method is presented in Figure 2.5. For each student, 4–9 lines are executed to detect the learning style, and it is continued until finding out the learning style for all.

```
Define MF and its range
Define rules using agent interaction model
for i = 1 to 35(student) do
    for j=1 to 130(rules) do
        Takes crisp value for every input agent
        Calculate fuzzy value for each rule using Fuzzification technique
    end
    Perform aggregation technique
    Calculate crisp value using Centroid defuzzification technique
end
Analysis of the result
Detect individual learning style
End
```

Figure 2.5: Algorithm for the proposed method.

Input: Static dataset
Output: Individual learning style.

The flowchart of the proposed method is presented in Figure 2.6.

2.4 EXPERIMENT, RESULT AND ANALYSIS

The method of implementation is described in this section. The way of gathering and processing data is represented and analyzed here.

We experiment with final-year students from the University of Barishal's computer science and engineering department to measure the effectiveness of our proposal. The attributes of students are collected using questionnaires. Thirty-five students participate in this study and fill out the questionnaires. Data are transformed into crisp values and subsequently into fuzzy values to assess each person's learning style. The dataset is applied to the proposed model and evaluates the learning styles of individuals. The experiment is simulated in the ANFIS toolbox (MATLAB) to evaluate the results. Figure 2.7 represents the simulation of rules. After putting all the inputs in, the simulation will be evaluated according to the rule that was set before. In Figure 2.7, the input is inserted where the mental load is 4.25 (high), the mental effort is 3.125 (medium), the motivation is 4.25 (high), the teaching style is 11.25 (group style), and the emotion is 2.5 (anxiety). Then the output value of 24 means that the learner is a visual learner.

To assess each individual's learning style, a pen-and-paper approach is employed in the dataset. It also helps to examine the accuracy of the proposed system. The learning style for individual students is calculated using the students' information and literature. This result and the output of the proposed model are compared to get an estimate of the accuracy of the proposed model. In this test, it is found that the accuracy is 91%. The percentages of learning styles for 35 students are given below. From Table 2.2, it can be seen that 5.7 percent of the people who participated in the survey are active learners. The majority of students are sequential and verbal learners. As the students who participate in questionnaires are engineering students, the number of students is highest in the visual, verbal, and sequential sections.

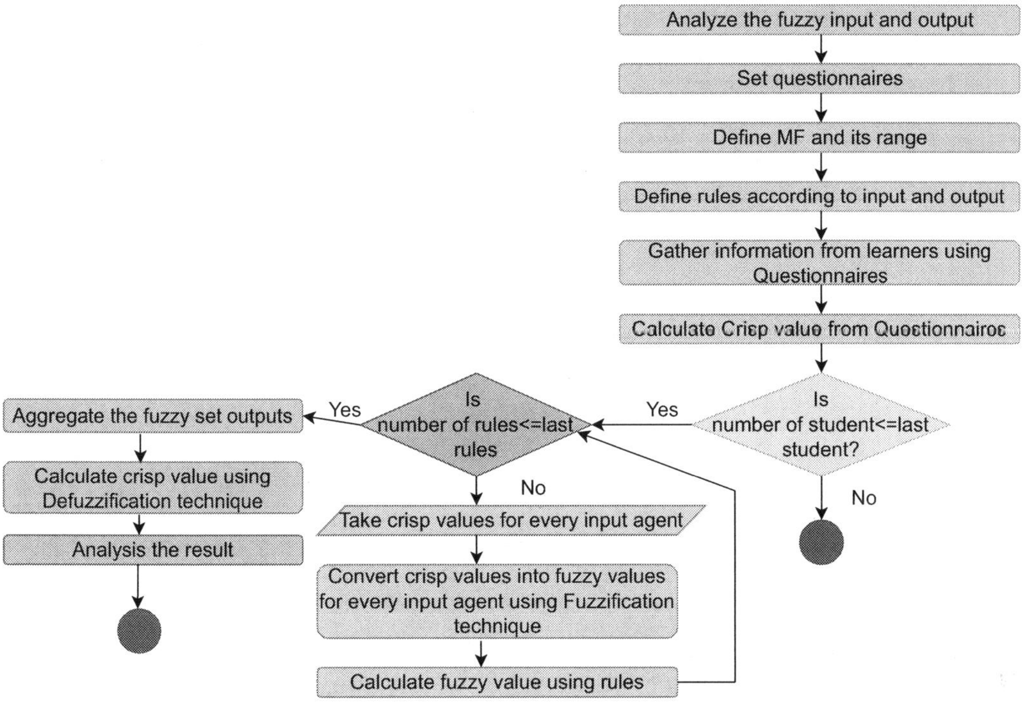

Figure 2.6: Flowchart for developing the proposed method.

This study proposed a model where datasets are collected from the students using questionnaires, and the information is converted into crisp and fuzzy values. The fuzzy values of the input agents are calculated using membership functions, and according to the rules, the learning style of the students is generated in the proposed model. Based on the literature, the methodology of this study is planned, and agent rules are established. In the proposed fuzzy agent-based method, intrinsic motivation is used as an input agent along with other input agents to get accurate results. Moreover, uncertainty and complexity are taken into account to correctly evaluate learning styles.

Table 2.2: Percentage of learning behavior for 35 students from the proposed model.

Learning style	**Percentage(%)**	**Learning style**	**Percentage(%)**
Active	5.7	Visual	40
Reflective	11.4	Verbal	37
Sensing	17	Sequential	34
Intuitive	29	Global	20
Total Accuracy	91		

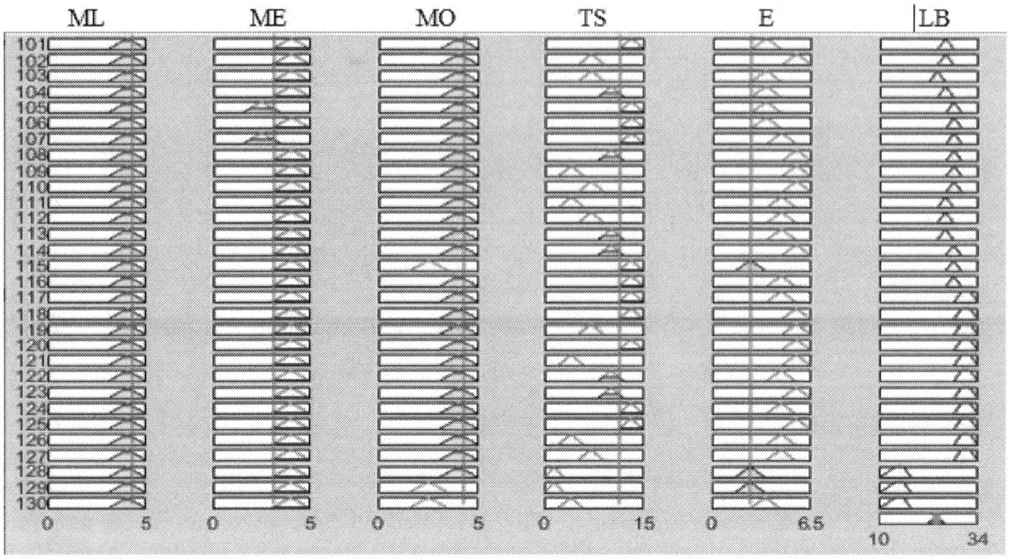

Figure 2.7: Simulation of fuzzy rules for the proposed model.

2.5 CONCLUSION AND FUTURE WORK

In this study, an automatic method is introduced to accurately evaluate the learning styles of individuals using a fuzzy agent-based model. At present, learners are increasing, but productive learning is not expanding. This study helps students understand their effective ways of learning and improve their performance in academic assessments. Human behavior is a complex concept, and it is hard to establish a model to evaluate the learning style. Many researchers persist in finding ways to precisely identify the learning styles of individuals. Fuzzy logic and agent-based modeling are two effective methods that are employed in this work to address the uncertainty and complexity of human behavior. Modeling human behavior as well as learning behavior is challenging because one needs to consider many factors to get an accurate result. Cognitive load, psychological, and sociological elements need to be considered when dealing with human behavior, and in this study, the link between cognitive load, motivation, emotion, teaching style, and learning behavior is taken into consideration. By using these connections, an agent interaction model is developed, which helps to generate fuzzy rules. A real database is used in this experiment, which is collected by the final-year CSE student. A set of questionnaires containing 40 questions is used to extract the required information.

A dynamic learning style detection model using a fuzzy agent-based model will be developed in the future. In future studies, more significant elements that affect the learning behaviors of learners will be considered.

2.6 ACKNOWLEDGEMENT

This work was supported by the Qatar National Library to fulfill the research activities.

2.7 APPENDICES

Questionnaires:

Appendix A. Mental Load

1. If I work in a group and need to answer difficult questions, I will use—to solve the questions?
   (a) little effort
   (b) Medium effort
   (c) High effort
2. In the course that I have taken, it is—to grasp new concepts.
   (a) Easy
   (b) Difficult
   (c) Very hard
3. When I read a book that has lots of written text, it is—to understand.
   (a) Easy
   (b) Difficult
   (c) Very hard
4. If there is a complex problem that I need to solve, I prefer to
   (a) Do easy part
   (b) Do difficult part
   (c) Do very difficult part
5. How many time(s) you will take to read the lecture note to understand any subject?
   (a) 1
   (b) 2
   (c) 3
6. If I see a diagram or sketch during classes, it is—to remember the pictures.
   (a) very easy
   (b) hard
   (c) very hard
7. When I need to answer a question by using some sequential steps, I found that—to solve it
   (a) It is easy
   (b) It is difficult
   (c) It is very difficult
8. When I am studying concrete material (facts, data), I am
   (a) easy to understand and memorize
   (b) easy to understand and difficult to memorize
   (c) difficult to understand and difficult to memorize

Bibliography

[1] Rasheed, F. and Wahid, A. Learning style detection in E-learning systems using machine learning techniques. *Expert Systems with Applications.* **174** p.114774 (2021).

[2] Sheeba, T. and Krishnan, R. Automatic detection of students learning style in learning management system. *In Smart Technologies and Innovation for a Sustainable Future.* (pp. 45–53). Springer, Cham (2019).

[3] Movchun, V., Lushkov, R. and Pronkin, N. Prediction of individual learning style in e-learning systems: opportunities and limitations in dental education. *Education and Information Technologies.* **26(3)**, pp.2523–25 (2021).

[4] Ding, Z., Hu, T., Li, M., Xu, X. and Zou, P.X. Agent-based model for simulating building energy management in student residences. *Energy and Buildings.* **198**, pp.11–27 (2019).

[5] Raoufi, M., and Fayek, A.R. Fuzzy Agent Based Modeling of Construction Crew Motivation and Performance, American Society of Civil Engineers, *Journal of Computing in Civil Engineering.* 32(5) (2018).

[6] Macal, C.M. Everything you need to know about agent-based modelling and simulation. *Journal of Simulation.* **10(2)**, pp.144–156 (2016).

[7] Lee, J.S., Filatova, T., Ligmann-Zielinska, A., Hassani-Mahmooei, B., Stonedahl, F., Lorscheid, I., Voinov, A., Polhill, J.G., Sun, Z. and Parker, D.C. 2015. The complexities of agent-based modeling output analysis. *Journal of Artificial Societies and Social Simulation.* **18(4)** (2015).

[8] Zadeh, L.A. Outline of a new approach to the analysis of complex systems and decision processes. *IEEE Transactions on Systems, Man, and Cybernetics.***(1)**, pp.28–44 (1973).

[9] Hellmann, M. Fuzzy logic introduction. Université de Rennes, 1(1) (2001).

[10] Vashishtha, S. and Susan, S. Fuzzy rule based unsupervised sentiment analysis from social media posts. *Expert Systems with Applications.* **138**, p.112834 (2019).

[11] Dumitrescu, C., Ciotirnae, P. and Vizitiu, C. Fuzzy logic for intelligent control system using soft computing applications. *Sensors.* **21(8)**, p.2617 (2021).

[12] Mittal, K., Jain, A., Vaisla, K.S., Castillo, O. and Kacprzyk, J. A comprehensive review on type 2 fuzzy logic applications: Past, present and future. *Engineering Applications of Artificial Intelligence.* **95**, p.103916 (2020).

[13] Felder, R.M. and Soloman, B.A. Learning styles and strategies. (2000).

[14] Honey, P. and Mumford, A. The manual of learning styles Maidenhead. P Honey. (1992).

[15] Felder, R.M. Matters of style. *ASEE Prism.* **6(4)**, pp.18–23 (1996).

[16] James, W.B. and Gardner, D.L. Learning styles: Implications for distance learning. *New directions for adult and continuing education.* **67**, pp.19–31 (1995).

[17] Dunn, R. Learning style and its relation to exceptionality at both ends of the spectrum. *Exceptional children.* **49(6)**, pp.496–506 (1983).

[18] Felder, R.M. and Silverman, L.K. Learning and teaching styles in engineering education. *Engineering education.* **78(7)**, pp.674–681 (1988).

[19] Graf, S., Liu, T.C., Chen, N.S. and Yang, S.J. Learning styles and cognitive traits–Their relationship and its benefits in web-based educational systems. *Computers in Human Behavior.* **25(6)**, pp.1280–1289 (2009).

[20] Van Zwanenberg, N., Wilkinson, L.J. and Anderson, A. Felder and Silverman's Index of Learning Styles and Honey and Mumford's Learning Styles Questionnaire: how do they compare and do they predict academic performance? *Educational Psychology.* **20(3)**, pp.365–380 (2000).

[21] El Aissaoui, O., El Alami El Madani, Y., Oughdir, L. and El Allioui, Y. A fuzzy classification approach for learning style prediction based on web mining technique in e-learning environments. *Education and Information Technologies.* **24**, pp.1943–1959 (2019).

[22] Zhang, H., Huang, T., Liu, S., Yin, H., Li, J., Yang, H. and Xia, Y. A learning style classification approach based on deep belief network for large-scale online education. *Journal of Cloud Computing.* **9**, pp.1–17 (2020).

[23] Li, L.X. and Abdul Rahman, S.S. Students' learning style detection using tree augmented naive Bayes. *Royal Society Open Science.* **5(7)**, p.172108 (2018).

[24] El Fazazi, H., Samadi, A., Qbadou, M., Mansouri, K. and Elgarej, M. A learning style identification approach in adaptive e-learning system. *In Information Systems and Technologies to Support Learning: Proceedings of EMENA-ISTL 2018.* **2**, pp. 82–89. Springer International Publishing. (2019).

[25] Brosso, I. and La Neve, A. Adaptive security policy using user behavior analysis and human elements of information security. INTECH Open Access Publisher. (2012).

[26] Ghosh, A. and Huang, S. April. Visual Search as a Probabilistic Sequential Decision Process in Software Autonomous System. *In SoutheastCon 2018.* (pp. 1–7). IEEE. (2018).

[27] Zadeh, L.A. Fuzzy logic—a personal perspective. Fuzzy sets and systems, **281**, pp.4–20 (2015).

[28] Sweller, J. Cognitive load theory. *In Psychology of learning and motivation.* **55**, pp.37–76. Academic Press.(2011).

[29] Krell, M. and Hui, S.K.F. Evaluating an instrument to measure mental load and mental effort considering different sources of validity evidence. *Cogent Education.* 4(1), (2017).

[30] Tokan, M.K. and Imakulata, M.M. The effect of motivation and learning behaviour on student achievement. *South African Journal of Education.* **39(1)** (2019).

[31] Shawer, S.F. Teacher-driven curriculum development at the classroom level: Implications for curriculum, pedagogy and teacher training. *Teaching and Teacher Education.* **63**, pp. 296–313 (2017).

[32] Tyng, C.M., Amin, H.U., Saad, M.N. and Malik, A.S. The influences of emotion on learning and memory. *Frontiers in Psychology.* p.1454 (2017).

[33] Plass, J.L. and Kalyuga, S. Four ways of considering emotion in cognitive load theory. *Educational Psychology Review.* **31(2)**, pp.339–359 (2019).

[34] Renou, J. A study of perceptual learning styles and achievement in a university-level foreign language course. *Criso Lenguas.* **1(2)**, pp.1–15 (2008).

[35] Sanchez-Martin, J. and Dávila-Acedo, M.A. Emotional Responses to Innovative Science Teaching Methods: Acquiring Emotional Data in a General Science Teacher Education Class. *Journal of Technology and Science Education.* **8(4)**, pp.346–359 (2018).

[36] Stiller, K.D. and Schworm, S. Game-based learning of the structure and functioning of body cells in a foreign language: effects on motivation, cognitive load, and performance. *In Frontiers in Education.* **4**, p. 18. Frontiers Media SA. (2019 March).

[37] Lee, M.K., Cheung, C.M. and Chen, Z. Acceptance of Internet-based learning medium: the role of extrinsic and intrinsic motivation. *Information & Management.* **42(8)**, pp.1095–1104 (2005).

CHAPTER 3

Quantifying Shifts in Word Contexts from Social Media Data

Kawshik Banerjee

Rajshahi University of Engineering & Technology, Dept. of Computer Science & Engineering, Rajshahi, Bangladesh. Email: k.banerjee2024@gmail.com ***(Corresponding Author)***

Azmain Yakin Srizon

Rajshahi University of Engineering & Technology, Dept. of Computer Science & Engineering, Rajshahi, Bangladesh. Email: azmainsrizon@gmail.com

Md. Farukuzzaman Faruk

Rajshahi University of Engineering & Technology, Dept. of Computer Science & Engineering, Rajshahi, Bangladesh. Email: faarukuzzaman@gmail.com

S. M. Mahedy Hasan

Rajshahi University of Engineering & Technology, Dept. of Computer Science & Engineering, Rajshahi, Bangladesh. Email: mahedycseruet@gmail.com

Md. Rakib Hossain

Rajshahi University of Engineering & Technology, Dept. of Electronics & Telecommunication Engineering, Rajshahi, Bangladesh. Email: rakib.ete12.ruet@gmail.com

CONTENTS

DOI: 10.1201/9781003605508-3

Words denote different things in different contexts. But measuring the differences and concluding whether such differences are statistically significant or not is a big computational challenge. Previous methods either needed larger datasets or were computationally intensive. But in this study, following a relatively simpler approach on a self-curated dataset of about 4 million tweets from the top 300 universities in the USA, such issues have been addressed. Using regression on embedded vectors of words, it is shown here that, after the mass protest defending the rights of the Black people in the USA in 2018, there has been a statistically significant ($p < 0.01$) change in the usage of racial and ethnic stereotypical terms like 'black' and 'race,' in comparison to trivial regular terms like 'excited' and 'welcome' in the social media. It is also shown how a word's association with other words can largely vary across time. Such methods can be used to address other problems in the field of computational linguistics in the future.

3.1 INTRODUCTION

Words, as we know them, can convey different meanings depending on the situation. Without a clear and similar understanding of words on both ends, proper communication cannot occur. As for humans the easier it seems to learn the meaning of a word or phrase, the more complex it is for machines to get a full grasp of it. We, humans, use combinations of symbols to form words and subsequently form sentences from words. Afterwards, these sentences are what we use. But machines are different. Machines do not use or incorporate these, the only thing they know is whether the electrical supply is enabled or disabled. Such states are expressed in the form of numbers. Thus, a potential solution to make computers to work with human language is to map human-understandable text data into numbers in some forms of mathematical modeling. Such was the idea behind the concept of word embedding [1, 2].

The core idea behind this was to assign a vector for each of the words [2]. As this was based on the idea of neural network-based language models from [1]. Initially the words were mapped into a multidimensional space, where each coordinate denoted a particular word. After successfully mapping each unique word into unique coordinates, each of the coordinate values was transformed into a vector. These vectors were used to work on later. Essentially this way, a large collection of words could be represented as a pool of 'word vectors'. Afterwards, these vectors and their relationship could be used to infer important meanings from the text, as shown in [3]. Although [1] worked efficiently, it required much larger datasets to proceed with.

Their success led other researchers to apply these methods for different tasks but was somewhat thwarted where enough data was not available [4] or the dataset was either too sparse or not feature-rich enough [5]. However, the primary interest of our present study is to measure how the discrepancy in word vector values would affect a difference in understanding or usage of the text. Additionally, we want to conclude how other factor would affect it. Using traditional word embedding methods alone we cannot conclude whether the numeric mapping of a certain term will remain similar for other instances of similar contexts and different for different contexts. To exemplify, in traditional word embedding, without considering the concept of 'context,' the vector of 'wood' in the sentence "Chopping off wood is not recommended" will be same as that in "The nearby wood is pretty dense." But the meaning of 'wood' is different in the sentences, and so should their numeric mappings be.

A significant difference in mapping is expected for a difference in context. As in real life, we have a difference in the meaning of the very same word in different contexts. A cricketer and a musician would differently interpret the term 'pitch'. While one would think of a flat surface on the ground, another would delve into an auditory level. Such different understandings or interpretations of the very same word lead to distinct vector representations across the corpus for different instances and such differences are dependent on the way context is classified [6].

To deal with contexts researchers had previously sought after topic modeling [7, 8]. The method shown in [7] works well for retrieving theme or the core idea behind a text. Along with themes, we can also find how mass people express their sentiments and views, as shown in [8]. But this method itself is not of much help when it comes to inferring word-level context and their differences [9, 10]. Other limitations include not being able to infer meanings when topics are correlated [11].

Following morphological [12] as well as contextual methods [13, 14], more recent studies have attempted to address this issue of word-level context using word embedding [15, 16, 17]. In [15] we see quantifying changes in the attitudes towards historically marginalized people in our community: women and ethnic minority people. Although their work is significant, it does not address whether this shifts are statistically significant or not. In [16] time-varying shifts are analyzed using relatively smaller data. Representations like these are helpful to extract meaningful information from the data. But contextual representations on their own are not much informative when it comes to statistical inference. For example, depending on the context, a certain focal term's representation will be different. But can we quantify how dissimilar these representations are? Also, we have no means to statistically conclude whether these dissimilarities are caused by any specific factor. Present embedding models are inherently non-interpretable, and other additional measures are needed to tackle these issues [18, 19].

In this study, these issues have been attempted to tackle using word embedding and regression analysis. Following the works of [17, 20], a regression equation is estimated on our self-curated social media corpus to test the statistical significance of differences in contexts affected by a confounding factor. A self-curated dataset of tweets has been used here for all the analysis. Current work conclusively shows that, after the mass protest in 2018 for the rights of racially discriminated Black people in the USA, namely the 'Black Lives Matter' movement, there has been a statistically significant ($p < 0.01$) change in the notion of ethnic stereotypical words such as 'black' and 'race,' in the social media compared to regular words; whereas previously such terms were more associated with color and competitions, now it is largely associated with community and heritage. Details of the study are explained in the following sections.

3.2 MATERIALS AND METHODS

We have incorporated a relatively less computationally extensive approach for the analysis of a large corpus. In this section, the data used and the methodology followed are discussed in detail.

3.2.1 Dataset Description

In this analysis, a novel self-curated dataset has been used. A total of 3,911,863 tweets were collected. All of these were tweeted from January 01, 2010, till December 31, 2022. All the tweets were from the official Twitter accounts of the top 300 universities in the USA. For the university list, US News 2022-23 ranking [21] has been followed. All the universities were classified as 'National University.' A Python package named 'snscrape' has been used to collect the tweets.

Our selected tweets were from 2010 to 2022. One reason for specifying this time frame for the collected tweets was to capture the tweeting trend before (2010-17), during (2018), and after (2019-22) the mass protest defending the rights of the Black community in the USA. By searching for tweets containing relevant keywords (like, 'race' and 'black') and plotting their frequency (Figure 3.1) throughout the years, we see a surge of tweets during and after the movement.

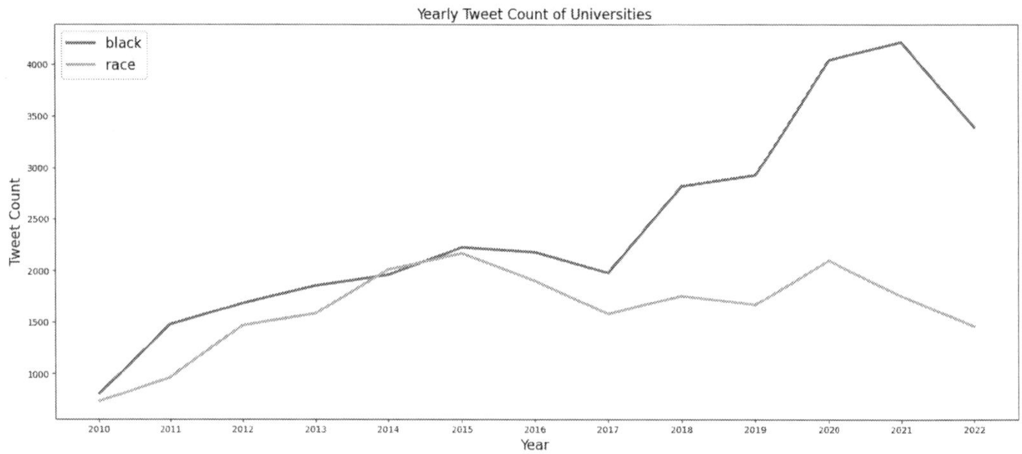

Figure 3.1: Yearly count of tweets containing keywords 'race' and 'black'

Before proceeding with the analysis, data was pre-processed. As part of the data pre-processing we have incorporated removing punctuation marks, non-alphanumeric characters, URLs, and special characters.

3.2.2 Methodology

In this study, the changes in the usage of some contextually relevant terms are analyzed with respect to changes in some other factor groups, and afterwards, it has been compared with the same for some other irrelevant terms for validation and comparison. Essentially, here it is shown how much the usage of racial and ethnic stereotypical terms (like 'race' and 'black') has changed by the administrative authorities in the social media because of the protest, in comparison to the usage of other regular words (like 'excited' and 'welcome').

For such comparison, two confounding groups have been selected: one that denotes whether a certain tweet is after the protest (year 2019 or later) or not, and the other denoting the category of the university from which it was tweeted (in our case, all universities are top-ranked national universities). The objective here is to measure the change in usage of words due to the protest, controlling for other confounding factor—category.

Following the promising leads of previous works [17, 20], a regression equation is estimated as follows:

$$Y = \beta_0 + \beta_1 \cdot Protest + \epsilon \tag{3.1}$$

where β_0 is the intercept term of the regression, β_1 is the regression coefficient, ϵ is the error term, and *Protest* is an independent variable. This *Protest* variable denotes '1' for all the mentions of the target terms in the tweets tweeted in 2019 and later years and '0' for all the other mentions of the target terms. Since in the dataset only time and university type are present except for the tweets, it can be evidently said from equation 3.1 that the outcome Y is solely dependent on time, controlling for the university category.

The idea behind such regression is somewhat straightforward, as first incorporated in [20] and extended by [17]: the use of simple linear regression on embedded vector representation of words. For example, assuming each mention of the target term's embedding vector v_{term_i} is of M dimension if all of such vectors are combined together, an $n \times M$ dimensional dependent variable Y can be measured, where n is the number of mentions of the target terms.

Now the question remains: how to obtain the initial embedding of any particular instance of the target term v_{term_i} and the embeddings of the terms around it, in other words, the embeddings of the 'context.' A pre-trained GloVe model [17] has been used to obtain a certain transformation matrix for further processing the corpus. Subsequently, this transformation matrix is used to down-weight the non-important and too frequent terms and terms that are not much meaningful at all [20]. Because of this additional layer of processing, a significant change than previous is observed in the embedding of the same words in different contexts. The overall workflow of this study is shown in Figure 3.2.

3.3 EXPERIMENT AND RESULT

In this section, we begin by outlining the environment specifications. This is followed by a discussion on feature co-occurrence and the measurement of similarity. Lastly, we delve into the explanation of regression analysis.

3.3.1 Environment Specification

Data collection and pre-processing steps were executed using Python (version 3.8) on an Intel®Core^{TM} i3-6006U machine running on Windows 10. All the other steps were

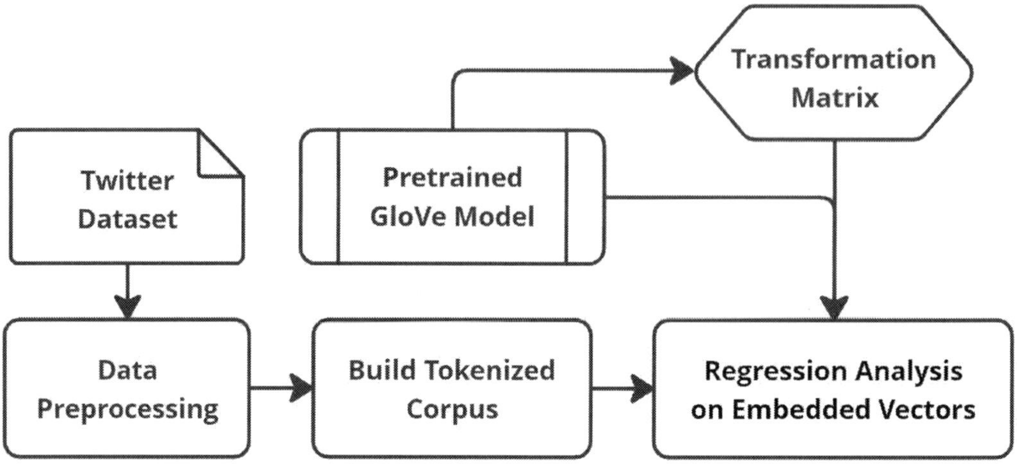

Figure 3.2: Research workflow

executed using the cloud platform freely provided by Kaggle without any hardware acceleration, pinned always on the latest notebook environment. All the used libraries were updated to the latest version at the time of the complete study.

3.3.2 Feature Co-occurrence

To observe how the protest affected the usage pattern of our target terms, a feasible and useful exploratory way could be to check the co-occurrence of the target terms with other co-occurring terms. The feature co-occurrence plots in Figures 3.3 and 3.4 show with other terms, the target term ('black') had been mostly associated and appeared together in tweets before (Figure 3.3) and after (Figure 3.4) the protest.

From Figures 3.3 and 3.4 it can be observed that in the last four years universities tweeted almost as much as they did in the previous nine years. Additionally, in later years, after the protest, we find new terms like 'anti-racism,' 'equity,' 'diversity,' 'voices,' 'together' and so on associated with the target word 'black.' Since we don't see these newer terms before the protest, we can conclusively say this positive addition of such terms is solely due to the movement.

3.3.3 Similarity Measurement

Here it is shown how much the embedding of our target terms, 'black' and 'race' are similar to that of 'color,' 'community,' and 'competition' when compared group-wise. From Table 3.1, it can be seen that before the protest, 'black' was more associated with the notion of 'color,' whereas after the protest the scenario reversed: 'black' was more similar to 'community' than it was for 'color.'

A similar trend is observed when 'race' was compared with 'competition' and 'community' showcased in Table 3.2. Although before the protest, the embedding of 'race' had almost equal similarity with that of 'community' and 'competition,' the

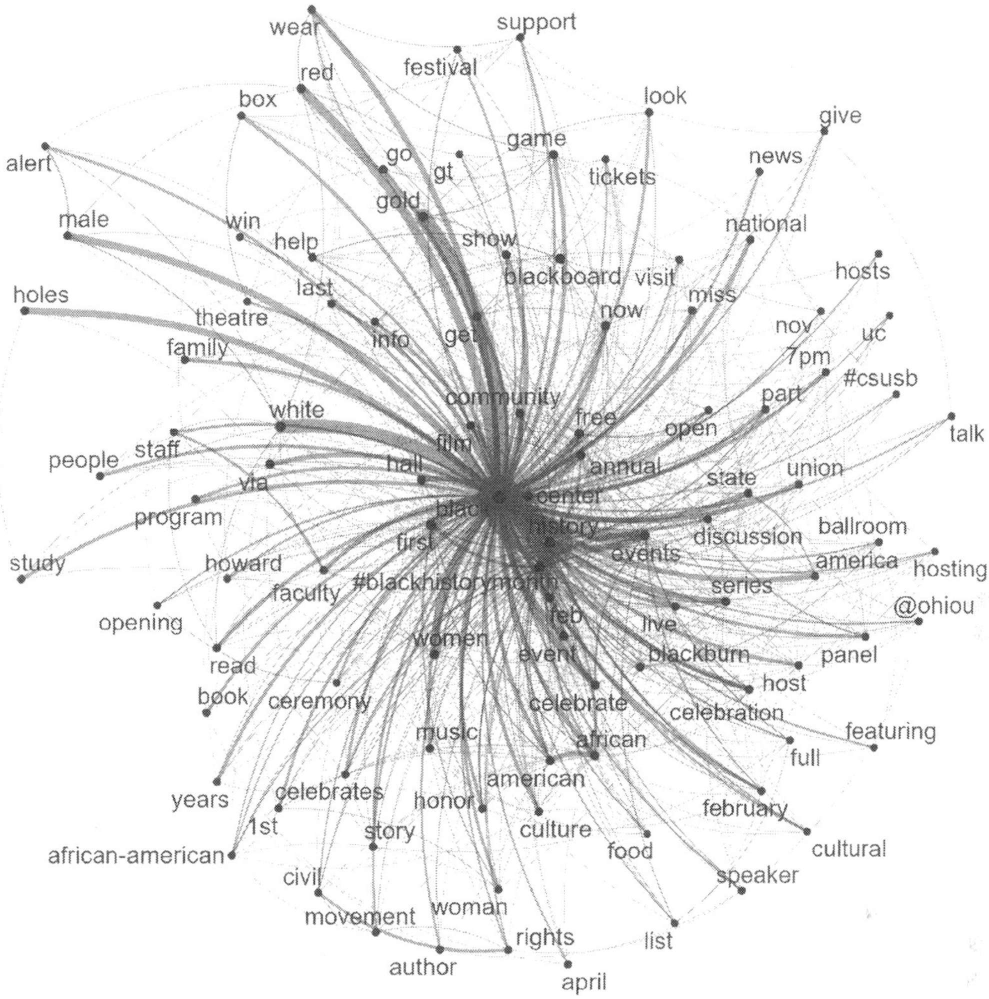

Figure 3.3: Before protest, year 2010-17: 17,628 tweets

difference becomes stark after the protest. Whereas initially, before the protest, the similarity difference was less than 1%, after the protest it rose to almost 16%. In other words, we can conclude, while the term 'race' was used in almost same degree of similarity as 'competition' and 'community,' the usage of 'race' became more and more similar to that of 'community' than it did to that of 'competition.'

For the similarity measurement, the cosine similarity method has been applied, with 500 bootstraps. Essentially, in the vector space, the cosine distance between the embedding of the target group (mentions of 'black' and 'race' in tweets before and after the movement) and that of the features ('color,' 'competition', and 'community') have been measured.

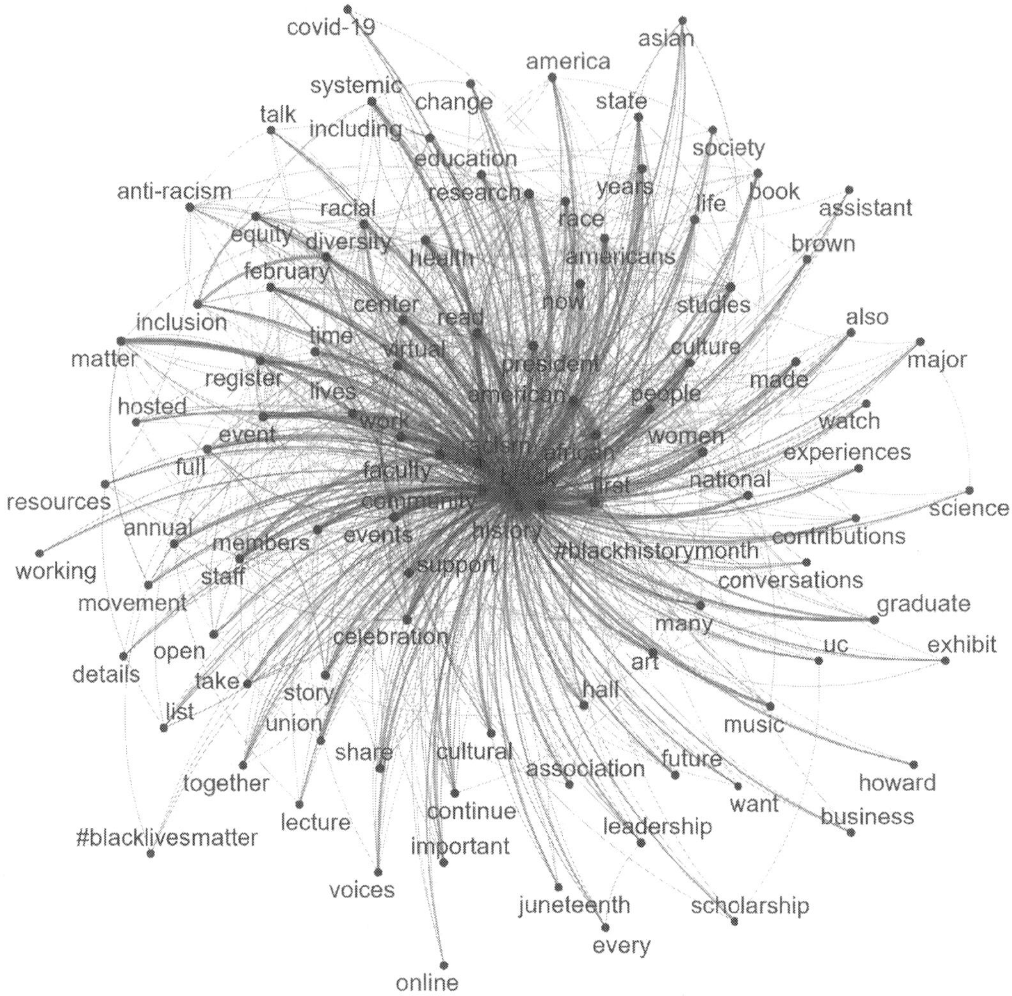

Figure 3.4: After protest, year 2019-22: 16,937 tweets

Table 3.1: Cosine similarity between each time-group's embedding for 'black' and certain features

Group	Feature	Similarity	Std. Error
Before protest	**color**	**25.4%**	0.0043
	community	22.1%	0.005
After protest	color	31%	0.0035
	community	**42.5%**	0.0039

Table 3.2: Cosine similarity between each time-group's embedding for 'race' and certain features

Group	Feature	Similarity	Std. Error
Before protest	competition	22.01%	0.0079
	community	**22.88%**	0.0081
After protest	competition	15.72%	0.0080
	community	**31.8%**	0.0079

3.3.4 Regression Analysis

Finally, it can be shown by incorporating equation 3.1 that in the tweets posted after the protest, in 2019 and later, the usage or context of racial and ethnic stereotypical terms, like 'black' and 'race' are significantly different (at 0.01 level) than those before or during 2018. When compared against regular trivial terms, like 'welcome' or 'excited,' a larger value in the norm of β was observed for 'black' and 'race,' relative to those for 'welcome' and 'excited.'

Figure 3.5 shows the norms of βs ($p < 0.01$) for our target terms - 'race,' 'black,' 'excited,' and 'welcome,' along with a 95% confidence interval.

The larger coefficient value for 'black' and 'race' essentially denotes a larger change in the context in the years 2019 onwards, whereas relatively much less of such difference is observed for trivial terms like 'excited' and 'welcome.'

3.4 CONCLUSION AND FUTURE SCOPE

Human language is constantly varying. Quantifying these variations across time and stating their significance level is a challenging task. Setting a regression framework on embedded vectors is an efficient approach, as shown in the literature [17]. In this study the very same is applied in a different setting: to quantify time-varying changes in the usage of words using social media data, which can be a powerful tool in these settings.

In this study we have used a self-curated dataset of tweets from the official accounts of the top 300 US universities, and incorporated a linear regression setting to test the effect of one confounding group controlling for another. Thus it is shown in this study that, after the mass protest in the year 2018 defending the rights of the Black community residing in the US, there has been a statistically significant change ($p < 0.01$) in the usage or context of the terms 'black' and 'race' in comparison with other trivial terms like 'excited' and 'welcome' in our day to day use in the social media.

Additionally, it is also shown how the protest has contributed to words' association over the years. In our study, we have observed that the 'black' word was more associated with the notion of 'color' before the protest, whereas afterward it was largely

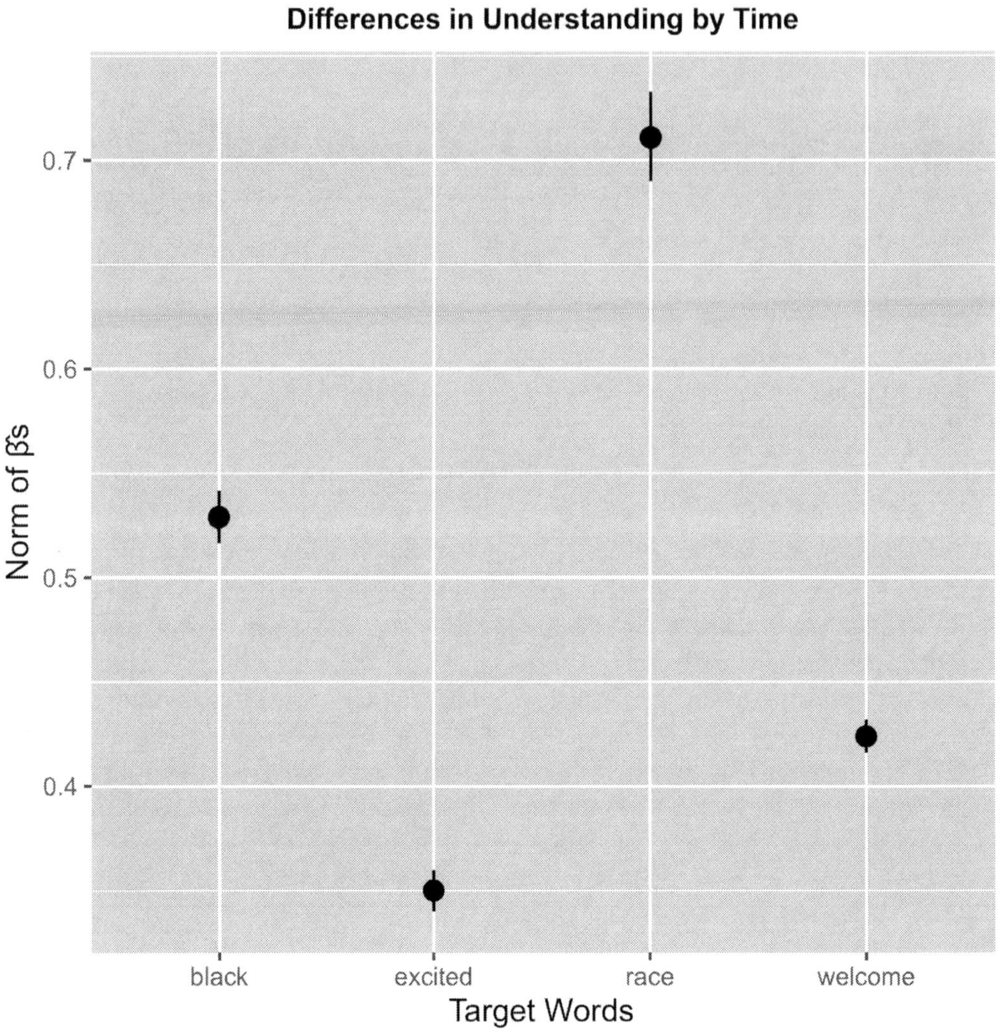

Figure 3.5: Difference in the understanding of words over the years

associated with 'community.' Similarly, 'race' became broadly more associated with 'community' than with 'competition' afterwards.

Detecting the meaning of a word in different contexts is a challenging task. This study aims to address this issue. Although efficient approaches such as this can be applied in the future to address other issues, including but not limited to word polysemy and homonymity in our regular language, there is much room for improvement yet.

Bibliography

[1] Bengio, Y., Ducharme, R. & Vincent, P. A neural probabilistic language model. *Advances In Neural Information Processing Systems.* **13** (2000)

[2] Mikolov, T., Chen, K., Corrado, G. & Dean, J. Efficient estimation of word representations in vector space. *ArXiv Preprint ArXiv:1301.3781.* (2013)

[3] Caliskan, A., Bryson, J. & Narayanan, A. Semantics derived automatically from language corpora contain human-like biases. *Science.* **356**, 183–186 (2017)

[4] Adams, O., Makarucha, A., Neubig, G., Bird, S. & Cohn, T. Cross-Lingual Word Embeddings for Low-Resource Language Modeling. *Proceedings Of The 15th Conference Of The European Chapter Of The Association For Computational Linguistics: Volume 1, Long Papers.* pp. 937–947 (2017,4), https://aclanthology.org/E17-1088

[5] Lazaridou, A., Marelli, M. & Baroni, M. Multimodal word meaning induction from minimal exposure to natural text. *Cognitive Science.* **41**, pp. 677–705 (2017)

[6] Soper, E. & Koenig, J. When Polysemy Matters: Modeling Semantic Categorization with Word Embeddings. *Proceedings Of The 11th Joint Conference On Lexical And Computational Semantics.* pp. 123–131 (2022,7), https://aclanthology.org/2022.starsem-1.10

[7] Blei, D. Probabilistic Topic Models. *Commun. ACM.* **55**, 77–84 (2012,4), https://doi.org/10.1145/2133806.2133826

[8] Satu, M., Khan, M., Mahmud, M., Uddin, S., Summers, M., Quinn, J. & Moni, M. TClustVID: A novel machine learning classification model to investigate topics and sentiment in COVID-19 tweets. *Knowledge-Based Systems.* **226**, 107126 (2021)

[9] Grimmer, J. & Stewart, B. Text as data: The promise and pitfalls of automatic content analysis methods for political texts. *Political Analysis.* **21**, 267–297 (2013)

[10] Roberts, M., Stewart, B. & Airoldi, E. A model of text for experimentation in the social sciences. *Journal Of The American Statistical Association.* **111**, 988–1003 (2016)

[11] Li, S., Pan, R., Luo, H., Liu, X. & Zhao, G. Adaptive cross-contextual word embedding for word polysemy with unsupervised topic modeling. *Knowledge-Based Systems.* **218** pp. 106827 (2021)

[12] Pad'o, S., Herbelot, A., Kisselew, M., Snajder, J. & Others Predictability of distributional semantics in derivational word formation. *Proceedings Of The 26th International Conference On Computational Linguistics (COLING 2016).* pp. 1285–1296 (2016)

[13] Herbelot, A. & Baroni, M. High-risk learning: acquiring new word vectors from tiny data. *ArXiv Preprint ArXiv:1707.06556.* (2017)

[14] Lazaridou, A., Marelli, M. & Baroni, M. Multimodal word meaning induction from minimal exposure to natural text. *Cognitive Science.* **41**, 677–705 (2017)

[15] Garg, N., Schiebinger, L., Jurafsky, D. & Zou, J. Word embeddings quantify 100 years of gender and ethnic stereotypes. *Proceedings Of The National Academy Of Sciences.* **115**, E3635-E3644 (2018)

[16] Rodman, E. A timely intervention: Tracking the changing meanings of political concepts with word vectors. *Political Analysis.* **28**, 87–111 (2020)

[17] Rodriguez, P., Spirling, A. & Stewart, B. Embedding regression: Models for context-specific description and inference. *American Political Science Review.* pp. 1–20 (2023)

[18] Kulkarni, V., Al-Rfou, R., Perozzi, B. & Skiena, S. Statistically Significant Detection of Linguistic Change. *Proceedings Of The 24th International Conference On World Wide Web.* pp. 625–635 (2015), https://doi.org/10.1145/2736277.2741627

[19] Lauretig, A. Identification, Interpretability, and Bayesian Word Embeddings. *Proceedings Of The Third Workshop On Natural Language Processing And Computational Social Science.* pp. 7–17 (2019,6), https://aclanthology.org/W19-2102

[20] Khodak, M., Saunshi, N., Liang, Y., Ma, T., Stewart, B. & Arora, S. A La Carte Embedding: Cheap but Effective Induction of Semantic Feature Vectors. *Proceedings Of The 56th Annual Meeting Of The Association For Computational Linguistics (Volume 1: Long Papers).* pp. 12–22 (2018,7), https://aclanthology.org/P18-1002

[21] U.S. News & World Report (n.d.). *Search U.S. News Best Colleges.* Retrieved April 25, 2023, from https://web.archive.org/web/20230425140921/https://www.usnews.com/best-colleges/search?_sort=rank&_sortDirection=asc

BanglaOngko: A New Dataset for Accurate Bengali Mathematical Expression Detection Utilizing YOLOv8 Architecture

Soumit Das

Rajshahi University of Engineering & Technology, Dept. of Computer Science & Engineering, Rajshahi, Bangladesh. Email: somudasctgbn@gmail.com, Corresponding author.

Ashraf Ul Alam

Rajshahi University of Engineering & Technology, Dept. of Computer Science & Engineering, Rajshahi, Bangladesh. Email: ashrafamit9227@gmail.com

Sudipta Progga Islam

Rajshahi University of Engineering & Technology, Dept. of Computer Science & Engineering, Rajshahi, Bangladesh. Email: proggasudipta0@gmail.com

Azmain Yakin Srizon

Rajshahi University of Engineering & Technology, Dept. of Computer Science & Engineering, Rajshahi, Bangladesh. Email: azmainsrizon@gmail.com

CONTENTS

DOI: 10.1201/9781003605508-4

This chapter proposes an efficient Optical Character Recognition (OCR) methodology that utilizes the latest object detection algorithm, You Only Look Once version 8 (YOLOv8), to identify handwritten Bengali digits and operators. A *new dataset*[1] called "BanglaOngko" has been introduced, containing 21 classes of Bengali digits and operators with annotated bounding box information for object localization. The proposed YOLOv8 model was trained on this dataset and achieved an average mAP (mean average precision) of 92.0%. However, YOLOv8 only provides localization data for detected objects, lacking a sorted list of bounding boxes. As a result, directly converting a handwritten line to a printed line poses a challenge. To address this, an efficient algorithm incorporating statistical concepts has been developed. With the proposed approach, precise localization of every character within an image can be achieved, and its position can be leveraged to identify the line where the digit or sign belongs. The proposed methodology can convert handwritten mathematical equations into digital text, improving accessibility and efficiency for tasks such as educational and scientific writing. The approach proposed in this research contributes significantly to the development of advanced OCR systems, particularly for Bengali language.

4.1 INTRODUCTION

The field of computer vision has undergone a revolution with the advancements in deep learning, particularly in Optical Character Recognition (OCR) which involves recognizing printed or handwritten text and converting it into machine-readable format [10]. This research article proposes an efficient OCR methodology that utilizes the latest object detection algorithm, You Only Look Once version 8 (YOLOv8), to identify handwritten Bengali digits and operators. YOLOv8 is an anchor-free model that directly predicts the center of an object rather than offsetting it from a predefined anchor box. With this approach, precise localization of every character within an image can be achieved, and its position can be leveraged to identify the line where the digit or sign belongs. However, it lacks a sorted list of bounding boxes, making the direct conversion from handwritten to printed lines challenging.

To overcome this challenge, we developed a statistical algorithm that efficiently converts handwritten expressions to printed ones, preserving their original structure. Leveraging YOLOv8's character position predictions, our algorithm overcomes YOLOv8's limitations, ensuring accurate and efficient detection and recognition of

[1]BanglaOngko Dataset: `https://github.com/ashraf-ul-alam-amit/BanglaOngko`

Bengali handwritten characters. This advancement in OCR technology marks a significant milestone in the development of advanced systems for recognizing and localizing Bengali handwritten characters.

In recent years, there has been significant research dedicated to recognizing handwritten Bengali characters and digits. This is because Bengali is the seventh most spoken language globally, with over 300 million speakers [13]. The development of a reliable system for recognizing and localizing Bengali handwritten characters is a challenging yet important task in enhancing OCR technology and facilitating communication and accessibility for Bengali speakers. In this work, Bengali digits, parentheses, and various mathematical operators such as +, -, x, /, etc. were considered. By analyzing the position of each character, the model is capable of converting handwritten mathematical equations into digital text. This task holds immense importance in several fields, including education, science, and engineering.

Previous researches have mostly focused on recognizing handwritten Bengali digits through object classification without considering localization information or multiple objects present in an image. To overcome these issues, a new dataset called "BanglaOngko" has been introduced, containing 21 classes of Bengali digits and operators with annotated bounding box information for object localization. This dataset can be used to train and test object detection models that more accurately recognize and locate handwritten Bengali digits and operators. It will eventually accelerate the advancement of OCR technology for Bengali literature. Details of the proposed methodology can be found in the materials and methodology section.

4.2 LITERATURE REVIEW AND BACKGROUND STUDY

This section is divided into three parts: previous works related to handwritten digit recognition, text localization, and mathematical expression recognition are discussed.

4.2.1 Previous Works on Handwritten Digit Recognition

Yuli Wu et al. utilized a Cascade R-CNN network for handwriting localization in their object detection-based approach [14]. However, the dataset consisted of both handwritten and printed data. Kavallieratou et al. developed a system for handwritten text localization in document images that involved skew angle correction and localization of handwritten areas based on shape and dimension regularity measures [7]. Another model was introduced that used a customized Faster-RCNN consisting of region of interest annotation, an improved Faster-RCNN with DenseNet-41, and regression and classification layers for digit localization and classification [1].

4.2.2 Previous Works on Handwritten Text Localization

Significant advancements have been made in localizing handwritten text in recent years. Dutta et al. utilized EAST [16] and Textboxes++ [8] methods for precise text localization in lecture video frames [4]. A modified region-growing approach was introduced to tackle text localization in video frames [9]. Zhou et al. [15] proposed an algorithm combining contour analysis, connected component analysis, and geometric properties for accurate identification and localization of characters in handwritten text. A system using the Wigner-Ville distribution successfully corrected skew angles and localized handwritten areas in documents [7]. Huang et al. [5] introduced a method combining stroke feature transform and text covariance descriptors for text localization.

4.2.3 Previous Works on Bengali Mathematical Expression Recognition

Previous research on Bengali handwritten digit recognition relied on conventional machine learning techniques such as k-nearest neighbors (KNN) [3] and support vector machines (SVM) [2]. BengaliNet, a cost-effective convolutional neural network architecture, was proposed for recognizing Bengali characters [12]. The model had a parameter range of 2.24 to 2.43 million and surpassed the performance of prior works on Bengali character datasets. A novel tri-level segmentation approach was proposed for Optical Character Recognition (OCR) of Indic scripts, with a specific focus on Bangla [11].

4.2.4 Advantages of Proposed Methodology Compared to Previous Works

Previous studies on Bengali handwritten character recognition have mainly focused on recognizing a single character or digit using various machine learning techniques along with morphological and graph-based methods. However, these methods are not suitable for detecting multiple digits and their positions from a single image. In contrast, multiple digits can be detected with their coordinates from an image by our proposed method, which eliminates the need for an additional line detection machine learning method. Based on these coordinates, the lines to which the digits belong can be identified, which allows accurate conversion of handwritten text into digital text.

4.3 MATERIALS AND METHODOLOGY

In this section, dataset preparation and proposed methodologies have been described.

4.3.1 Data Collection

Previous research on recognizing handwritten Bengali digits has predominantly focused on object classification without incorporating localization information or considering the presence of multiple objects in a single image, which limits the development of OCR technology for Bengali. Therefore, a new dataset was required for such advancement. In this research, the images were captured using mobile-phone camera and obtained from diverse demographics and sources, including undergraduate students, general people of Bangladesh, and professionals. The dataset consists of 267 images containing 21 classes of Bengali digits, operators, and parentheses. The distribution of the objects in the dataset is shown in Table 4.1, where the total number of objects is 7648.

Table 4.1: The distribution of the samples in the dataset for each class

Class Label	No. of Samples	Class Label	No. of Samples
0	258	=	368
1	487	(	375
2	434	)	362
3	258	-	528
4	348	×	386
5	310	।	427
6	401	{	308
7	410	}	275
8	358	[	309
9	307	]	309
÷	430		

4.3.2 Dataset Preparation

In this subsection, techniques utilized for data preparation have been explained.

4.3.2.1 *Image Binarization and Resizing*

The RGB images were converted to grayscale using the BGR to Gray conversion method. Gaussian blur was then applied to remove any noise present in the image using a kernel size of 3×3. The processed image was finally binarized by setting the color of the digits to white and the background to black. For the YOLO object detection algorithm, downsampling occurs during the CNN architecture, which requires a minimum image size that is a multiple of 32. Downsampling reduces spatial dimensions of the feature map and enables detection of objects at different scales. To prevent rounding errors during downsampling, it is recommended to use image sizes that are multiples of 32 (e.g. 320×320, 416×416, 640×640) when training a YOLO model. Hence, all the images were resized to 640×640 pixels to ensure consistency in image sizes for further processing and analysis. The binarization process has been illustrated in the second column of Table 4.2.

4.3.2.2 *Image Annotation*

To detect multiple objects in an image, the first step is to localize them. To achieve this, manual annotation of images was done using *Roboflow*[2]. Each object (digits, operators, and parentheses) was annotated by drawing a bounding box around it and assigning it to one of the 21 different classes: 0 to 9, addition, subtraction, multiplication, division, equal, and brackets. The third column of Table 4.2 illustrates the process of image annotation.

[2] `Roboflow : https://roboflow.com/`

Table 4.2: Two steps of data preparation: image binarization and image annotation

Original	Inverse Binarized	Annotated
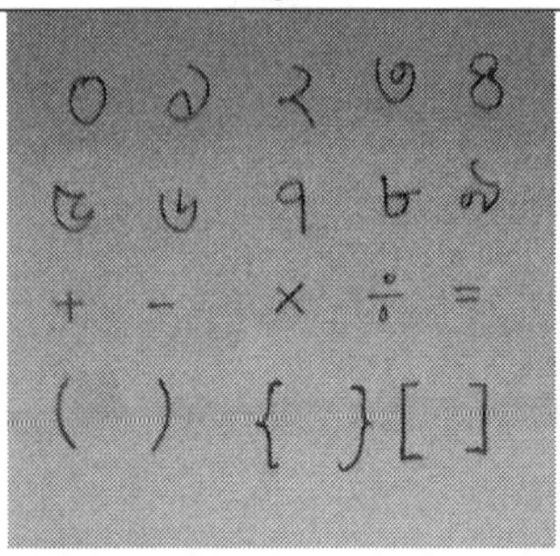	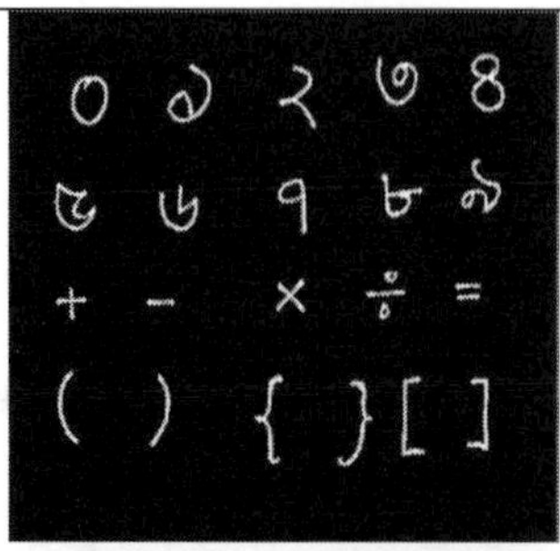	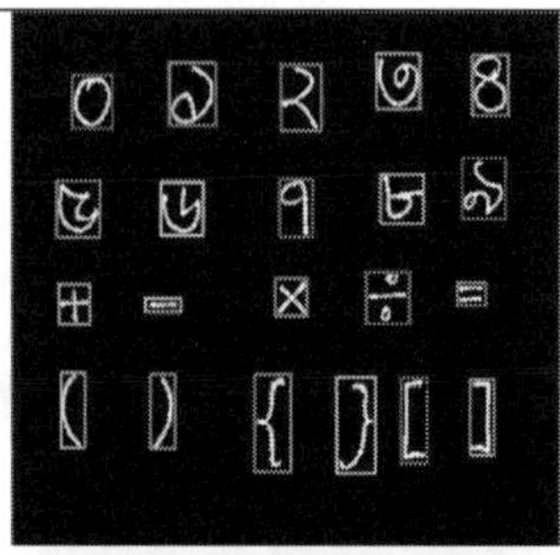
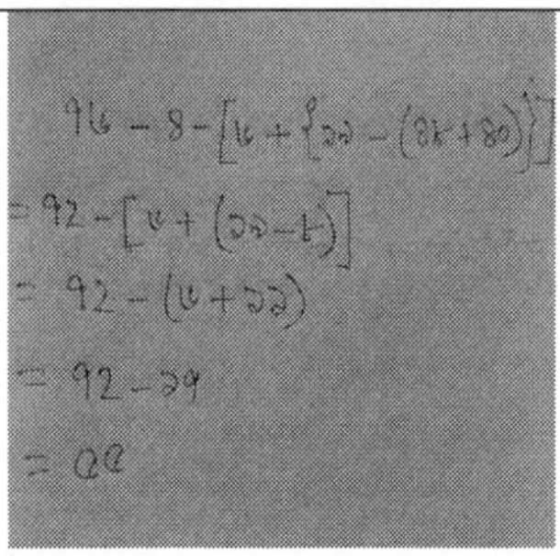	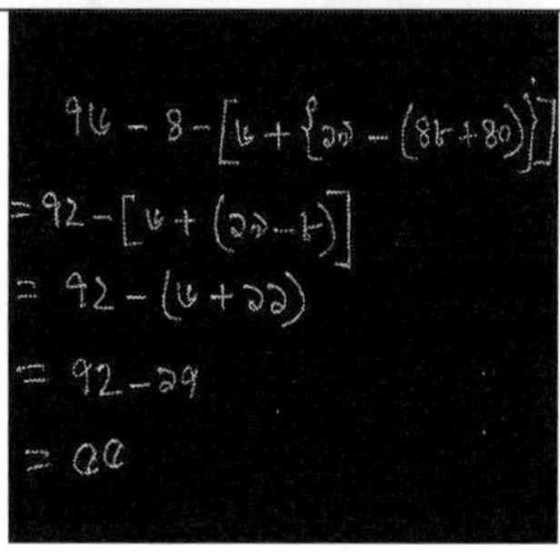	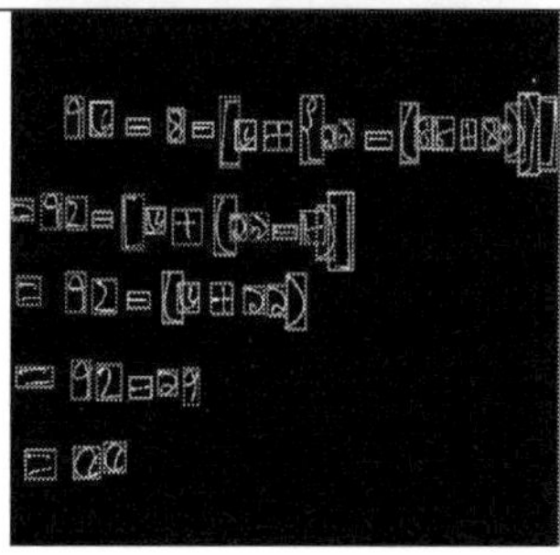

4.3.3 YOLOv8 Architecture

Convolutional neural networks (CNN) are known for their effectiveness in object detection algorithms, with several models such as R-CNN, Fast R-CNN, Faster R-CNN, SSD, RetinaNet, and YOLO being developed over the years. Unlike the previous models that employed a two-stage detection process involving selective search or regional proposal network to identify interesting regions, YOLO handles detection as a regression problem and performs both detection and classification with a single neural network, making it the fastest one-stage detector available.

The most recent version of the YOLO object detection algorithm is YOLOv8, which emphasizes speed, size, and accuracy. It incorporates new features such as a new backbone network, an anchor-free split head, and loss functions that enhance its performance while maintaining a compact size and exceptional speed. YOLOv8 is suitable for various computer vision tasks, including detection, segmentation, pose estimation, tracking, and classification. It offers superior accuracy and performance, making it an invaluable tool for object detection and image segmentation. In the YOLOv8 model, the input image is first passed through a backbone network, which generates a feature map:

$$f_{\text{backbone}}(x) = \{f_{\text{backbone}}(x)_1, f_{\text{backbone}}(x)_2, \ldots, f_{\text{backbone}}(x)_n\} \tag{4.1}$$

This feature map is then processed by a split head, which generates predictions for objectness, class probabilities, and bounding box coordinates:

$$f_{\text{head}}(f_{\text{backbone}}(x)) = \{(o_1, p_{1,1}, p_{1,2}, \ldots, p_{1,c}), \ldots, (o_n, p_{n,1}, p_{n,2}, \ldots, p_{n,c})\} \tag{4.2}$$

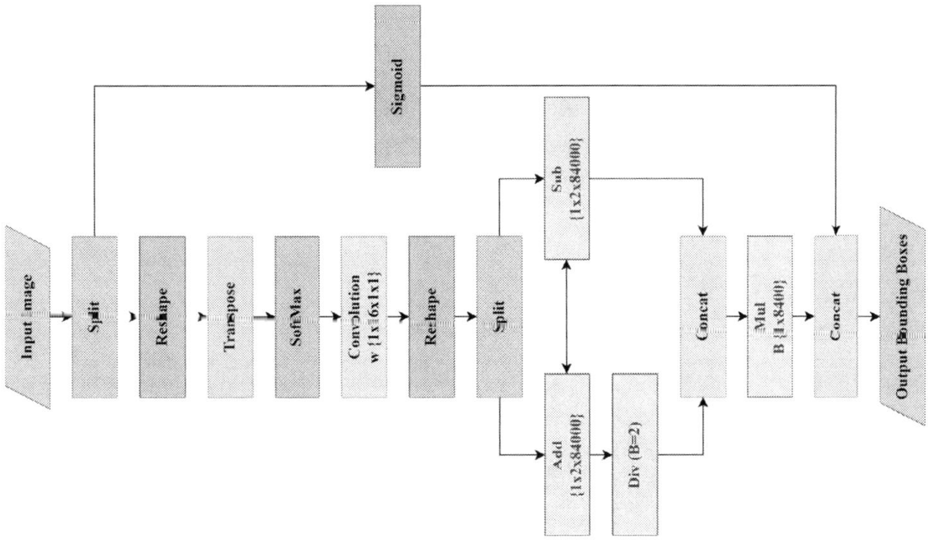

Figure 4.1: YOLOv8 Architecture [6]

where o_i is the objectness score for the ith anchor box, $p_{i,j}$ is the probability of the ith anchor box belonging to class j, and c is the number of classes. The predicted bounding box coordinates are normalized with respect to the width and height of the input image:

$$b_i = \left\{ \frac{x}{w}, \frac{y}{h}, \frac{w}{w}, \frac{h}{h} \right\} \tag{4.3}$$

where (x, y) are the coordinates of the center of the predicted bounding box, and (w, h) are its width and height. Finally, the objectness score and class probabilities are combined to obtain the final confidence scores:

$$c_{i,j} = o_i \times p_{i,j} \tag{4.4}$$

The object with the highest confidence score is selected as the prediction for that region. Fig. 4.1 provides an illustration of the architecture of YOLOv8 based on [6].

4.3.4 Proposed Workflow

This subsection showcases a detailed description of the proposed workflow.

4.3.4.1 Preprocessing of Input Image

To achieve accurate recognition of handwritten characters, the proposed YOLOv8 model requires preprocessed images. This involves converting the raw image to grayscale and then binarizing it, resulting in a black background with white characters. This preprocessing step is crucial to ensure optimal results when feeding the images into the model.

4.3.4.2 Obtaining Output from the Proposed YOLOv8 Model

The YOLOv8 model is capable of detecting and localizing objects within an image. Once the model processes an image, it returns a list of boundary box information for each detected object. This information includes the coordinates of the top-left and bottom-right corners of the bounding box, as well as the class and confidence rate of the detected object. With this information, it is possible to accurately identify the location and class of each object within the image.

4.3.4.3 Feeding the Output Array to Localization Algorithm

The output generated by the proposed YOLOv8 model was utilized as input for the proposed localization algorithm. The algorithm was able to process the data and produce an array of arrays. Each sub-array contained the indices of bounding boxes that belong to a particular line within the image. This approach enabled accurate and efficient localization of the handwritten characters and operators within the image, contributing to the development of advanced OCR technology. Algorithm 1 showcases the steps of the proposed localization algorithm.

Algorithm 1 Proposed Localization Algorithm

Input: detection_result: an array of tuples where each tuple contains (class_label, center_point, $[x_1, y_1, x_2, y_2]$)
Output: Line: an array of arrays containing the indices of bounding boxes that form a line

Sort detection_result according to y-value of center_point and put them inside sorted_detection_result
$d\_arr \leftarrow []$
for each element in sorted_detection_result **do**
    $d_i \leftarrow y_i - y_{i-1}$
    Insert d_i into d_arr
end for
th $\leftarrow$ Threshold(d_arr)
$Line \leftarrow []$
for each element in sorted_detection_result **do**
    $d_i \leftarrow y_i - y_{i-1}$
    if $d_i <$ th **then**
        Insert i^{th} bounding box into $Line$
    else
        Sort the elements of $Line$ according to x-value (in non-descending order)
        Print $Line$
        Clear $Line$
        Insert i^{th} bounding box into $Line$
    end if
end for

4.3.4.4 Setting Threshold for Line Splitting

The center point coordinates of all objects were derived by analyzing the collected bounding box information. However, there was no image-processing technique found in the literature that could be used for line detection, and therefore, a mathematical approach was developed to solve this problem, which calculated the distance between neighboring objects based on their center y-values, and if the distance exceeded a certain threshold value, it was concluded that the objects belonged to different lines. Firstly, a list was created where bounding boxes were sorted according to their center y value. Then, from the sorted array, the vertical distance between every i^{th} and $i+1^{th}$ bounding box was measured and put in a list. Next, the outliers are measured from the distances where the mean value of the outliers is the threshold. Suppose Q_i is the ith quartile and n is the number of data points. Then, thresholds can be calculated using the following equations:

$$Q_i = \left[\frac{i \cdot (n+1)}{4}\right]^{\text{th}} \tag{4.5}$$

$$\text{IQR} = Q_3 - Q_1 \tag{4.6}$$

$$\text{Outlier Threshold} = Q_3 + 1.5 \times \text{IQR} \tag{4.7}$$

$$\text{drr\_out} = \text{Values in } d\_arr \text{ greater than or equal to Outlier Threshold} \tag{4.8}$$

$$\text{Threshold} = \frac{1}{n}\sum_{i=1}^{n} \text{drr\_out}_i \tag{4.9}$$

Here, IQR refers to interquartile range and d_arr refers to the list of distances between the center y value of two adjacent bounding boxes.

4.3.4.5 Bounding Box Overlap Calculation and Selection Algorithm

During the object detection process, it was observed that multiple bounding boxes were sometimes detected on a single object. To overcome this issue, we implemented a solution based on calculating the overlap between two bounding boxes. This overlap was determined by measuring the percentage of one box's area that is enclosed within the other. If the overlap percentage exceeded 80%, we selected the bounding box with the higher confidence rate as the accurate representation of the object. Through trial and error analysis, we found this threshold percentage to be feasible. Mathematical steps for finding overlap between two bounding boxes have been mentioned below.

Suppose, $P_1 = (x_{A1}, y_{A1})$ and $P_2 = (x_{A2}, y_{A2})$ be the top left and bottom right points of bbox1 and also, $Q_1 = (x_{B1}, y_{B1})$ and $Q_2 = (x_{B2}, y_{B2})$ be the top left and bottom right points of bbox2 as shown in Fig. 4.2. Then,

$$LX = \begin{cases} x_{A2} & \text{if } x_{A2} < x_{B2} \\ x_{B2} & \text{otherwise} \end{cases} \tag{4.10}$$

$$RX = \begin{cases} x_{A1} & \text{if } x_{A2} > x_{B2} \\ x_{B1} & \text{otherwise} \end{cases} \tag{4.11}$$

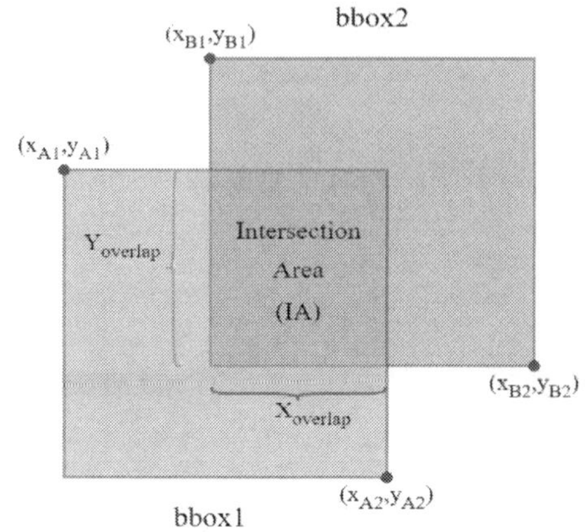

Figure 4.2: Overlapping of two adjacent boxes

$$LY = \begin{cases} y_{A2} & \text{if } x_{A2} < x_{B2} \\ y_{B2} & \text{otherwise} \end{cases} \tag{4.12}$$

$$RY = \begin{cases} y_{A1} & \text{if } x_{A2} > x_{B2} \\ y_{B1} & \text{otherwise} \end{cases} \tag{4.13}$$

where LX and LY are the top-left coordinates of the bounding box. Similarly, RX and RY are the bottom-right coordinates of the bounding box. Therefore,

$$\text{Overlap on X-axis, } X_{overlap} = \begin{cases} RX - LX & \text{if } RX - LX > 0 \\ 0 & \text{otherwise} \end{cases} \tag{4.14}$$

$$\text{Overlap on Y-axis, } Y_{overlap} = \begin{cases} RY - LY & \text{if } RY - LY > 0 \\ 0 & \text{otherwise} \end{cases} \tag{4.15}$$

Considering $X_{overlap}$ and $Y_{overlap}$, the intersection area (IA) and area of bounding box1 ($area_A$) can be calculated.

$$IA = X_{overlap} \times Y_{overlap} \tag{4.16}$$

$$area_A = (x_{A2} - x_{A1}) \times (y_{A2} - y_{A1}) \tag{4.17}$$

Then, the overlapping can be calculated by,

$$\text{Overlapping} = \begin{cases} 0 & \text{if } area_A = 0 \\ \frac{IA}{area_A} & \text{otherwise} \end{cases} \tag{4.18}$$

4.4 EXPERIMENTAL ANALYSIS

In this section, experimental setup and obtained results have been presented.

4.4.1 Pre-processing and Experimental Setup

As mentioned earlier, in the research, a two-step preprocessing approach was applied to reduce the dimensionality of the images. Firstly, the raw images were converted to grayscale, and then they were binarized into black and white before feeding them into the model. An attempt was made to train the model using raw images as well, but it resulted in poor predictions. Next, the dataset was split into train, validation and test set. There were a total of 267 images. Among them, 208 images were kept in the train set, 36 images were kept in the validation set and the rest of the 23 images were utilized for the testing. There exist several training settings for the proposed YOLOv8 model, which refer to the various hyperparameters and configurations used during the training process to optimize the model's performance on a given dataset. These settings are usually fine-tuned through careful experimentation and analysis to achieve the best possible results. Table 4.3 presents the training results of different setups, evaluated using mAP.

Table 4.3: Obtained mAP for different setups. Here, LR denotes learning rate.

Serial	**Model**	**No. of Epochs**	**Batch Size**	**Initial LR**	**Final LR**	**mAP**
Setup 1	yolov8s.pt	300	16	0.0001	0.001	86.5
Setup 2	Pretrained Model	300	16	0.0001	0.0001	86.0
Setup 3	yolov8s.pt	300	16	0.0001	0.0001	85.8
Setup 4	yolov8s.pt	300	16	0.001	0.001	91.7
Setup 5	Pretrained Model	200	32	0.01	0.01	91.6
Setup 6	yolov8s.pt	200	16	0.01	0.01	92.0

4.4.2 Obtained Results while Training

The mAP was achieved from Setup 6, as shown in Table 4.3. Table 4.4 displays the complete training results, which offer a deeper understanding of the model's performance during the training process with Setup 6 with box probability, recall score, mAP at 50% overlap, and mAP across 50% to 95% overlap. The model was trained for 200 epochs using YOLOv8, with a batch size of 16 images. The training process utilized an SGD optimizer with an initial learning rate and a final learning rate of 0.01, an SGD momentum of 0.937, and an optimizer weight decay of 0.0005.

4.4.3 Test Results and Analysis

After training the model, the proposed workflow was applied to the test images to obtain the results. The obtained results for some test images have been presented in Table 4.5. Fig. 4.3 illustrates the confusion matrix (here, for better illustration, we have used simplified class labels such as eq for '=', div for '÷' etc.), and Fig. 4.4 illustrates the precision-recall curve. The average mAP was determined to be 92.0%. While there were some instances of duplicate or double detection of a single object during testing, this issue was largely resolved using the technique described in the methodology section. It can be observed from Table 4.5 that in sets 1 and 2,

Table 4.4: Results for setup 6. Here, BoxP, mAP50, mAP50-95 denote box probability, mean average precision at 50% and across 50% to 95% overlap respectively

Class Label	Images	Instances	BoxP	Recall	mAP50	mAP50-95
0	36	41	1.000	0.934	0.993	0.730
1	36	85	0.864	0.835	0.886	0.602
2	36	72	0.844	0.899	0.880	0.644
3	36	43	0.947	0.977	0.993	0.729
4	36	59	0.886	0.966	0.967	0.689
5	36	48	0.974	0.938	0.981	0.703
6	36	62	0.878	0.952	0.923	0.631
7	36	71	0.979	1.000	0.993	0.744
8	36	56	0.921	0.875	0.863	0.621
9	36	51	0.933	0.941	0.970	0.730
÷	36	43	0.862	0.884	0.837	0.627
=	36	47	0.879	0.830	0.816	0.488
(	36	49	0.902	0.878	0.934	0.654
)	36	42	0.765	0.776	0.850	0.510
-	36	75	0.953	0.810	0.941	0.356
×	36	43	0.928	0.907	0.894	0.625
+	36	50	0.914	1.000	0.980	0.704
{	36	46	0.759	0.935	0.877	0.648
}	36	47	0.800	0.767	0.864	0.646
[	36	40	0.890	0.950	0.928	0.617
]	36	41	0.915	0.902	0.942	0.675
All	**36**	**1111**	**0.895**	**0.903**	**0.920**	**0.637**

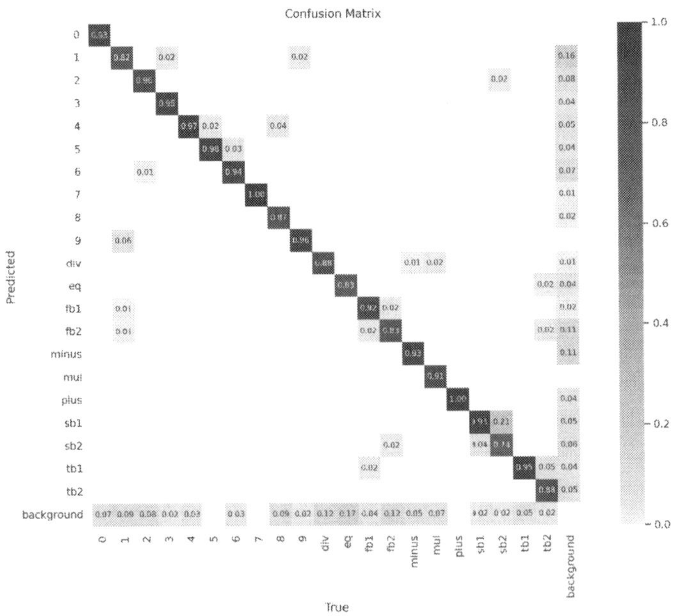

Figure 4.3: Confusion matrix for the test instances

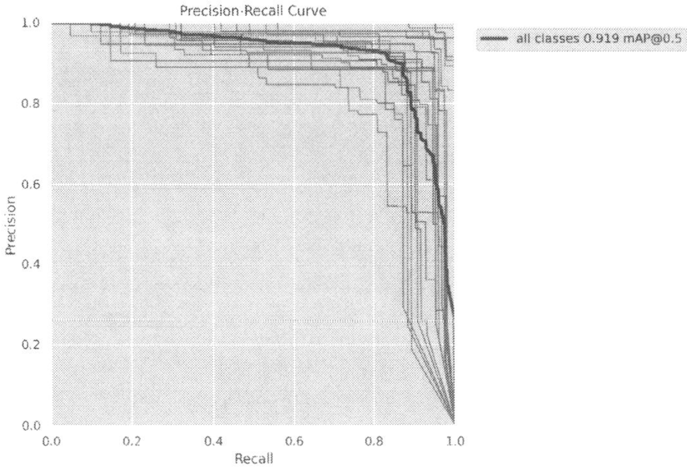

Figure 4.4: Precision-recall curve for the test instances

Table 4.5: Image Processing and Output Text

Set No.	Original Image	Binarized Image	Detected Image	Output
1	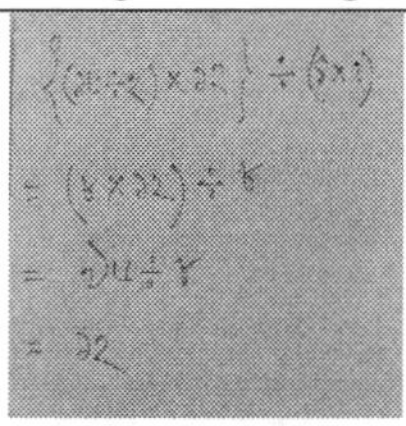	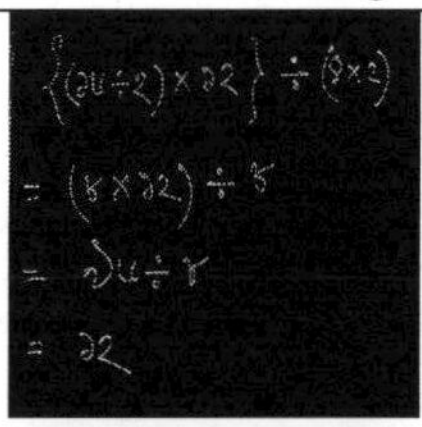	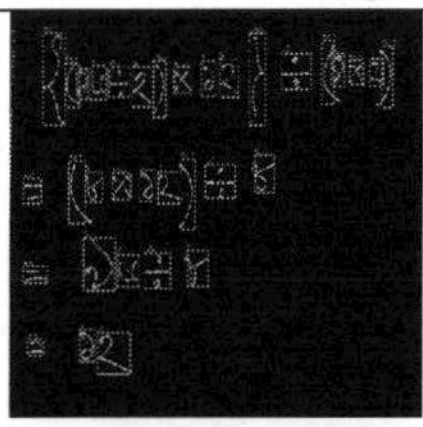	Thresold: 65.32 Converted Text: {(১৬/২)x১২}/(৪x২) =(৮x১২)/৮ =৯৬/৮ =১২
2	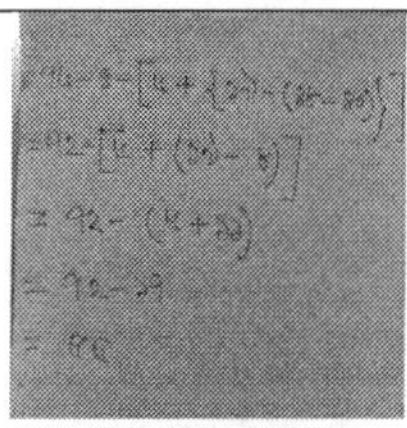	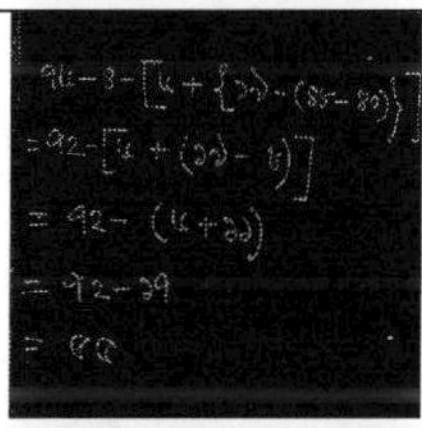	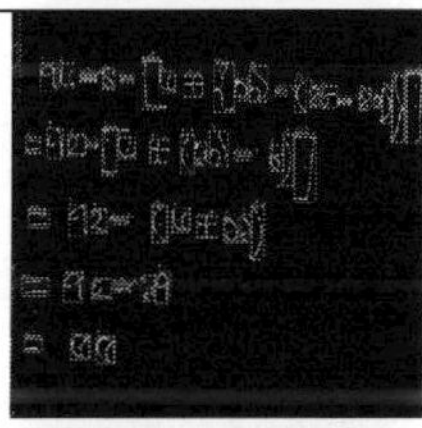	Thresold: 46.61 Converted Text: ৭৬-৪-[৬+{১৯-(৪৮-৪০)}] =৭২-[৬+(১৯-৮)] =৭২-(৬+১১) =৭২-১৭ =৫৫
3	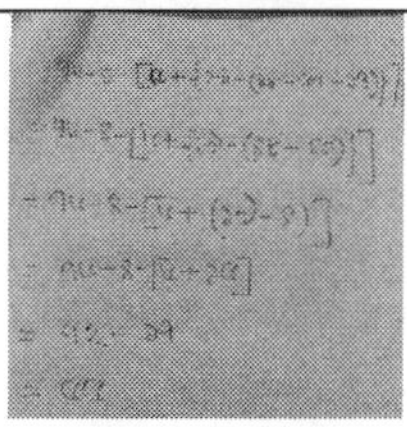	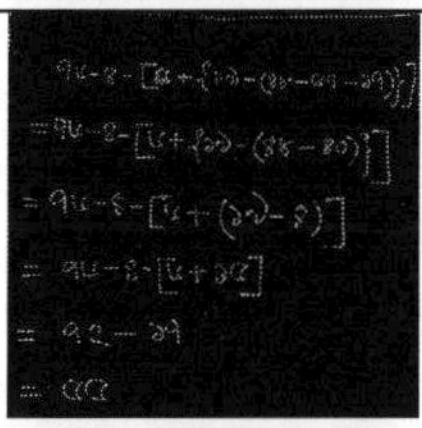	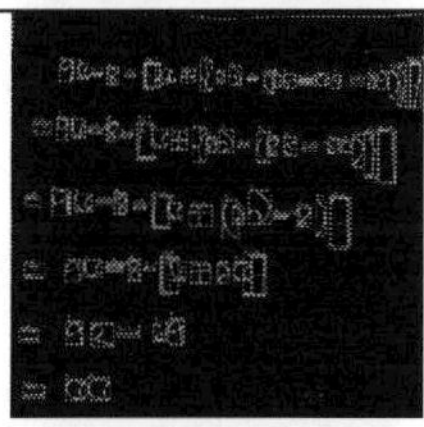	Thresold: 35.8 Converted Text: ৭৬-৪-[+{১৯-(৪৮-৫৭-১৭)}] =৭৬-৪-[৬+{১৯-(৪৮-৪০)}] =৭৬-৪-[৬+(১৯-৪)] =৭৬-৪-[৬+১৫] =৭২-১৭ =৫৫
4	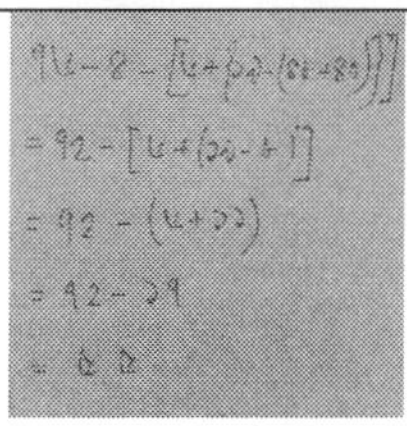	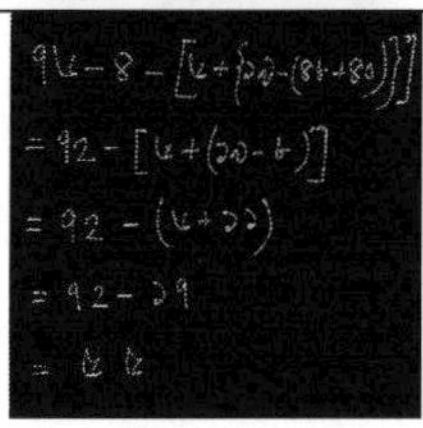	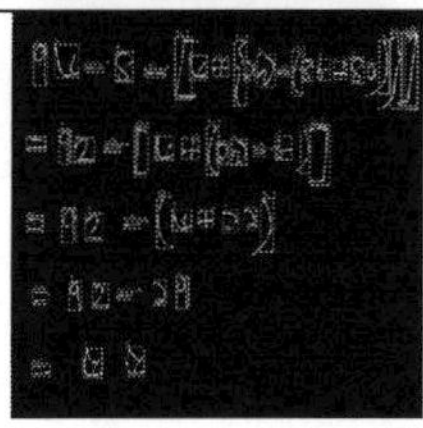	Thresold: 39.97 Converted Text: ৭৬-৪-[৬+{১৯-(৪৮+৪০)}] =৭২-[৬+(১৯-৮)] =৭২-(৬+১১) =৭২-১৭ =৫৫

predictions were fully accurate with no redundant or false objects detected. However, in sets 3 and 4, more than one bounding box was detected on a single object. The method to resolve this issue is described elaborately in the methodology section.

4.5 CONCLUSION

Developing a reliable system for recognizing and localizing Bengali handwritten characters is a challenging yet important task for enhancing OCR technology and facilitating communication and accessibility for Bengali speakers. The proposed approach in this study utilizes the YOLOv8 object detection algorithm for accurately detecting and localizing multiple Bengali digits and operators in an image. The use of YOLOv8 ensures precise localization of every character within an image, allowing for identification of the line where the digit or sign belongs. By analyzing the position of each character, the proposed method can convert handwritten mathematical equations into digital text, improving accessibility and efficiency for tasks such as educational and scientific writing. An average mAP of 92.0% was achieved, indicating promising results. Further improvement in performance and accuracy can be achieved through training the model with additional datasets. In the future, the objective will be to apply the proposed approach to all characters present in Bengali literature.

Bibliography

[1] Albahli, S., Nawaz, M., Javed, A., & Irtaza, A. (2021). An improved faster-RCNN model for handwritten character recognition. Arabian Journal for Science and Engineering, 46(9), 8509–8523.

[2] Azim, R., Rahman, W., & Karim, M. F. (2016). Bangla hand-written character recognition using support vector machine. International Journal of Engineering Works, 3(6), 36–46.

[3] Aziz, T. I., Rubel, A. S., Salekin, M. S., & Kushol, R. (2017, December). Bangla handwritten numeral character recognition using directional pattern. In 2017 20th International Conference of Computer and Information Technology (ICCIT) (pp. 1–5). IEEE.

[4] Dutta, K., Mathew, M., Krishnan, P., & Jawahar, C. V. (2018, August). Localizing and recognizing text in lecture videos. In 2018 16th international conference on frontiers in handwriting recognition (ICFHR) (pp. 235–240). IEEE.

[5] Huang, W., Lin, Z., Yang, J., & Wang, J. (2013). Text localization in natural images using stroke feature transform and text covariance descriptors. In Proceedings of the IEEE international conference on computer vision (pp. 1241–1248).

[6] Jacob Solawetz, Francesco. (Oct 23, 2024). What is YOLOv8? A Complete Guide.. Roboflow Blog: https://blog.roboflow.com/what-is-yolov8/.

[7] Kavallieratou, E., Balcan, D. C., Popa, M. F., & Fakotakis, N. (2001, October). Handwritten text localization in skewed documents. In Proceedings 2001 International Conference on Image Processing (Vol. 1, pp. 1102–1105). IEEE.

[8] Liao, M., Shi, B., & Bai, X. (2018). Textboxes++: A single-shot oriented scene text detector. IEEE Transactions on Image Processing, 27(8), 3676–3690.

[9] Medida, L. H., & Ramani, K. (2021). Impact of Deep Learning on Localizing and Recognizing Handwritten Text in Lecture Videos. International journal of Advanced Computer Science and Applications, 12(4).

[10] Mori, S., Nishida, H., & Yamada, H. (1999). Optical character recognition. John Wiley & Sons, Inc..

[11] Rakshit, P., Halder, C., Ghosh, S., & Roy, K. (2018). Line, word, and character segmentation from Bangla handwritten text—a precursor toward Bangla HOCR. Advanced Computing and Systems for Security: Volume 5, 109–120.

[12] Sayeed, A., Shin, J., Hasan, M. A. M., Srizon, A. Y., & Hasan, M. M. (2021). Bengalinet: A low-cost novel convolutional neural network for bengali handwritten characters recognition. Applied Sciences, 11(15), 6845.

[13] Wikipedia contributors. (n.d.). Bengali language. Wikipedia, The Free Encyclopedia. Retrieved April 26, 2023, from https://simple.wikipedia.org/wiki/Bengali_language.

[14] Wu, Y., Hu, Y., & Miao, S. (2021). Object Detection Based Handwriting Localization. In Document Analysis and Recognition–ICDAR 2021 Workshops: Lausanne, Switzerland, September 5–10, 2021, Proceedings, Part II 16 (pp. 225–239). Springer International Publishing.

[15] Zahedi, M., & Eslami, S. (2012). Improvement of random forest classifier through localization of persian handwritten OCR. ACEEE Int. J. Inf. Technol, 1(2), 31–36.

[16] Zhou, X., Yao, C., Wen, H., Wang, Y., Zhou, S., He, W., & Liang, J. (2017). East: an efficient and accurate scene text detector. In Proceedings of the IEEE Conference on Computer Vision and Pattern Recognition (pp. 5551–5560).

CHAPTER 5

Pandemic Prediction and Prevention in Bangladesh by Data Mining Approach

Md. Nazmus Sakib

Institute of Information Technology, Jahangirnagar University.

Fahima Tabassum

Institute of Information Technology, Jahangirnagar University.

CONTENTS

In today's world, the word 'pandemic' has been made so normalized because of the vast amount of industrialization, chemical uses and climate change. Since Bangladesh is not a well-developed country, it is important for the people to be aware of the upcoming pandemic situation and take proper measures to prevent as much as possible. We have seen the emergence of COVID-19 disease in the recent past and then the variants of SARS COV-2 'Delta', 'Omicron' are coming consecutively. Here, it is aimed to develop a system which will predict the imminent threats. It will also assist the affected ones with their do's and don'ts during the illness. The application might include visualization of pandemic information, breakdown of the epidemic risk, tracking of confirmed cases through Google API and adequate assessment of pandemic

DOI: 10.1201/9781003605508-5

prevention and control. The system will fetch accurate information from authentic web sources through web scrapping structured keyword analysis. Data mining will be used for the related information of Government facilities and vaccine issues for the concern disease. Weather forecasting will also be in operation for location tracking of the user to observe the weather conditions. Finally, optimum solutions will be provided by the system after getting some information from the user on the basis of the symptom analysis. Overall, this application will attempt to minimize the associated risks of Covid dissemination and will effectively curtail the harmful effects of future pandemics. After testing, using Bernoulli Naive Bayes Classifier, it is found that this system has 82.40% accuracy in prediction function which outperforms SVM and DTC model.

5.1 INTRODUCTION

The rapid multi-point transmission of a pandemic can be caused by the geographic spread of emergent infectious illnesses caused by large-scale human migration [1]. The analysis of the pandemic scenario based on multi-source urban data is an effective technique at the early stages of the COVID-19 outbreak. Due to the large transit in the run-up to the Chinese Spring Festival, tracking human migration from Wuhan and other cities in Hubeii province is very essential. Asymptomatic and mild-symptomatic cases are difficult to detect within the infected community.

In the recent past, we've seen the introduction of COVID-19 disease, followed by the SARS COV-2 'Delta' and 'Omicron' forms. Here, I aim to create a system that can predict impending hazards. It will also advise those who are afflicted on what to do and what not to do during their illness. Visualization of pandemic data, breakdown of epidemic risk, tracking of confirmed cases via Google API, and proper assessment of pandemic prevention and control could all be included in the app. Through web scraping and structured keyword analysis, the system will get correct data from reliable web sources.

Despite there are some works available about recent covid prediction, the projection for other pandemics has not been introduced so frequently. The pathogenic issue's demography in any specific area is also rare in other works which can be addressed as a vital issue.

The main objectives of this project are mentioned below:

- To predict the propagation of corona-related pathology along with similar kind of epi/pandemics.
- To visualize – with easily digestible graphical aids – the demography of pathogenic occurrences in any particular area, updated information.
- To offer customized prevention guidelines by taking specific inputs from users.

In section 2, a literature review on related works is given including problems and explanations for the most important terms used in this research project on the basic concept and technological tools of prediction, how other researcher have done their tasks, what is their limitations etc. section 3 introduces the system methods including system mathematical expressions, formulations and system diagrams and processes of transmission on channels, etc. section 4 concludes the proposed system.

5.2 RELATED WORK

M Rohini et al (2021) represented a model where KNN, SVM, Decision Tree, and Random Forest models were used to construct supervised Machine Learning classification models for the COVID-19 virus. Various types of method have ensured the accuracy level of this work. [2] Kaiser et al (2021) introduced a phone version of smart transportable healthcare technology, to help businesses identify potential COVID-19 contamination possibilities among their workforce who could require preventive care where the best possible outcome can be retrieved. [3] Liu [4] developed a strategy to help medical practitioners and the government deal with the COVID-19 issue and supplement distribution in terms of the overall accuracy of deep learning-based models which make the operating process more sustainable. [4] Tubre and Rodeghero [5] investigated the reasons for cloud installations, as well as the challenges that software engineers confront. Dina Fitria Murad et al (2020) recommended web-based learning models should have a minimum degree of proficiency. [6] Morse et al (2012) mentioned over the last few decades, the foundation on which to develop a worldwide pandemic prevention plan has shifted dramatically. [7] Xian-Xian Liu et al [8] presented like SIR and SEIR models, which were widely employed during the SARS outbreak, may fall short in COVID-19 prediction. Kumar et al (2020) carried out a disease analysis and prediction study utilizing ARIMA and FBProphet, two extensively used forecasting models. [9] Joshua Ritterman et al (2009) demonstrated that adding features concerning the historical context of the current day's feature counts has a markedly beneficial effect on the forecast accuracy after news about H1N1 broke out. So, we can get the accuracy of forecast just getting the news of the dissemination of that particular virus. [10] M. Shur et al (2020) explained the novel Pandemic Equation, which is the generalized rate equation describing the pandemic curve hollowing and budgeting for the preventive actions.This kind of equation will help for taking the measures which can be a solution concern for this specific issue [11].

5.3 METHODOLOGY

In this project, the system will retrieve data from public web resources and info sites by data mining framework where through web scrapping data will be fetched and categorized for the pre-processing unit. The categorisation and sanitization of data will ensure the expected output of the system. Dual layer filtering, feature extraction and tokenization process have been introduced here for the successful operation. Structured keyword analysis is introduced for the expected result which will be shown through the android application. Google API and weather forecasting will assist to indicate the locations and humidity of the particular areas.

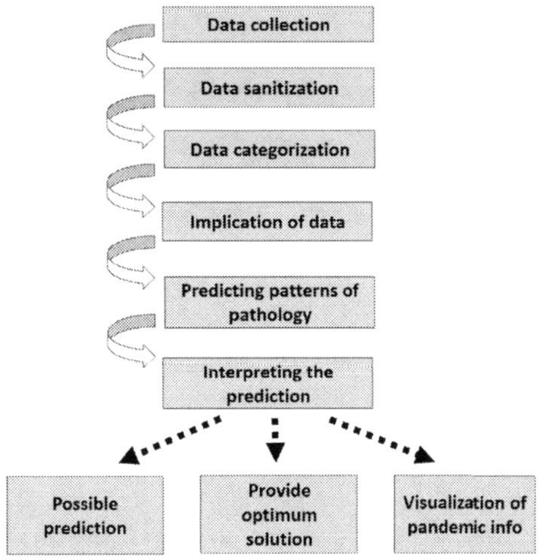

Figure 5.1: Block diagram on pandemic prediction and prevention app

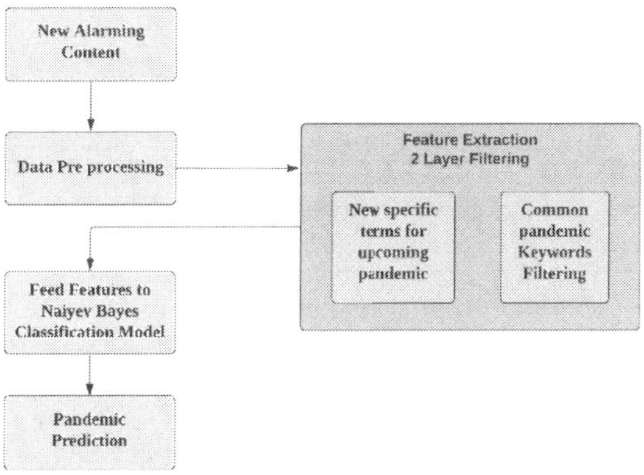

Figure 5.2: Model For Pandemic Prediction

5.3.1 Data Collection

One of the key jobs in data analysis is data collecting. For this work, matched data collecting has been used. Since a huge number of content is stored on relevant sites and thus the whole system works. The scrapper will fetch data for the basis of structured keyword analysis. Also Gathering data is one of the most important tasks in data analysis. Matching data collection has been utilized for this project. Because a vast amount of content is kept on pertinent websites, the system functions as a whole. Data for the foundation of structured keyword analysis will be retrieved by the scraper.

5.3.2 Data Pre-Processing

Data Pre-processing is a basic task for doing such operation in mining. There are many extra things which disturbs to gain of the main data. That's why we use a

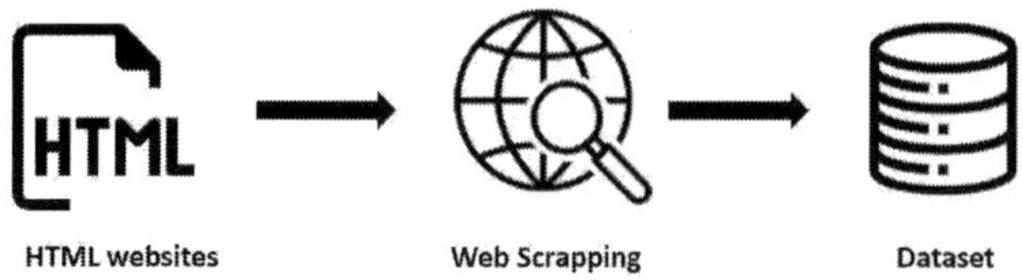

Figure 5.3: Web scrapping process

process like following which is essential for furthermore activities. There are several types of unnecessary information, including commas, punctuation, emojis, spaces, and more. In addition to eliminating these extraneous components, pre-processing reduces lengthy paragraphs into phrases, words, and finally word roots. Stopwords are ignored since they are insignificant for retrieving anything useful from data.

5.3.3 Dual Layer Filtering

Dual Layer Filtering extracts from a post the frequency of words that are ubiquitous and pandemic event-specific. The whole first layer looks for common Pandemic words, whereas the second tier looks for epidemic event-specific words. Every sort of word's appearance is noted.

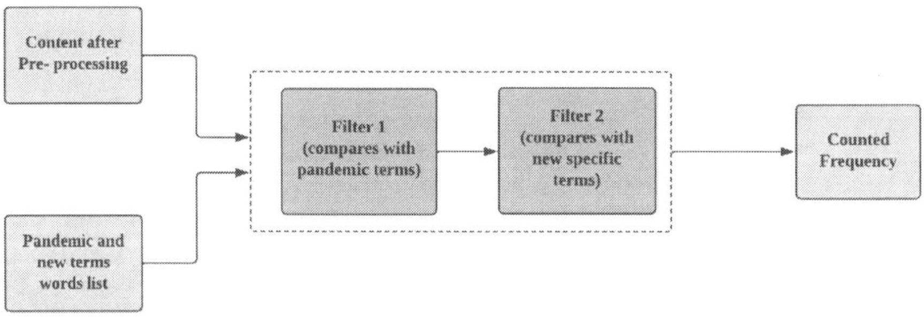

Figure 5.4: Dual Layered Filtering

5.3.4 Common and Novel Pandemic Words and Phrases

There are some findings for words and phrases that could be efficient for predicting the new upcoming pandemic or epidemic. These will help to scrap by structured keyword input. The overall theme depends upon the significant properties that have been utilised here to make on. So, the data collection method is here also involved and those words and phrases are mentioned in the following Table 5.1.

Table 5.1: Pandemic Structured Keywords: common pandemic words for layer-1 filtering, celebrating, novel pandemic words for layer-2 filtering purpose.

Common Pandemic Words	Novel Pandemic Words	Common Pandemic Phrases	Novel Pandemic Phrases
transmission, out- break, contagious, strain, viral, flu, threat-vector, severe, acute, viral, threat	deadliest, strains, vaccine, vaccination, variant, delta, syndrome, novel-virus, rH-antigen	public health emergency, initial outbreak, infection fatality, transmitted via, death range, mortality rate, influenza epidemic, disease surveillance	social distance, quantify mortality, C-reactive protein, epidemic threshold, hypothetical strain

5.3.5 Model Training and Prediction

The model of Naive Bayes classification obeys the 'Bayes theorem' and those are given in no. 1 and no.2 equation. Here, 'Pr' is an acronym of 'probability'.

$$Pr(class|property) = \frac{Pr(property|class) * Pr(class)}{Pr(property)} \tag{5.1}$$

$$Pr(class|property) = Pr(property_1|class) \times Pr(property_2|class) \times ... \\ Pr(property_n|class) \times Pr(class) \tag{5.2}$$

Pr(class|property): It means the posterior probability of class(aim) that was given.

Pr(class): It is previous probability of that class.
Pr(property|class): This represents the likelihood which is the probability of estimator where there are given class.
Pr(property): Finally this is the prior to the probability of final predictor.

5.3.6 Flowchart

A flowchart refers to a pictorial visuals process which is systematic for a specific task. The show of an algorithm is meant by the flowchart or in common saying, a chronological approach for solving complex problems is meant to be the flowchart.

There are some findings for words and phrases that could be efficient for predicting the new upcoming pandemic or epidemic. The tokenization procedure in this task is accompanied by the stopwords removal procedure. Stopwords in the post are replaced and eliminated.

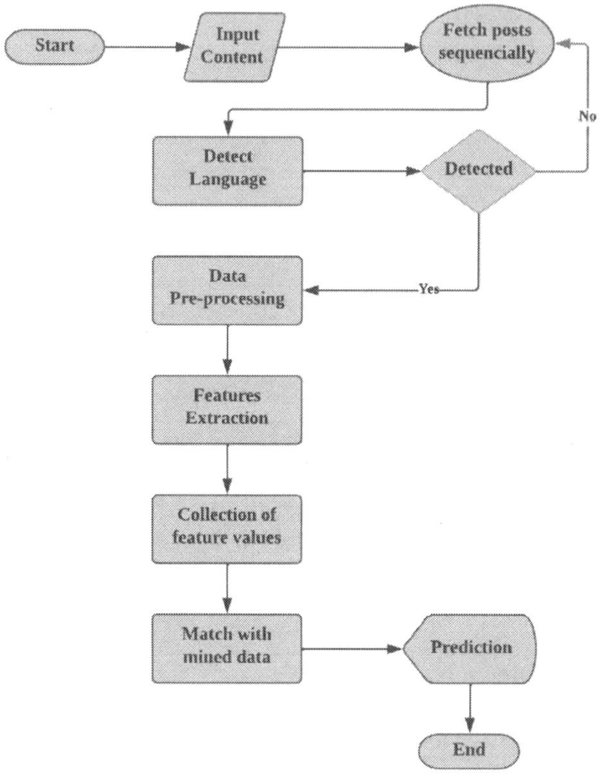

Figure 5.5: Flowchart of Pandemic Prediction Process

Material will be retrieved from a few carefully chosen websites, where an internal system will scrape the articles or contents. In this way, new terminology and the epidemic are found. Certain words and phrases have been found to be effective in anticipating the next pandemic or epidemic. In this task, the stopword elimination approach goes hand in hand with the tokenization process. The post replaces and gets rid of stopwords. It can be carried out the content scraping process by browsing some of the articles that are displayed below. Pre-processing includes the tokenization to process payments procedure and stopword removal.

5.4 IMPLEMENTATION

Data will be fetched from some selective web sources from where the articles or contents would be scrapped by some internal algorithm. In this way, the pandemic and novel terms are detected. The scrapping of the content can be executed by surfing some where articles which are shown below. Tokenization process and stopword elimination are included in pre-processing.

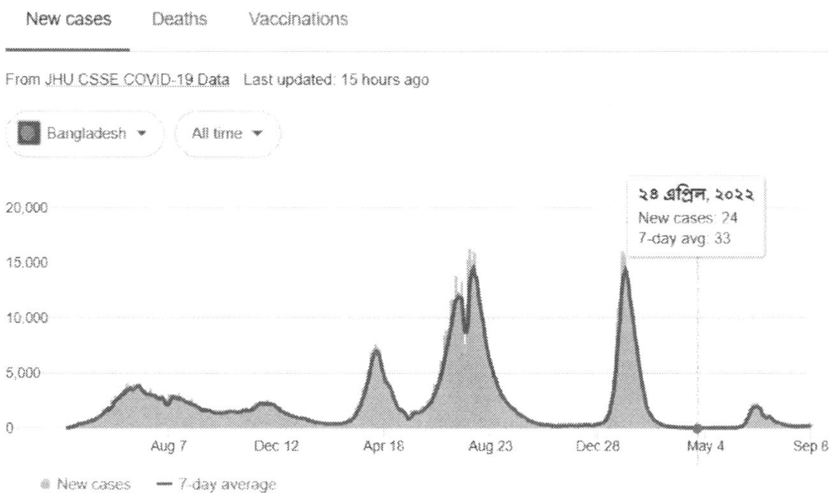

Figure 5.6: Stats from Native Sites

5.5 RESULT AND DISCUSSION

Bernoulli Naive Bayes Classification Model has been used in this pandemic prediction which represents 82.4% accuracy in predicting events. This task has been operated with the binary feature values which shows the better result.

The proposed model has been examined with more than twenty reliable websites through the web scrapping process. There are less prediction about the current pandemic and represents based on common pandemic words, novel pandemic words, common pandemic phrases, and novel pandemic phrases.

The model has been tested with two other classification models. They are: Support-Vector Machines (SVM) and Decision Tree Classifier (DTC). Bernoulli Naive Bayes Classification Model has shown the better performance than the other two: the SVM model and the DTC models in this work. SVM and DTC models show an accuracy of 79.62% and 78.66% respectively in this task whereas Bernoulli Naive Bayes Classification shows 82.4% accuracy that is quite better in comparison with two models.

Table 1.2 represents the recall, precision and f1-score of Bernoulli Naive Bayes, SVM and DTC models respectively.

Table 5.2: Performances of Bernoulli Naive Bayes, SVM and DTC models in pandemic prediction. [Here, CPW = Common Pandemic Words; NPW = Novel Pandemic Words; CPP = Common Pandemic Phrases; NPP = Novel Pandemic Phrases]

Method	keyword Type	Precision	Recall	f1-score	Accuracy
Bernoulli Naive Bayes	CPW	0.57	0.57	0.58	0.8240
	NPW	0.82	0.89	0.91	
	CPP	1.00	0.82	0.88	
	NPP	0.880	0.81	0.86	
Support Vector Machines (SVM)	CPW	0.58	0.58	0..58	0.7962
	NPW	1.00	0.67	0.81	
	CPP	.88	0.69	0.80	
	NPP	0.80	0.93	0.85	
Decision Tree Classifier(DTC)	CPW	0.52	0.78	0.60	0.7866
	NPW	0.90	0.73	0.81	
	CPP	1.00	0.68	0.84	
	NPP	0.77	0.93	0.84	

Table 5.3: True and false PANDEMIC prediction performances of Bernoulli Naive Bayes, SVM and DTC model on basis of output models.

Method	CPW		NPW		CPP		NPP	
	True CPW	False CPW	True NPW	False NPW	True CPP	False CPP	True NPP	False NPP
Bernoulli Naive Bayes	0.84	0.09	0.54	0.35	0.75	0.21	0.81	0.15
SVM	0.75	0.28	0.55	0.38	0.69	0.25	0.85	0.07
DTC	0.78	0.24	0.81	0.18	0.68	0.27	0.88	0.11

Table 5.3 shows the performance of Bernoulli Naive Bayes, SVM and DTC models on each output class of this work. Bernoulli Naive Bayes shows relatively less accuracy in case of CPW but it tasks good on the others. SVM model performs with average accuracy in the NPP class but so less accuracy in CPW, NPW and CPP classes. DTC model visualises well and average accuracy in NPW and NPP classes but low in CPP and NPP phenomenons. Overall, the Bernoulli Naive Bayes Classification Model shows well and approximately likewise accuracy in each class of outputs there.

It is seen that the Bernoulli Naive Bayes Classification Model displays an accuracy of 82.40% in prediction function. This model outperforms the other two type of models: SVM and DTC Model. Accuracy might have been refined but this model does not succeed to detect some structured keywords and phrases on several websites. A trio of objectives have been integrated and attempted to be realized in this work. Since scrapping is still a questionable technology, there may occasionally be problems that cause it to produce ineffective results.

In this work, the integration of three objectives has been tried to be fulfilled. The infographics are also fetched by this technology as per people interactions concerned. And the expert system will give the output based on the given inputs of the users. So, it will up to the users who use it in right or wrong way. As there is integration about three phenomenons, it is rare to have it all in same platform. This process or system can be run short in case of a single objective comparison with others but will surely effective as a whole of three prime pandemic concern. As the scrapping is not out of questioned technology, there can be some issues where it can show some inefficient result sometimes.

5.6 CONCLUSION AND FUTURE WORK

Prediction of the propagation corona-related pathology and similar other pandemics is the prominent outcome of this project. The process works basically on web scrapping where naiyev bayes has been there for classifying the data perfectly. It can be seen that the accuracy of the prediction function of the Bernoulli naive Bayes classification model is 82.40%. Since scrapping is still in its early stages, there may sometimes be issues that cause scrapping to produce poor results. This model has a higher accuracy than the other two types of models: the SVM model and the DTC model. The accuracy of the model may have been improved. The system will use structured keyword analysis and web scraping to get precise data from reliable online sources. Data mining will be performed to obtain relevant information on government facilities and vaccination concerns pertaining to the disease in question. In order to track the user's location and monitor the weather, weather forecasting will also be used. Ultimately, the system will suggest the best options based on the symptom analysis once it has collected some data from the user. By creating specified algorithms the additional tasks are also to be executed. Along with visualising the demography of risks in particular area and updated information Using multi-source urban data is also a concern here. Consequently, the examination will be in operation of the relationship between several probable parameters and daily new confirmed cases. Finally, the system will offer guidelines by taking logical inputs from the user end.

Since, The development of this system through an app that can predict and prevent pandemics in Bangladesh. One of the most important thing is: By visualizing pandemic data and breaking down epidemic risk, the app could help policymakers make informed decisions about pandemic prevention and control. Eventually, The development of this Android application could serve as a model for other countries looking to develop similar systems for predicting and preventing pandemics.

5.7 ACKNOWLEDGEMENT

I would like to express my special gratitude to the Information and Communication Technology (ICT) Division, Government of the People's Republic of Bangladesh for the fellowship they have provided me to implement this research project.

Bibliography

[1] I. Ahmed, M. Ahmad, G. Jeon, and F. Piccialli, "A framework for pandemic prediction using big data analytics," Big Data Research, vol. 25, p. 100190, 2021.

[2] M. Rohini, K. Naveena, G. Jothipriya, S. Kameshwaran, and M. Jagadeeswari, "A comparative approach to predict corona virus using machine learning," in 2022 International Conference on Artificial Intelligence and Smart Systems (ICAIS), pp. 331–337, IEEE, 2021.

[3] M. S. Kaiser et al., "iWorksafe: Towards Healthy Workplaces During COVID-19 With an Intelligent Phealth App for Industrial Settings," in IEEE Access, vol. 9, pp. 13814–13828, 2021, doi: 10.1109/ACCESS.2021.3050193.

[4] T. Liu, "Us pandemic prediction using regression and neural network models," in 2020 International Conference on Intelligent Computing and Human-Computer Interaction (ICHCI), pp. 351–354, IEEE, 2020.

[5] B. Tubre and P. Rodeghero, "Exploring the challenges of cloud migrations during a global pandemic," in 2020 IEEE International Conference on Software Maintenance and Evolution (ICSME), pp. 784–785, IEEE, 2020.

[6] L. Yan, H.-T. Zhang, J. Goncalves, Y. Xiao, M. Wang, Y. Guo, C. Sun, X. Tang, L. Jing, M. Zhang, et al., "An interpretable mortality prediction model for covid-19 patients," Nature machine intelligence, vol. 2, no. 5, pp. 283–288, 2020.

[7] S. S. Morse, J. A. Mazet, M. Woolhouse, C. R. Parrish, D. Carroll, W. B. Karesh, C. Zambrana-Torrelio, W. I. Lipkin, and P. Daszak, "Prediction and prevention of the next pandemic zoonosis," The Lancet, vol. 380, no. 9857, pp. 1956–1965, 2012.

[8] X.-X. Liu, S. J. Fong, N. Dey, R. G. Crespo, and E. Herrera-Viedma, "A new seaird pandemic prediction model with clinical and epidemiological data analysis on covid-19 outbreak," Applied Intelligence, vol. 51, no. 7, pp. 4162–4198, 2021.

[9] N. Kumar and S. Susan, "Covid-19 pandemic prediction using time series forecasting models," in 2020 11th International Conference on Computing, Communication and Networking Technologies (ICCCNT), pp. 1–7, IEEE, 2020.

[10] J. Ritterman, M. Osborne, and E. Klein, "Using prediction markets and twitter to predict a swine flu pandemic," in 1st international workshop on mining social media, vol. 9, pp. 9–17, 2009.

[11] M. Shur, "Describing and predicting covid19 evolution using pandemic equation," in 2020 IEEE International Conference on Big Data (Big Data), pp. 5822–5824, 2020.

CHAPTER 6

Impacts of Passenger Request Trends on Ride-sharing System Performance

Md Sayem Ahamed

Deptartment of Computer Science, United International University, Bangladesh.

Md Shahzamal

Independent Researcher

CONTENTS

Ride-sharing is extremely popular in every country and covers a major proportion of commuting services nowadays. This has attracted the attention of the research community to find ways to optimize the performance of ride-sharing systems and hence user satisfaction and revenue can be uplifted. Our research investigated how trends in passenger ride requests across geographical and temporal contexts impact the performances of a ride-sharing system. Uber pickup data collected from New Your City was used as a proxy of ride requests to a ride-sharing simulation model, which was built based on the taxi movement data from the same city. The performances achieved by operating cars with strategies where: 1) cars allocated to a geographical area proportionate to the request generation intensity of that area, and 2) cars are operated during the hours when request generation was high are compared with the performances of random car operations to understand the impacts of request trends.

DOI: 10.1201/9781003605508-6

Random car operations show very poor request acceptance rates even though fleet size (number of cars) is equal to the number of ride requests. The request trend-based car allocation and searching passengers made acceptance of all requests with the same fleet size and low request acceptance delay. However, car utilization (duration when the car carries passengers) drops significantly when fleet size increases. With the fleet size when car utilization is 50% in the trend-based approach, five times more ride requests were accepted and drivers have three times more revenue than random car operations. The passenger searching effort was also two times more in the random-based ride-sharing system.

6.1 INTRODUCTION

Ride-sharing is the practice of using an automobile vehicle to offer rides to other passengers with the fare. This is very often organized through a ride-sharing application. It provides a low-cost and flexible transportation option for individuals who do not have access to a car or decided not to drive. Ride-sharing facilities were started in the 1940s when carpooling was used to reduce fuel during World War Two. However, the flourishing of ride-sharing applications like Uber and Lyft make this a widely used transportation choice. Recent studies show that ride-sharing services are responsible for almost 5% of all daily trips in the United States.

The research communities and industries have put immense effort into improving the performance of the ride-sharing system with the aim of increasing user satisfaction and high revenue generation. Recently, the availability of the data has made it possible to understand the factors that underpin the system's performance [1] [9]. Most of the studies have focused on two key areas: i) passengers do not want to wait for a long time to get the ride, and ii) cars do not want to be underutilized which leads to less income for drivers [2] [6]. The underutilization may come from the situation where cars spend a long time searching for passengers in proximity.

Our research studied using real-world datasets how the intensity of ride-share requests over: i) the hours of the day and ii) across the various locations affect the ride-sharing system performance. To conduct our research, a simulation model was built where cars move from one location spot to another location according to rules derived from real-world taxi trip data. In this model, the cars move within the 400 location boxes created in the GPS coordinates of New York City. The car travels to a destination box which is defined by the lognormal distribution fitting parameters from real-world data. A speed is also assigned to a car derived from the log-normal distribution.

We conducted three experiments to capture the impact of ride request trends. In the first experiment (baseline strategy), cars start their service in a random location box and at a random time of the day. In the second experiment (area-based strategy), the cars start services in the location boxes with high priority where more rider requests were made. The intensity of request trends at destination boxes and neighboring boxes are considered in the car's movement when cars move to destination and neighboring search boxes. In the third experiment (area and time-based strategy), cars start servicing during the hours when more ride requests are made. The simulation is run for car fleet sizes up to 10K and various metrics are collected

The analysis results indicate that the ride request trend strongly impacts the system's performance. The random movement of cars wastes fleet resources. Low car utilization and request acceptance rate (46%) are found even though the fleet size was made equal to the number of ride requests. However, the other two experiments indicated that if the car moves according to the spatial and temporal trends of ride requests, the car utilization and percentage of request acceptance are increased significantly in some situations almost 5 times. The acceptance delay and passenger searching efforts are significantly low. The research output clearly shows that ride request trends across locations and time are critical to system performance.

This paper is organized as follows. In section 2, the related works are reviewed where we discuss the methods and techniques used to study ride-sharing system performances. Then our methodology is discussed in section 3 which details the data and simulation technique. The obtained results are presented in section 4. The discussion on the research outcome and limitations are described in section 5.

6.2 RELATED WORK

There has been a wide variety of research on ride-sharing system performance. The work of [4] discussed a method for analyzing the relationship between spatiotemporal patterns of traffic flow and the structure of urban culture. The study shows that the spatial concentration of taxi services varies over time and is impacted by the urban design, institutions, and lifestyle of Shanghai. The study also found a high availability of taxi services in the CBD area, a ring-like service zone with stochastic distribution, and emerging regional business centers and external cargo hubs [8]. The Clustering Method proposes a density-based spatial clustering method employed to identify pick-up and drop-off hotspots [6]. They argued that the study results can be useful for taxi operation management and user travel pattern analysis.

A data mining algorithm is proposed to discover latent knowledge and a triplet (Time, Location, Strategy) to find feature patterns of passenger-finding strategies. The experimental results show that trip frequency and urban hot spot distribution play a major role in the income levels of the drivers and taxicab companies [10]. They analyze real taxi data from Changsha and mine some efficient strategies that give insights into taxi operations. A method to discover trip hot routes using large-scale taxi trajectory data [7]. Then hot routes were extracted using large-scale taxi trajectory data. They argue that our experiment shows that the trajectory clustering algorithm and hot route extraction method are effective [3].

As we can see, researchers used many different techniques to do their research. Some use weight matrices but try to find spatial concentrations of taxi services that fluctuate over time. Some of them use density-based clustering methods to indicate pickup and drop-off hotspots and use weight matrix to identify pickup hotspots [2] [5]. Some of them use time and location strategies to find passengers and find that trip frequency and urban hotspot distribution play a major role in the income levels of drivers and taxi companies [10]. However, there is no research to quantify how ride request trends impact system performances compared to random car operations. Our research focused on addressing this gap in the literature.

6.3 METHODOLOGY

This section describes the methodology of the research.

6.3.1 Dataset

Two real-world datasets relevant to ride-sharing systems were used in this study. The first dataset contains 14.3 million Uber pickup requests from New York City from April to September 2014. This data provides information about the date and time of the Uber pickup and the pickup location coordinates – longitude and latitude. This dataset was used as the Uber ride requests which need to be served by the car fleet of the ride-sharing system under investigation. The second dataset is the yellow taxi trip records collected from New York City which includes capturing pick-up and drop-off dates/times, pick-up and drop-off locations, and trip distances. This data is used to design the car movement simulator that is used to investigate the impact of Uber request trends found in the first dataset.

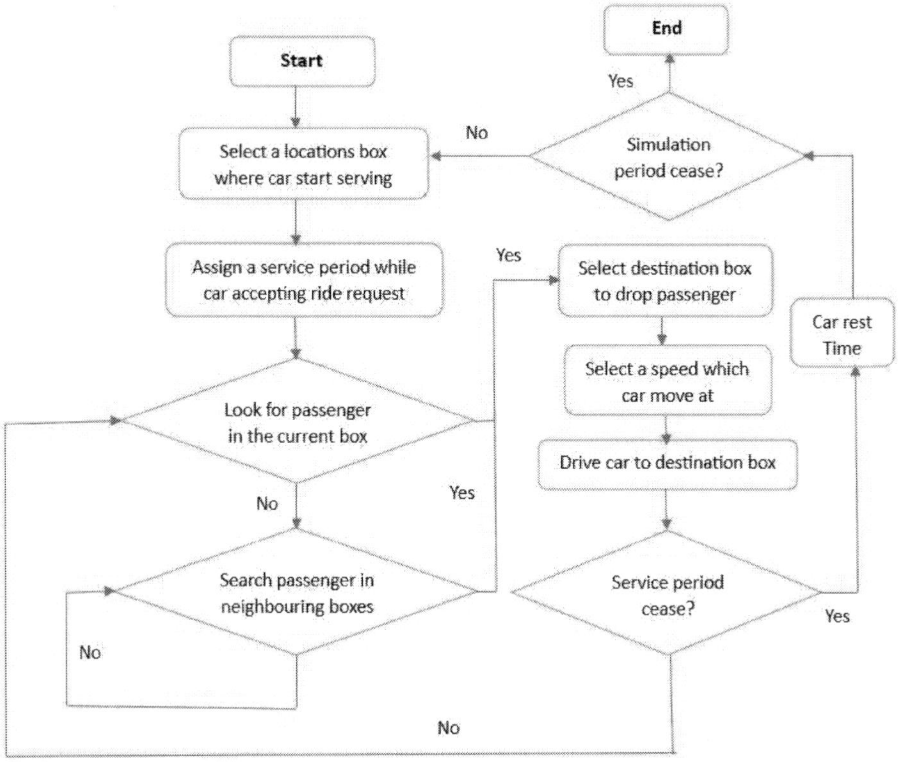

Figure 6.1: Flowchart of car movement in the developed simulation model. The spot in the flowchart represents the location box discussed in this paper

6.3.2 Car Movement

The used car movement model is presented in Figure 6.1. In this model, a geographical area is divided into small location boxes, and cars move from one box to the other box over time. When a car starts serving (available to offer a ride), it is assigned a location box according to the experiment strategy and starts looking for passengers. The car is also provided a start time and a stop time to indicate the operation period while it is available to carry passengers. After this period, the car will rest up for at least 8 hours before starting the next operating period. During the operating period, the car scans for the ride request from its current location box. If there are requests from the current location box, it accepts the oldest request made within 2 hours. Then, a destination location box is assigned based on the experimentation strategy. It is also assigned a speed drawn from a lognormal distribution and estimates pick up and drop off time. So that we can determine when the car is available for the next ride in the destination location box. If the car does not find the ride request in the current location box within a search time period, it moves to the neighboring location box which is chosen according to the experiment strategy and searches for ride requests.

6.3.3 Experiment Setup

The Uber pick-up data from New York City is used as the proxy of ride requests in this analysis. Therefore, the number of ride requests generated during the experimentation period of one month is constant. But, the number of cars (fleet size) is varied to understand the system performance in various scenarios. Performances are initially analyzed for the fleet sizes of 200, 400, 600, 800, and 1000 cars and then up to 10K at each 1k interval. The simulation is run for the geographical area of New York City with a longitude of -73:3 to -74.0 and a latitude of 40 to 40.4. This area is divided into 400 boxes of 0.01 coordinate width in both longitude and latitude.

A car movement simulator is run to track car status and request status over each time step of 5 min during an observation period of 30 days. At each step, the status of every car is inspected. First, the availability and ride-service status of the car is checked. No status update is needed if the car is not operated. The car status registry records information about the number of trips the car made, the duration of time the car was carrying the passengers, and the number of locations the car searched to find the passengers during an operating period. When a ride request is accepted, we save the delay between request generation and acceptance. We also marked the request as expired if its generation time is more than two hours.

The strategy of assigning car operating periods and choosing starting boxes is varied to explore the impact of ride request trends. In the first strategy, a car assumes any time as the car service start time. Then the stop time is determined by adding 10 hours to the start time. However, the start time must be 8 hours away from the previous service stop time. The starting box and next destination are also chosen randomly but follow the distance selection method from the lognormal distribution which is fitted with mean $= 2$ and standard deviation $= 0.3$. The analysis of pickup and drop-off records within the same longitude and latitude coordinates provided these parameters. The speed is also chosen based on the lognormal distribution with mean $= 0.05$ and standard deviation $= 0.2$. The generated distribution of distance and speed is shown in Figure 6.2.

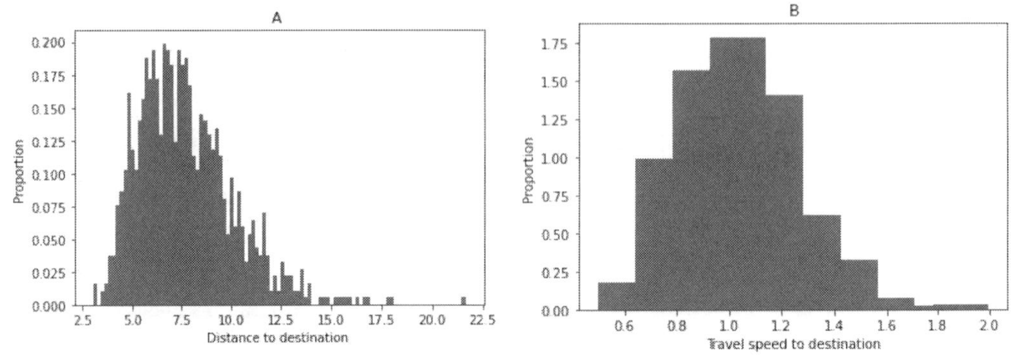

Figure 6.2: Density of log-normal distribution: a) distance to destination box where the passenger will be dropped and b) travel speed to destination box

The analysis indicated (see Figure 6.3) that the intensity of Uber requests varies over the location boxes. We have created a weight matrix for each box which represents the proportion of requests made from these location boxes. The proportion of

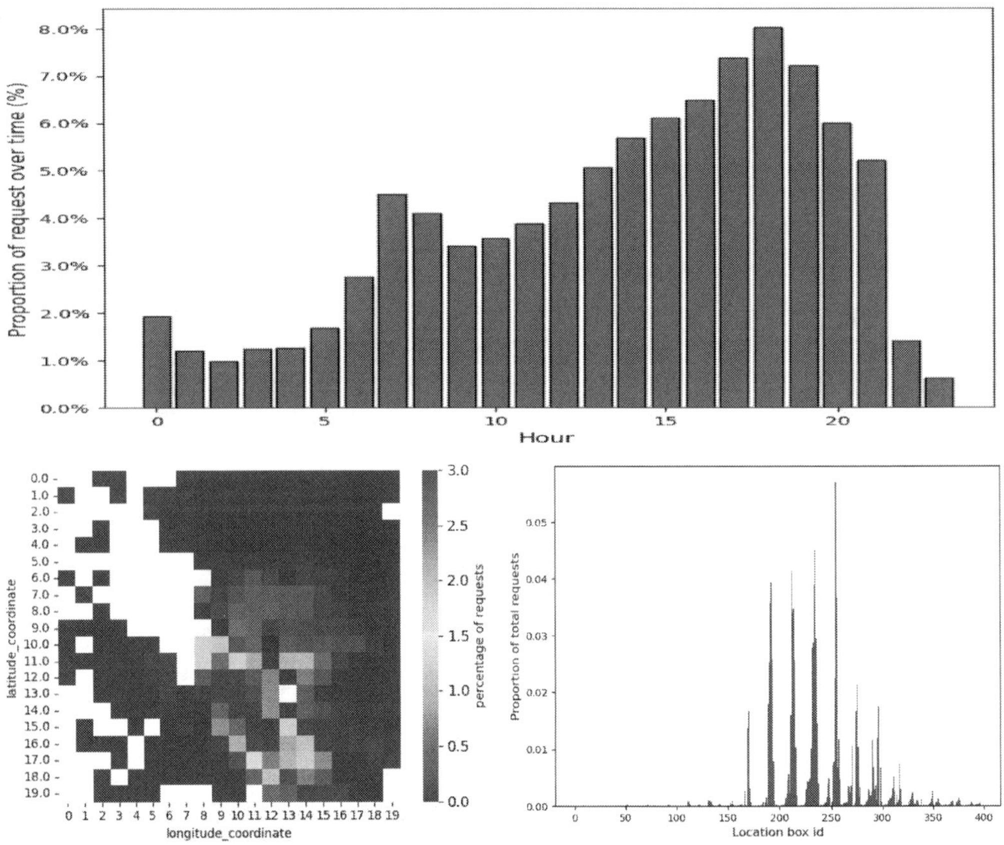

Figure 6.3: a) Trend of ride requests over the hours of the day, b) intensity of ride request over geographically divided location boxes, c) trend of ride request over location box coordinate

cars starting within a box matched the Uber request intensity from this box. Similarly, when a car searches a neighbor's boxes to find the passenger, it moves to the neighbor's boxes that have a higher weight. The same principle is applied to select the destination boxes. We first select the distance from a log-normal distribution to indicate how far a car goes to drop the passenger. Then, we look at the boxes with that distance and select the boxes that have a higher weight. We also find the time weight - the proportion of requests over time for each box. Then, the multiplication of weights gives the weight for location boxes which capture both the request intensity from the area and over time.

6.4 RESULTS AND EVALUATION

The simulation was run for 30 days over the defined geographical areas of 400 location boxes. The key metric of system performance is the number of ride requests that were accepted for the observation period. When the location boxes are chosen randomly (baseline strategy) by cars, it is observed that the ride request acceptance rates do not change significantly as the car fleet size increases (see Figure 6.4). However, the request acceptance rates increased significantly as the car fleet size increased in the second scenario (area-based strategy) where the location boxes were chosen based on the intensity of request generation from the boxes With the smallest fleet size of 200 cars, about 17% of ride requests are accepted, which is much higher than the acceptance rate of baseline strategy at a fleet size of 1000 cars (see Figure 6.4). A maximum 51% acceptance rate was achieved with 1000 cars in the area-based strategy. Further improvement in acceptance rate is achieved if the time parameter (time-area strategy) is included in the location box selection where time weights are

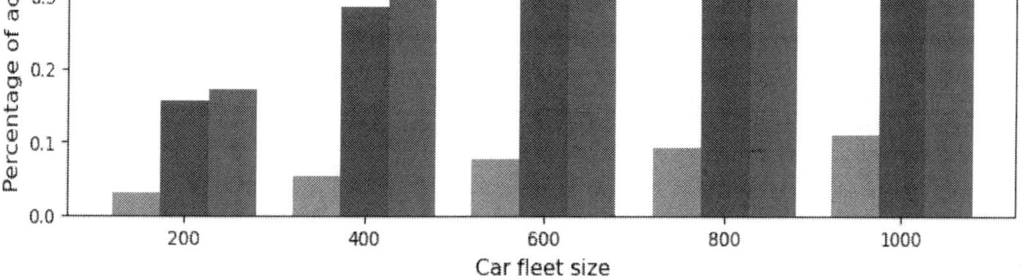

Figure 6.4: Percentage of ride requests were accepted in various strategies

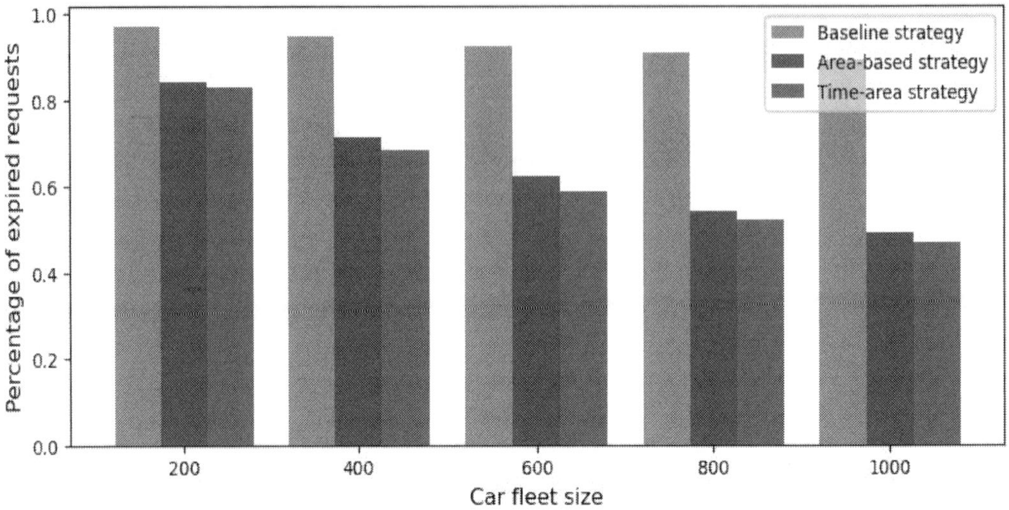

Figure 6.5: Percentage of ride request expired for different strategy

added to the area weights of boxes. The maximum 54% request acceptance rate is achieved with a fleet size of 1000 cars.

In our simulation, requests expired by 2 hours after it was generated. It is important to understand how much percentage of requests expire over the car fleet size. The results are presented in Figure 6.5. As it is expected, the baseline strategy has very high expiry rates while the other two strategies have low expiry rates. The expiry rates drop from 76% to 40% for changing the fleet size from 200 cars to 1000 cars in the time-area strategy. However, decreasing rates got slower when the fleet size was changed from 800 cars to 1000 cars.

Figure 6.6 shows the delay for each strategy for various fleet sizes. For the baseline strategy, we can see that the rate of delay is decreased when we increase the car fleet size. This is because when the system chooses a location box randomly, the rate of request acceptance is low. In the area-based strategy, we see the delay has slightly increased compared to the first scenario. This is because, for this request acceptance system, the rate of accepted requests has increased. For a high rate of request acceptance, the delay has increased slightly as the old request is accepted first. Figure 6B indicates that the number of requests with low delay decreases as the fleet size increases because more old requests are expected and recent requests wait further.

System performances are analyzed based on car utilization and presented in Figure 6.7 and Figure 6.8. In the baseline strategy, we can see that the average trips that a car makes per day are not significant. We also see that when we increase the car fleet size the rate of trips is also decreasing. In the area-based strategy (see Figure 6.7), we can see cars made good numbers of trips on average. So, we can say that if we distribute cars based on the intensity of request generation from location boxes, average trips made by car increase. The number of trips was further increased for

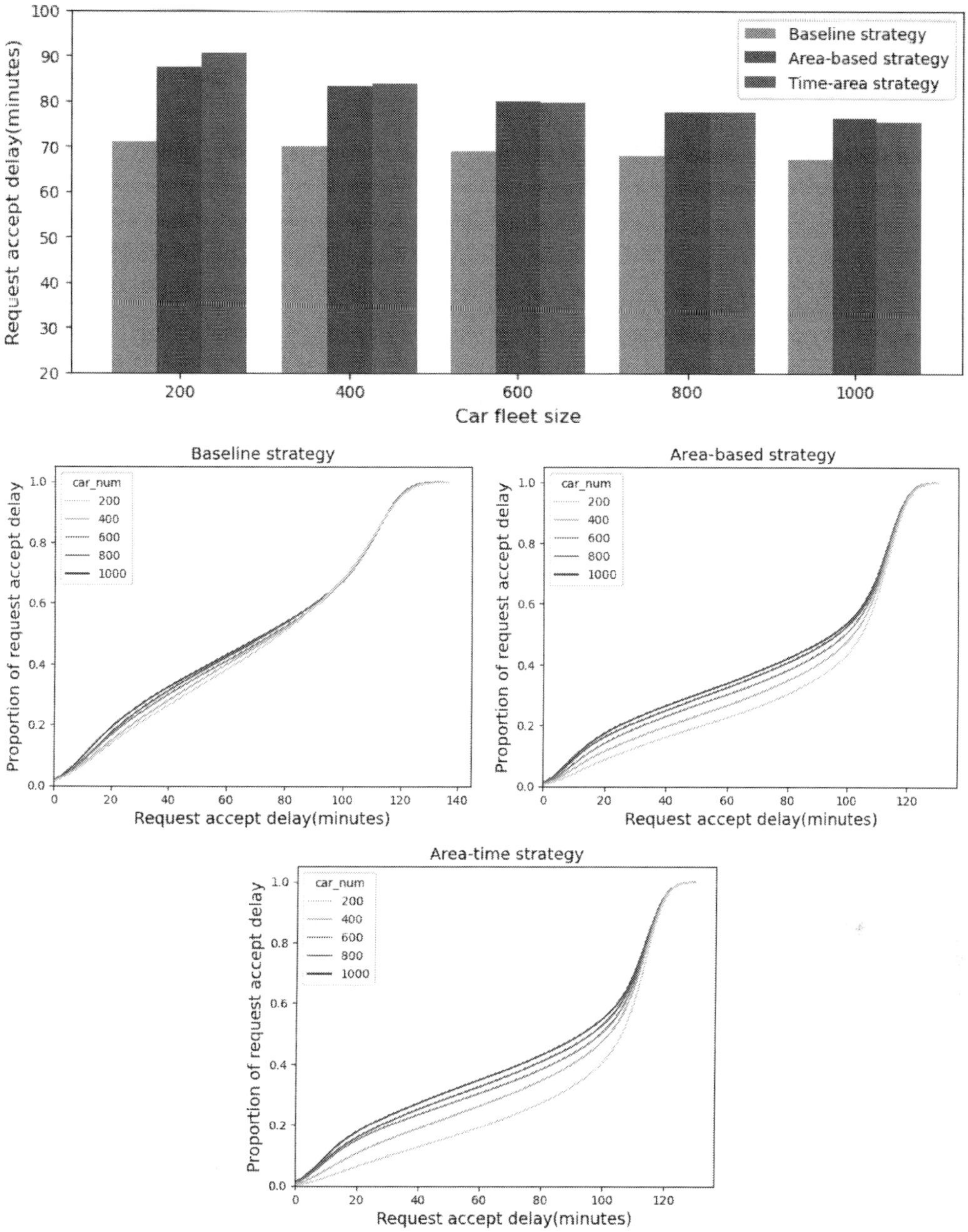

Figure 6.6: Delay to accept the request across various strategies: a) average delay for each fleet size and b) indicates cumulative distribution of delay for each fleet size

the time-area strategy as they now know when requests are made more. Figure 6.7 (time-area strategy) shows a further increase in average trips. The impact of request trends over time should be considered to designing a ride-sharing system.

Figure 6.8 shows the daily average time cars were engaged in carrying the passengers. This is the revenue generation time which excludes the passenger searching

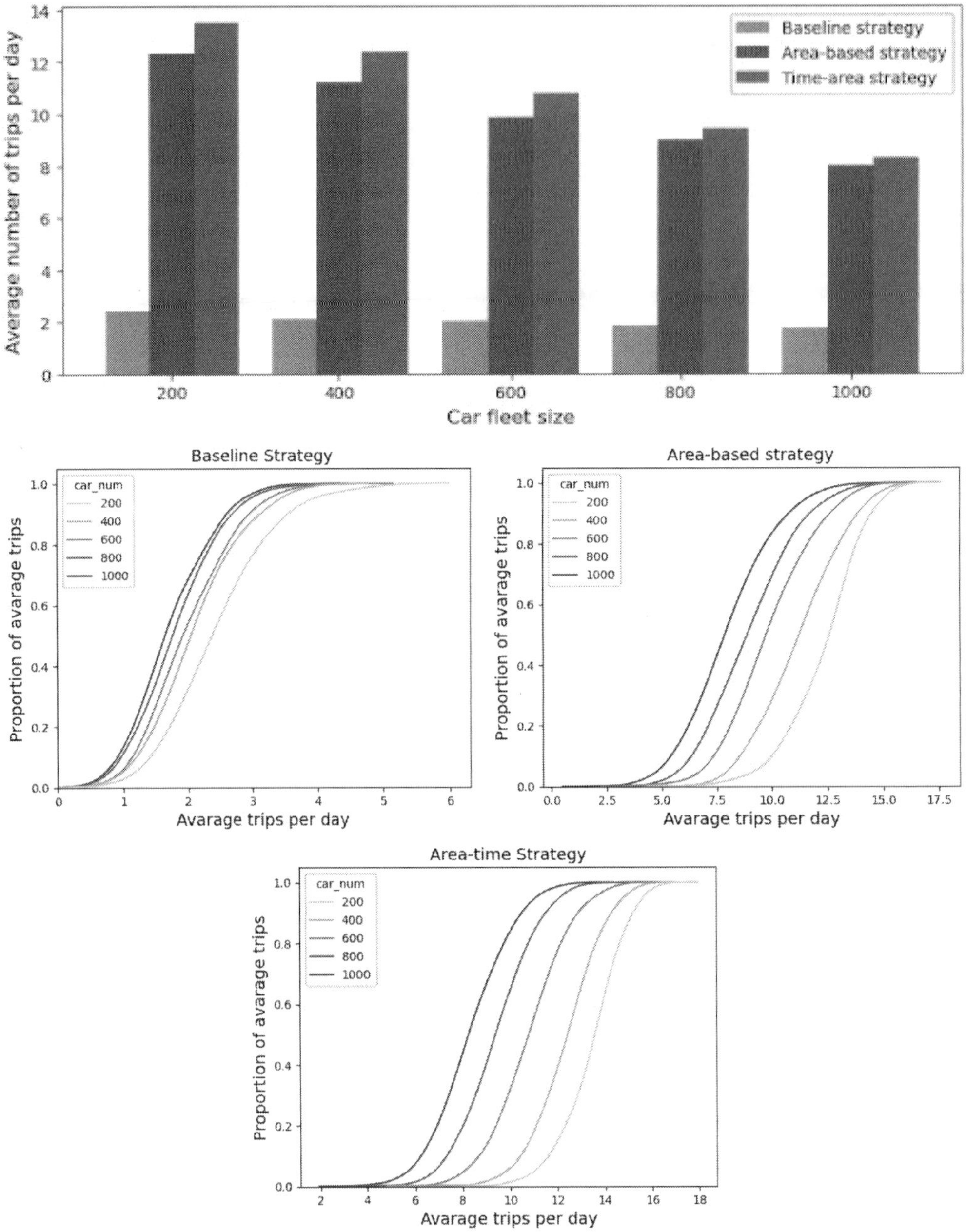

Figure 6.7: The number of trips cars made over different fleet sizes: a) mean and b) cumulative value

time. Very poor car utilization performance is observed in the baseline strategy while it is significantly improved in the other two strategies. Area – time-based strategy shows the best performance. However, car utilization also decreased as the fleet size increased.

Car drivers could be more satisfied if they spent less time searching for the next passenger. Figure 6.9 presented how ride request trends affect the next passenger

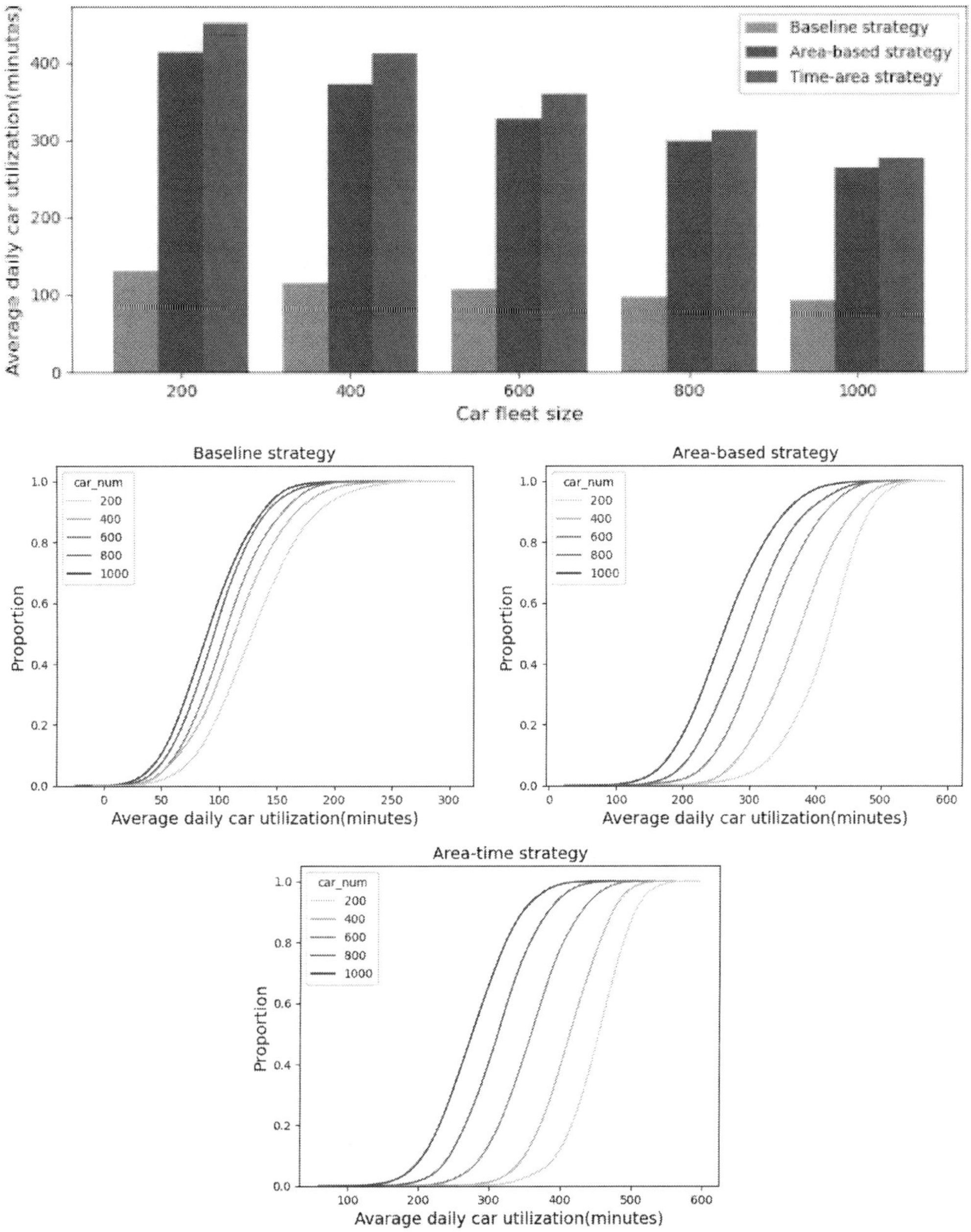

Figure 6.8: Average daily car utilization: average summary and cumulative distribution

searching time. Cars in the first strategy have higher searching time and increase as the fleet size increases. The other two strategies have low searching time. The small fleet has always had low searching time as there is not much competition in the system to accept requests.

The experiments are now conducted to achieve 100% ride request acceptance with increasing fleet sizes. However, other performance metrics such as acceptance delay,

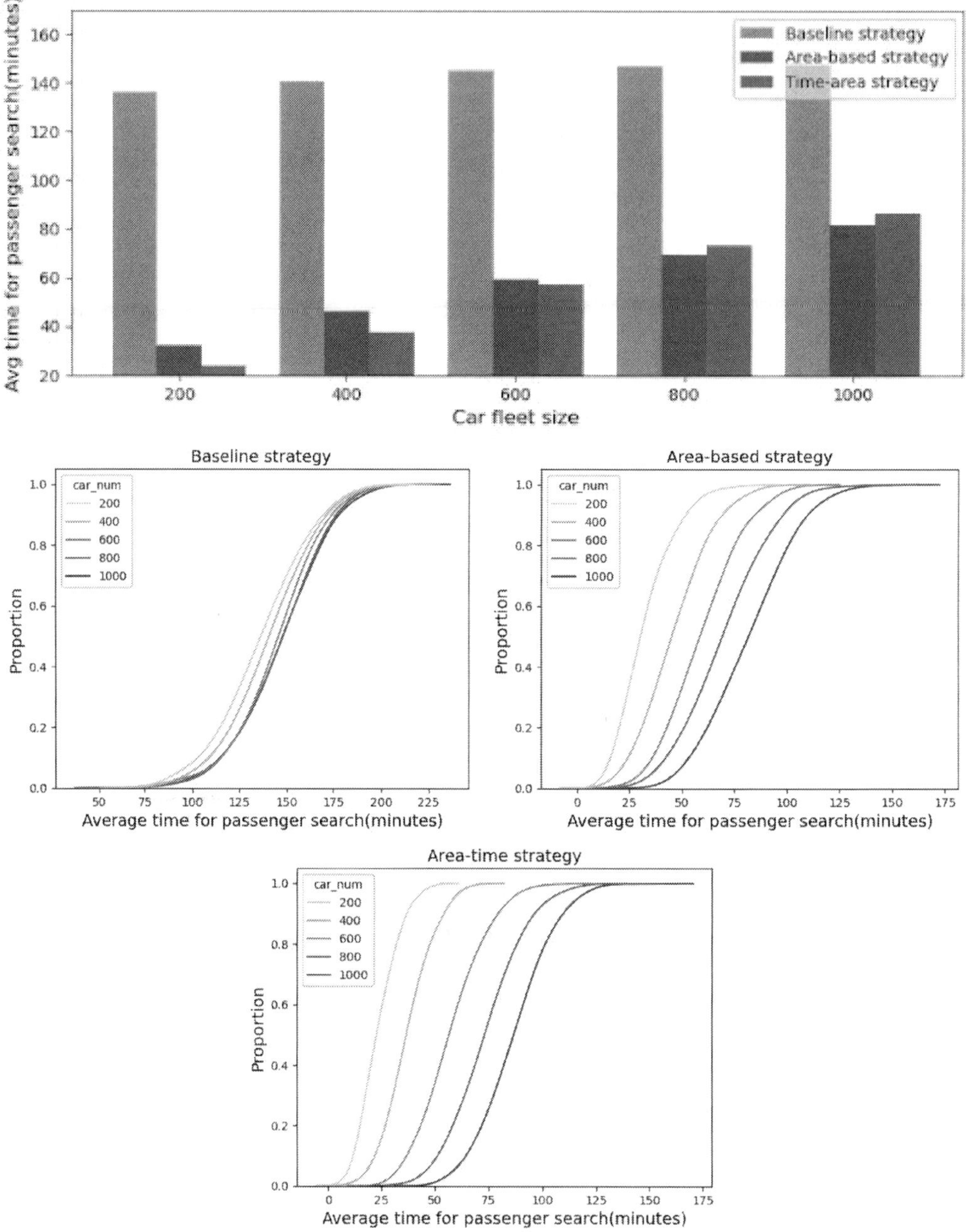

Figure 6.9: Daily passenger searching time: average summary and cumulative distribution

car utilization, and search number may deteriorate as fleet size increases. The following figure shows the variations of metrics for various fleet sizes. With the fleet size of 10K, random car operations achieved only 45% request acceptance while intensity-based car operations achieved 100% request acceptance. However, car utilization drops as fleet size increases in both methods. We assumed that 300 minutes

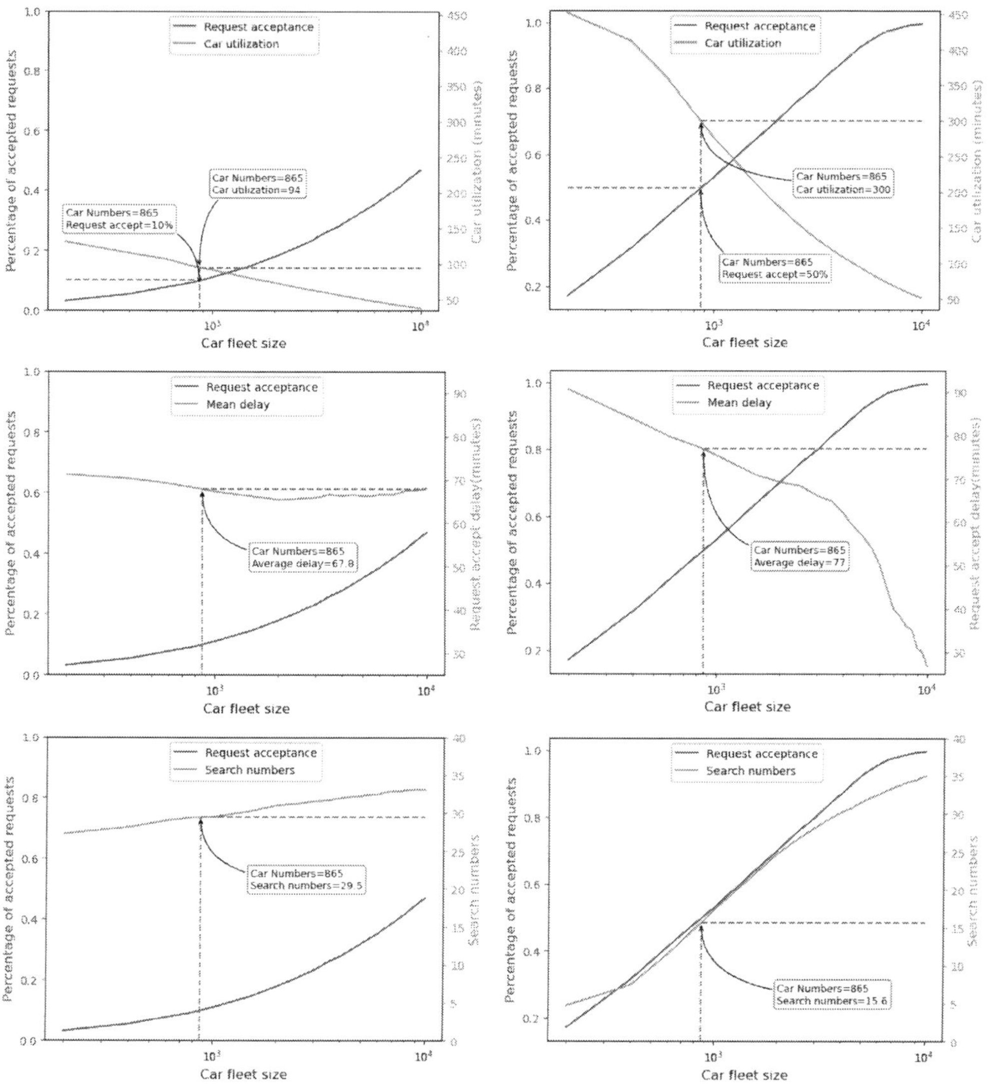

Figure 6.10: Maximum reachability for both baseline and trend-based system and comparison of performance metrics: car utilization, acceptance delay and searching number

of car utilization (50%) will be the standard from the diver's perspective. If the fleet size is set to 865, 50% utilization is possible with about 50% request acceptance. With this fleet size, five times acceptance is achieved compared to random car operations, half search effort is reburied and three times more car utilization is achieved compared to random car operations.

6.5 CONCLUSION AND FUTURE WORK

The intensive simulation indicates that ride request trends are very crucial to designing a ride-sharing system. The average acceptance request increased significantly from the baseline strategy to the area-time strategy by 5 times. Maximum 40% of ride requests were accepted in random car operations with the fleet size equal to number of ride requests. The care utilization increased 3 times in the area-time strategy than in the baseline strategy. We also found that passengers' searching time also decreased by 50%. However, there is a trade-off between request acceptance rates and driver incomes in both approaches. This study shows that optimization systems should consider these impacts of ride request trends across geographical and temporal contexts.

This research has several limitations and improvement opportunities. We have only considered the limited geographical area in our research for one month. It would be interesting for future work to look at the problem in a large geographical area for a longer observation period. The experiment can be done in other cities. It could also be interesting to find ways to get 100% acceptance rates without losing car utilization. The movement model can be more realistic by applying contextual information.

Bibliography

[1] Rong, Huigui, et al. "Mining efficient taxi operation strategies from large scale geolocation data." IEEE Access 5 (2017): 25623–25634

[2] Zhou, Dong, Rongrong Hong, and Jingxin Xia. "Identification of taxi pick-up and drop-off hotspots using the density-based spatial clustering method." CICTP 2017: Transportation Reform and Change—Equity, Inclusiveness, Sharing, and Innovation. Reston, VA: American Society of Civil Engineers, 2018. 196–204

[3] Moreira-Matias, Luis, et al. "Online predictive model for taxi services." Advances in Intelligent Data Analysis XI: 11th International Symposium, IDA 2012, Helsinki, Finland, October 25–27, 2012. Proceedings 11. Springer Berlin Heidelberg, 2012

[4] Deng, Zhongwei, and Minhe Ji. "Spatiotemporal structure of taxi services in Shanghai: Using exploratory spatial data analysis." 2011 19th Int Conference on Geoinformatics. IEEE, 2011

[5] Jamil, Mohammad Sabar, and Saiful Akbar. "Taxi passenger hotspot prediction using automatic ARIMA model." 2017 3rd International Conference on Science in Information Technology (ICSITech). IEEE, 2017

[6] Zheng, Linjiang, et al. "Discovering trip hot routes using large scale taxi trajectory data." Advanced Data Mining and Applications: 12th International Conference, ADMA 2016, Gold Coast, QLD, Australia, December 12–15, 2016, Proc 12. Springer International Publishing, 2016

[7] Li, Bin, et al. "Hunting or waiting? Discovering passenger-finding strategies from a large-scale real-world taxi dataset." 2011 IEEE International Conference on Pervasive Computing and Communications Workshops (PERCOM Workshops). IEEE, 2011

[8] Lalos, Petros, et al. "A framework for dynamic car and taxi pools with the use of positioning systems." 2009 Computation World: Future Computing, Service Computation, Cognitive, Adaptive, Content, Patterns. IEEE, 2009

[9] Altshuler, Tal, et al. "Modeling and prediction of ride-sharing utilization dynamics." Journal of Advanced Transportation 2019 (2019): 1–18

[10] Li, Jacky PK, Sandjai Bhulai, and Theresia van Essen. "Optimization of the revenue of the New York City taxi service using Markov decision processes." 6th International Conference on Data Analytics, Barcelona (Spain), November, 2017

CHAPTER 7

Protein Structure Prediction Using Feature Selective Support Vector Machines

Md. Golam Morshed Raju

Khulna University of Engineering and Technology, Dept. of CSE, Khulna, Bangladesh.

Pintu Chandra Shill

Khulna University of Engineering and Technology, Dept. of CSE, Khulna, Bangladesh.

CONTENTS

Numerous biological processes depend on proteins, one of the most fundamental and important substances. However, in order to construct novel proteins with the appropriate activity, secondary structure prediction is an essential and vital task.

 DOI: 10.1201/9781003605508-7

In this study, we introduce the SVM and feature selection as new approach to determining protein's secondary structures. The proposed framework consists of the following steps: The messy data can be managed by first pre-processing the dataset with the primary sequence. Each sequence of primary and secondary data in the dataset is orthogonally encoded to provide a distinct embedding that is a feature of the model. Using the window-padding method, which considers the sequence and sliding window's size in addition to the features of the windowing outcomes where a feature is constructed from the basic sequence of the data set. During the feature selection process, the most important and relevant characteristics are picked out of the set of developed features. The secondary protein structural states are determined by first preprocessing the dataset and then the support vector machine is trained with the appropriate features. The simulation results on the ccPDB 2.0 dataset demonstrate a promising 91.73 percent accuracy. The research demonstrates that the suggested method outperforms the majority of the alternatives and accomplishes the primary goal of improving prediction accuracy and decreasing error.

7.1 INTRODUCTION

In the contemporary era of scientific study, protein structure prediction is a significant topic. The relative abundance of each amino acid component affects how a protein is predicted to fold. In molecular biology, predicting protein secondary structures is a major obstacle. In between its primary and tertiary structures, proteins adopt a secondary structure. Treating an illness shortly after discovering its protein's secondary structure is a major breakthrough in molecular biology. These problems can be solved with the help of neural networks. Biological neural networks have a hard time being applied to this problem. So that these issues can be resolved the amino acid chains that make up proteins are held together by peptide bonds. Each polar amino acid in the chain possesses an NH group that can act as a hydrogen bond donor and a free carbonyl group that can accept hydrogen bonds. Therefore, these subunits are able to communicate with one another inside the protein framework. Protein macromolecules are built from the 20 amino acids that make up their structural components. Peptide bonds connect each subsequent amino acid in these linear sequences. One hydrogen atom and one carboxyl atom make up each individual amino acid. Building blocks of proteins can be thought of as the "bricks and mortar" of the body. Protein structures can be figured out with the use of spectroscopy and crystallography. This is the rudimentary framework for all proteins. The next step in protein folding is determined by the sequence of amino acids at this location. The primary structure is folded and twisted to produce the secondary. Ternary structure refers to a polypeptide chains the geometry of three dimensional. The ultimate protein structural class, a quaternary structure of three dimensional, is formed by additional folding and twisting of polypeptide chains. A protein's secondary structure is an intermediate protein sequence structure that may be utilized to foretell a tertiary protein's structure. Coil, sheet, and helix are the three types of secondary structures. The function of a protein can be deduced from its structure. Assumptions about how to control, affect, or alter it may be made. New treatments can't be developed without first understanding the fundamental processes that underpin the wide variety of species. Protein secondary structure prediction sheds light on protein function, interactions, and applications. Due to their extreme complexity, tertiary structures are not optimal for efficient medication delivery. Thus, it has been concluded that predicting secondary structures is the best approach to take when creating new medicines. Support vector machine, sometimes known as SVM, is a cutting-edge approach of robust classification that has its roots in machine learning [1]. Natural language processing, robotics, facial and speech recognition, and a host of other applications also rely on it [2, 3]. Bioinformatics, social networks, picture/audio authentication, text analysis, facial recognition, and faulty card identification are just some of the areas where this has been put to use [7, 8]. Predicting a protein's secondary structure [9] requires using sequence alignment, a method for comparing and ordering biological sequences.

This is how the research is designed: In Section 3 we list the relevant research being done in many different areas of protein structure prediction. In Section 5, we compare and contrast the results with existing methods for predicting protein shapes. In Section 4, we examine the features and capabilities of the best protein structure prediction algorithms in detail.

7.2 RELATED WORK

Several researchers have looked into the challenge of Predicting Protein Secondary Structure by utilizing various computational approaches. The ability of a system to learn from its mistakes and alter its behavior in response to new knowledge is what is meant by the term "Machine Learning" [4]. The basic principle behind the SVM is that an ideal hyperplane can be used to classify patterns [1]. By analyzing the amino acid sequences surrounding it, machine learning systems can predict its secondary structure [9]. The most precise methods, when applied to unknown protein sequences, achieved an average accuracy of about 77% [7]. Although SVM was initially used to predict protein secondary structures, its accuracy was not much higher than that of the previous technique [5]. Modern SVM algorithms, such as binary classifiers that use two layers, complex techniques, and PSI-BLAST profiles, can significantly outperform older SVM discoveries. This is true despite the fact that PSIPRED and contemporary SVM algorithms perform similarly in terms of performance [6]. The techniques which are most accurate, like PSIPRED, have a 77 percent average accuracy in predicting a residue. Finding a novel PSSP strategy is necessary to attain the needed level of prediction accuracy. Big data and protein data tend to slow down algorithms, reducing their prediction accuracy [11]. Historically, SVM has been used mostly in academic settings. The cross-validation accuracy for the RS126 set was 71.3% using this method, which used multiple types of alignment data. An innovative method taken by Chandonia and Karplus in training their NNs was to employ a dataset consisting of 681 proteins that are not identical to each other. The plan was 74.8 percent effective in achieving its goals [15]. The use of frequency profiles encoded with evolutionary information is one of the more recent attempts to put this SVM learning engine to use in the prediction of secondary structures [3]. SVM-based protein secondary structure prediction was found to be more accurate after a feature selection strategy using mutual information and genetic algorithms was implemented. Several studies have shown that SVM-based protein secondary structure prediction can be made more accurately by employing feature selection algorithms [21]. Linear discriminant analysis (LDA) and the maximum relevance, minimum redundancy (m-RMR) criterion were proposed as a means of selecting features [22]. To boost the precision of SVM-based structure prediction of secondary protein [23], extra parameters such as solvent accessibility, secondary structure propensity, and dihedral angles were used. This approach combined data on structural evolution with information on evolutionary history. The underlying issue with these models is that they do not take into consideration everything that could possibly occur. Even if these older models are more effective, it is probable that they are lacking key aspects of protein's secondary structure. These aspects are vital for understanding the protein. In addition to being unable of ensuring the finding of the best possible protein sequence, a considerable number of approaches struggle when presented with big protein datasets. These models have a significant flaw in that, in order to calculate their estimations, they rely far too heavily on probability values. This is the model's primary problem. Using this method may not be the most effective strategy when trying to uncover the secondary structures of proteins; as a result, the findings that are obtained using this method may not be accurate. Developing a method that would lower the present restrictions while still providing predictions about the structure of proteins is both a time-consuming and important task. This was proved by the preceding analysis of the conventional and combinatorial methodologies that are now in use. Data mining is incomplete without the ability to make predictions.

According to the findings of the research on conventional and combinatorial approaches that were described earlier, developing a method that would remove the constraints that are now placed on protein structure prediction is both essential and time-consuming. This is because the current constraints are placed on protein structure prediction.

7.3 PROPOSED METHOD

Predicting protein secondary structure using a model that incorporates data preprocessing and feature selection is the desired strategy. The PSSP approach, which integrates feature selection with support vector machines, will be described in greater depth below (SVMs). The traditional orthogonal matrix or patterns of successive amino acids found in protein sequences were among the encoding methods considered for this purpose. The protein's encoded structure allowed us to use Radial Basis Function (RBF) kernel functions in SVM to assess the protein's fundamental structure. In order to achieve the best possible outcomes, the SVM's regularization and its kernel function settings were both optimized.

A simplified version of the method we present is shown in (Fig. 7.1).

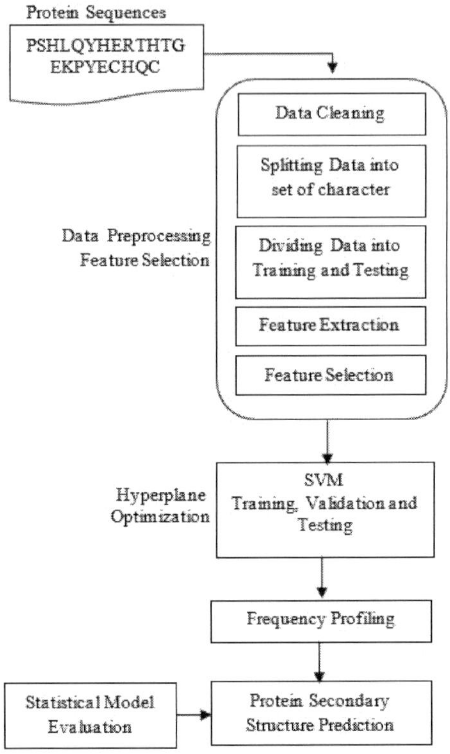

Figure 7.1: Block schematic of method.

7.3.1 Data Preprocessing

The technique uses orthogonal encoding and window sliding to organize data. All the information about the primary and secondary structures is first orthogonally segmented before being encoded. The string sequence was converted into an array of characters. A 2D array with a string in each row constitutes the input sequence. Convert each row into a character array. After the primary structure of the input sequence has been recovered, the secondary structure can then be broken down into its component parts of the target sequences.

As a result, Separate orthogonal segments are created for each primary and also secondary structure's data before encoding. Both primary and secondary protein structures can be broken down, and this can result in orthogonal encoding and the target labeling of the protein. Despite how challenging it is, every residue needs to be assigned a secondary state.

The Dictionary of Secondary Structure Prediction then labeled it based on the tertiary structure that was determined experimentally. The letters H, G, I, E, B, C, S, and - are used to denote the eight distinct forms of its secondary structure. The respective meanings of these letters are helices, strand, isolated bridges, turns, twists, loop, and bends (rest).

To simplify them into three categories, one of the following approaches is typically used:

(1) H for G, H, and I; E for E; and C for rest

(2) H for G, H; E for E, B; and C for remainder

(3) H for G, H; E for E; and C for remaining

(4) H for H; E for E, B; and C for reminder

(5) E for E; H for H; C for remaining

For the sake of a balanced comparison, we used procedure (2) even though it decreases prediction accuracy more often than method (5) boosts it. The precision of a prediction can be measured using a number of established KPIs. Amino acid information is gathered by using Windows padding as a new model feature [10].

7.3.2 Feature Selection

Additional features are extracted from the protein sequences utilizing the ccPDB 2.0 dataset [24] by utilizing evolutionary data from multiple sequence alignment (MSA) of homologous proteins obtained via PSI-BLAST. Structure-related characteristics such as solubility in solvents, propensity for secondary structure, and dihedral angles are also desirable. To calculate these features, the DSSP method is used. Feature-based selection follows technique, which is Window sliding that recovers characteristics from encoded sequences by using the machine learning classification, across a large number of feature subsets. Each window uses a residue feature vector. Statistics-based optimization takes in observations of potentially correlated variables and transforms them into values for linearly uncorrelated variables based on all of the characteristics selected. Both major and secondary structures must be broken down into orthogonal components before encoding. Protein primary and secondary structural dissection yields orthogonal encoding and target labeling for each amino acid. Each protein secondary structure disruption will be labeled or targeted. The dataset will have at least some of the required information using the total number of labels.

7.3.3 Classification using SVM

As a popular supervised learning model, the support vector machine (SVM) makes use of statistically-informed learning strategies [1]. The SVM technique, developed by Vapnik and Cortes, is a modern learning system [16]. The regularized hinge loss function minimization is analogous. Training the computer involves utilizing a feature with high-dimensional space containing linear hypotheses and a learning method based on optimization theory [17]. Because of its ability to reduce structural risk, this machine is particularly useful for preventing overfitting. A convex QP formulation is possible, and training unambiguously yields a global optimum [18]. The proposed method relies on the labeled data for both training and prediction. Thus, SVM is used by supervised classification. SVM classifies binaries. As a way to train SVM to distinguish between the three structures, three input files are generated. The input files can be built using one of three different windowing strategies: C/C', H/H', or E/E'. It was observed that 17 of the window sizes tested performed well. Support vector machines (SVMs) construct hyperplanes in n-dimensional spaces to make classifications between groups of data easier. The classifier uses the hyperplane with the highest margin to make the partitions. To better classify new data points, SVM calculates the optimal hyperplane. To classify new data points, SVM locates an appropriate hyperplane. A kernel function and regularization parameter must be selected for each Binary Classifier. A kernel is a piece of software that raises the dimensionality of data that has previously been stored in lower ones. Pattern analysis and the mapping of nonlinear to linear spaces are two common applications of kernel techniques. The kernel approach is an effective and useful tool. We found that the Radial Basis Function (RBF) would be the best kernel given the non-linear data specification of our SVM. Radial basis function kernels provide not just prediction on datasets but also analysis of reports. There are a total of 24747 occurrences across three groups in this data set. The results can be linearly divided into a number of different dimensional spaces. The kernel of the radial basis is stated as follows:

$$k(x_1, x_2) = \exp\left(-\frac{\|x_1 - x_2\|}{2\sigma^2}\right) \tag{7.1}$$

Where, Distance of Euclidean (L2-norm) which is distance between X1 and X2 is denoted by (X1 - X2) and our hyper-parameter is also defined by (X1 - X2).

The RBF kernel has great local features and a large convergence region, making it applicable to data sets of varying sizes. By just retaining the support vectors during training rather than the complete dataset, RBF Kernel Support Vector Machines get around the problem of space complexity. This allows the machines to learn more efficiently.

7.4 RESULTS & ANALYSIS

Protein Databank (PDB) source was used in the compilation and creation of the data set; PDB is a repository of datasets produced by the DSSP. The machine is trained using a dataset that makes predictions for these three scenarios. We tested our technique on the ccPDB 2.0 dataset [24], which contains protein sequences and structural class labels. The protein primary structure is shown on the odd-numbered line of data, while the secondary structure is shown on the even-numbered line. This allows us to determine whether or not the primary and secondary structures have distinct character counts. The SVM is derived from a linearly separable optimum classification decision surface [14], which is created by the support vector of the training sample. The SVM classifier requires a high-performing RBF kernel.

7.4.1 Confusion Matrix for Multi-Class Classifier

The effectiveness of different classification methods can be graphically shown with the use of confusion matrices. After testing and analyzing the model, the percentage of events successfully detected and mistakenly identified in the test data is displayed (see Table 7.1) [13].

Table 7.1: Confusion Matrix for Multi-Class classifier.

Actual label	Predicted		
	H	**C**	**E**
H	3162	633	9
C	427	17134	236
E	14	774	2358

According to the execution schema, where the value zero denotes the H structural class, one denotes the C class, and two denotes the E class for the classification.

7.4.2 Correctly Classified Instances (CCI)

The following abbreviations will be used: TP is denoted for True Positive, TN is denoted for True Negative, FP is denoted for False Positive and FN is denoted for False Negative.

$$True\ Positive\ Rate\ (TPR) = \frac{TP}{(TP + FN)} \tag{7.2}$$

$$False\ Positive\ Rate\ (FPR) = \frac{FP}{(FP + TN)} \tag{7.3}$$

The degree to which one can have faith in the model, is determined by how well it performs in terms of the accuracy value [13].

$$Accuracy\ Value = \frac{(TP + TN)}{(TP + TN + FP + FN)} \tag{7.4}$$

7.4.3 Specificity/True-Negetive Rate

It denotes the proportion of non-infected individuals who received a negative test result [13].

$$Specificity = \frac{TN}{(FP + TN)} \tag{7.5}$$

$$FPR = 1 \breve{}\, Specificity \tag{7.6}$$

7.4.4 Precision (PRE)

Scientists characterize it by [13].

$$Precision\ (PRE) = \frac{TP}{(TP + FP)} \tag{7.7}$$

7.4.5 Sensitivity or Recall (REC)

The academic community settled on the following definition for this variable in [13].

$$Recall\ (REC) = \frac{TP}{(TP\ +\ FN)} \tag{7.8}$$

7.4.6 Kappa Statistics Value (KS)

Kappa is a statistic for evaluating the discordance between a dataset's expected and actual arrangements [13].

$$Kappa\ (K) = \frac{Ro\ -\ Re}{1\ -\ Re} \tag{7.9}$$

In this formula, R_e represents the probability of a random assertion, and Ro represents the extent to which raters generally agreed with the statement being made.

7.4.7 F-Measure (FM) or f1-Score

Harmonic mean recall and accuracy are computed using the F1 score (harmonic mean since accuracy and recall are rates). If the F-Measure indicator is FM, then

$$F\breve{}Measure\ (FM) = \frac{2\ \times\ TP}{(2\ \times\ TP + FP + FN)} \tag{7.10}$$

Table 7.2: Statistically Measured Value.

Measured Value	H	C	E
Sensitivity(Sn)	0.8776	0.9241	0.9059
Specificity(Sp)	0.9696	0.8932	0.9640
Precision(Pr)	0.8312	0.9627	0.7495
Negative Predictive Rate (NPR)	0.9789	0.7976	0.9887
False Positive Rate (FPR)	0.0304	0.1068	0.0356
False Discovery Rate (FDR)	0.1688	0.0373	0.2505
False negative Rate (FNR)	0.1224	0.0759	0.0941
Accuracy value (ACC)	0.9562	0.9164	0.9583
F- measure (FM)	0.8538	0.9430	0.8203
Matthews Correlation Coefficient (MCC)	0.8285	0.7883	0.8015

The statistically measured value which is shown in (Table 7.2) defining the high outperform of the proposed method than the previous.

7.4.8 Mean Absolute Error (MAE)

IIt is common practice to disregard the magnitude of errors without first determining their sign [13].

$$MAE = \frac{(| p1 - b1 | + ... + | pn - bn |)}{n} \tag{7.11}$$

7.4.9 Relative Absolute Error (RAE)

In the same sense of normalization, it is the sum of all possible errors [13].

$$RAE = \frac{| p1 - b1 | + \cdots + | pn - bn |}{| b1 - b | + \cdots + | bn - \bar{b} |} \tag{7.12}$$

7.4.10 Matthews Correlation Coefficient (MCC)

PRE and REC coherence are assessed for MCC [13].

Table 7.3: Simulation Result at a glance.

Lebel	Recall	Precision	F1 Score	Support
H	0.88	0.83	0.85	3804
C	0.92	0.96	0.94	17797
E	0.91	0.75	0.82	3146
Average of Macro value	0.90	0.85	0.87	24747
Average of Weighted value	0.91	0.92	0.91	24747
Accuracy score				0.917
Kappa Statistical Value				0.801
MAE				0.084
RAE				0.288

Table 7.3 shows the structural accuracy and error measures, which demonstrate that the suggested method outperforms.

7.4.11 ROC & Recall vs Precision curve

In light of the test data, it is the likelihood that the classification system would assign a randomly chosen positive instance a higher rating than a randomly chosen negative instance [13]. This probability can be expressed as a percentage. In order to determine the precision of a binary result, Receiver Operating Characteristic (ROC) curves can be plotted and shown (Fig. 7.2).

In Fig. 7.3, we track the effects of threshold values of different candidates, which range from 0.0 to 1.0, on both the TP rate and the FP rate.

Against put it another way, it compares the rate of hit to the rate of false alarm. By dividing the good outcome's entire number by the negative outcomes total number, one may get the genuine positive rate. Results that are positive demonstrate how well the model predicted the success class. Talented models are represented by the point at (1,1), whereas those without it are portrayed as the straight line superimposed on

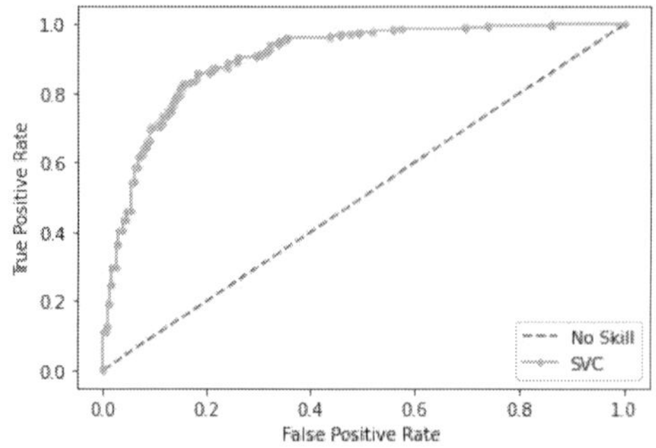

Figure 7.2: Graphical view of ROC.

the curved representation of the talentless model at (1,1). At each stage, the most advanced model is presented at (1,1).

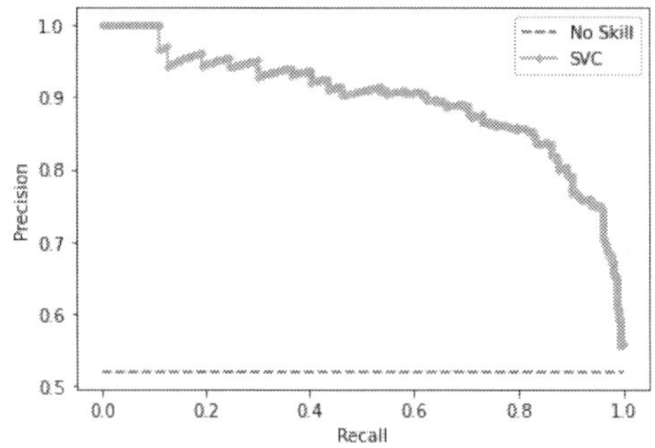

Figure 7.3: Recall v/s Precision Graph.

Precision-recall curves, which show the relationship between accuracy (y-axis) and recall (x-axis) at different cutoffs, are analogous to the ROC curve shown in (Fig. 7.2). In this scenario, the no-skill classifier is unable to differentiate between the classes and always makes the same prediction. The line denoting the absence of talent could move in either direction based on the ratio of positive to negative ratings. The positive case ratio for the dataset is shown by a horizontal line that is accompanied by a numerical value. For a perfectly even sample, this would be 0.5. In this case, the model is represented by a dot at (1,1). A model with a curve that slopes toward (1,1) is more competent than one whose curve is flat across the board.

7.4.12 Comparative Analysis

To evaluate how well the proposed strategy stacks up against other data-set and models approaches than the previous which is shown(see Table 7.4), we used CB513 [6] and RS126 [6] datasets. The RCSB Protein Data Bank was consulted in order to acquire information regarding the secondary structures of RS126 [15, 17] Of the 23,411 residues here, helices make up 45.63 percent, sheets 31.16 percent, and coils 31.23 percent. Manufacturing the CB513 [6]. Out of a total of 84,034 residues, 34.30% are in helices, 22.54% in sheets, and 43.16% in coils, according to the ccPDB 2.0. It consists of 65.78% coil, 15.37% helix, and 12.71% sheet across its 24,747 residues [24]. Our primary objective is to predict sequences with similarity to the protein data set. It's promising that our method outperforms a variety of established prediction methods [6].

Table 7.4: Comparative Analysis.

Structural Class	Predicted Accuracy (%)		
	RS126	CB513	ccPDB 2.0
C	84.083%	87.564%	91.635%
H	89.639%	90.718%	95.624%
E	90.223%	92.651%	95.826%

The RBF kernel function is the best outperformer function of the Support Vector Machine (SVM) compared to other simulation results, providing the most outstanding prediction accuracy with equivalent simulation time, outperforming certain other previous prediction elements.

7.5 CONCLUSION AND FUTURE WORK

In this study, Protein secondary structure can be predicted with greater accuracy using SVM learning technology with feature selective method. For this aim, testing with various window lengths is used to identify the optimal binary classifier window size for support vector machines. Features are selected and data is pretreated to improve encoding methods. Statistics played a vital role in this approach. In order to ascertain whether or not our method of prediction is useful, the first step was to determine how accurate it is. The outcomes are promising because our method outperforms a number of traditional prediction methods. The nonlinearly separable data is successfully implemented using the machine learning technique SVM. When it comes to binary classification tasks, SVM is quite generalizable. Since the kernel function is responsible for determining the output, we are forced to the conclusion that the selection method is extremely important. This is an exceptionally challenging issue that has not yet been resolved to anyone's satisfaction. The accuracy of the experimental results was determined to be 91.73 percent when applied to the ccPDB 2.0 datasets. In comparison to the other solutions, support vector machine (SVM) based on RBF is the most optimal and beneficial proposed prediction method for the aspect of a large dataset. Our method outperforms other prediction methods, resulting in intriguing outcomes.

Bibliography

[1] Mayuri, P. & Hitesh, S. "Protein secondary structure prediction using support vector machines (SVMs)," *in Proc. Int. Conf. on Machine Intelligence Research and Advancement*, ICMIRA 2013, pp. 594–598.

[2] Sasan, K., Shahidan, M. A. Mehran, H., Jafar S. & Mohammad J. R. "Advantage and Drawback of Support Vector Machine Functionality," *IEEE Int. Conf. on Computer, Communication and Control Technology (I4CT)*, 2014, pp. 63–65.

[3] Sujun, H. & Zhirong S. "A Novel Method of Protein Secondary Structure Prediction with High Segment Overlap Measure: Support Vector Machine Approach," *J. Mol. Biol.*, vol.308, pp.397–407, 2001.

[4] Latil, M., Janmejay P., Priyanka, S.& Arvind K. "Support Vector Machine Accuracy Improvement with Classification," *12th Int. Conf. on Computational Intelligence and Communication Networks (CICN)*, 2020, pp. 1–5.

[5] Yanchun, W., Jinyong, C., Yihui, L. & Yehong, C. "Prediction of Protein Secondary Structure using Support Vector Machine with PSSM Profiles," *in Proc. IEEE Information Technology, Networking, Electronic and Automation Control Conference*, 2016, pp. 502–505.

[6] Mohammad, S., Syed, M. B., Taskeed, J., Firoz, A. & Haseena K. "Protein secondary structure prediction with high accuracy using Support Vector Machine," *10th Int. Conf. Comput. Inf. Technol. ICCIT*, 2007, pp. 9–12.

[7] Jianyi Y., Ivan A., Hahnbeo P., & David B. "Improved protein structure prediction using predicted inter residue orientations," *Biophysics and Computational Biology*, vol.117, no. 3, pp. 1496–1503, 2020.

[8] Yang, Z. & Jeffrey S. "The protein structure prediction problem could be solved using the current PDB library," *Proc Natl Acad Sci*, vol.102, no.4, pp.1029–1034.

[9] Yun, Z., Bingru, Y., Lijun W. & Bing A. "New trend of protein secondary structure prediction," *Int. Symp. Intell. Ubiquitous Comput. Educ. IUCE 2009*, pp. 121–124.

[10] Yehong, C., Yihui, L., Jinyong, C. & Yanchun, W. "Prediction of Protein Secondary Structure using SVM-PSSM Classifier Combined by Sequence Features,"*IEEEAdvanced Information Management, Communicates, Electronic and Automation Control Conference (IMCEC)*, 2016, pp. 103–106.

[11] Heba, M., Afify, M. B., Abdelhalim, Mai, S. M. & Ahmed Y. S. "Protein secondary structure prediction (PSS) using different machine algorithms," *Egypt. J. Med. Hum. Genet.*, vol.22, no.54, 2021.

[12] Mingshun, Y., Zijiang, Y., Guangzao, H. & Guoli, J. "A Novel Feature Selection Method to Predict Protein Structural Class," *Comput. Biol. Chem.*, vol. 76, pp.118–129, 2018.

[13] Burkhard, R.& Chris, S. "Prediction of protein secondary structure at better than 70% accuracy," *J. Mol. Biol.*, vol. 232, pp. 584–599, 1993.

[14] Wenlei, S.& Xinhai, F. "Speech classification based on cuckoo algorithm and support vector machines," *2017 2nd IEEE International Conference on Computational Intelligence and Applications (ICCIA)*, Beijing, 2017, pp. 98–102,

[15] John, M. C. & Martin, K. "New methods for accurate prediction of protein secondary structure," *Proteins*, vol. 35, pp. 293–306, 1999.

[16] Søren, K. S.& Ole, W. "Protein secondary structure prediction with long short term memory networks", 2014.

[17] Nello, C. & John, S. T. An Introduction to Support Vector Machines. Cambridge, U.K.: Cambridge Univ. Press, 2000.

[18] Ishrat, K., Hossain, M. A. F. & Pintu, C. S. "A hybrid kernel support vector machine for predicting protein structure", In: *2021 International Conference on Automation, Control and Mechatronics for Industry 4.0, IEEE*, 2021, pp. 1–5.

[19] James, C. "Protein secondary structure prediction with support vector machines," M.Sc. thesis, Univ. Sussex, Brighton, U.K., 2002.

[20] Hyunsoo, K. & Haesun, P. "Protein secondary structure prediction based on an improved support machines approach," *Dept. Comput. Sci. Eng., Univ. Minnesota*, Minneapolis, 2003.

[21] Yonge, F., Hao, L. & Liaofu, L. "Prediction of protein secondary structure using feature selection and analysis approach", *Acta biotheoretica*, vol. 62, no. 1, pp. 1–14, 2014.

[22] Arti, P. & Deepak, S. C. "Svm kernel functions for classification", 2013 International Conference on Advances in Technology and Engineering (ICATE), pp. 1–9, 2013.

[23] Jiyun, Z., Hongpeng, W., Zhishan, Z., Ruifeng, Xu. & Qin, L. "Cnnhpss: Protein 8-class secondary structure prediction by convolutional neural network with highway", *BMC Bioinformatics*, vol. 19, 2018.

[24] Piyush, A., Sumeet, P., Rajesh, K., Vinod, K., Harinder, S., Pawan, K. R. & Gajendra, P. S. R. "ccPDB 2.0: an updated version of datasets created and compiled from Protein Data Bank", *Database*, vol. 2019.

CHAPTER 8

Sentiment Analysis of Bangla Text Using Transformer Based Model

S.M. Shadman Sakib

Khulna University of Engineering & Technology, Dept. of CSE, Khulna, Bangladesh.

Md. Jannatun Nayeem

Khulna University of Engineering & Technology, Dept. of CSE, Khulna, Bangladesh.

Abdul Aziz

Khulna University of Engineering & Technology, Dept. of CSE, Khulna, Bangladesh.

Abu Zafar Md Nuruzzaman Abir

Tata Consultancy Services Limited, United States of America

Jakaria Rabbi

Khulna University of Engineering & Technology, Dept. of CSE, Khulna, Bangladesh.

CONTENTS

SEntiment analysis falls under Natural Language Processing (NLP) and involves identifying and extracting the emotions or feelings a given text conveys. The main target of sentiment analysis is to identify the polarity of a given sentiment expressed in the form of a text, which can be positive, negative, or neither positive

DOI: 10.1201/9781003605508-8

nor negative. The highly inflected properties of the Bengali language makes sentiment analysis a challenging task. The absence of adequately annotated datasets in the Bangla language domain presents an additional difficulty for sentiment analysis. Several models based on "Transformer" have been developed for various languages, indicating substantial advancements in this field of Natural Language processing. This study examines the sentiment of Bangla text documents using the latest transformer-based model, Bidirectional Encoder Representations from Transformers (BERT). On top of BERT, Bi-directional long-short-term memory (BiLSTM) and Bi-directional gated recurrent unit (BiGRU) have been integrated. We have created a benchmark dataset of approximately 5000 sentences. The dataset is classified into 2 categories (positive and negative). The pre-trained model called "Bert-base-multilingual-cased" was adjusted or fine-tuned, and as a result, it achieved an accuracy of **90%** when applied to our collected dataset. Our proposed BERT-BiGRU and BERT-BiLSTM model achieves 85% and 82% respectively. We hope that this study could function as a valuable standard for assessing sentiment analysis in the Bangla language.

8.1 INTRODUCTION

In the modern era of the Internet, there has been a significant surge in the sharing of information, knowledge, and opinions compared to the previous decade. Users on the Internet actively participate in reviewing various things such as restaurants, dramas, books, videos, e-commerce products, and more. This user-generated content plays a crucial role in providing upfront feedback to subsequent users, enabling them to make informed decisions. This type of text content is particularly relevant for `sentiment analysis`[1], a task that involves classifying the polarity of a given text as positive, negative, or neutral.

Bangla, also known as Bengali, is a significant Indo-Iranian language and holds the sixth position in terms of global popularity. It is spoken by a population of over 250 million people. Bangla serves as the primary language in Bangladesh and is widely spoken as a second language in India. While sentiment analysis has been extensively researched for various languages like English [1], Chinese [2], and Arabic [3], the situation is different for Bangla. Sentiment analysis remains an unresolved research challenge for Bangla due to limited resources and the inherent complexity of the language, resulting in a scarcity of research work in this domain.

In this study, we created a benchmark dataset by collecting 5000 comments having positive and negative sentiments. We developed a hybrid BERT-BiGRU and a hybrid BERT-BiLSTM model. Also, we fine-tuned the pre-trained multilingual BERT model using three datasets, one is our own collected, another from [4] which is a drama dataset and another is the book review dataset from [5]. All these three datasets are also evaluated on the previous two hybrid models. Finally, in the comparison Table 8.5, we can see that the BERT fine-tuned model has better test accuracy for all three datasets.

[1] `https://aws.amazon.com/what-is/sentiment-analysis/`

8.2 RELATED WORK

In progressive years, sentiment analysis has drawn much interest from various linguistic viewpoints. For sentiment analysis, a sizable number of datasets in various languages are accessible. One may analyze a viewpoint from that language by employing a phrase or sentence. There are few datasets available in a low-resource language like Bangla. We looked at some of the previous studies and tried to figure out which approach they did. Recently, deep learning has been vastly used in such as sentiment analysis, natural language processing [6], recommendation systems [7], providing security using face detection [8], danger detection [9, 10], etc. In this study, we discussed some related works of sentiment analysis thoroughly that use several deep learning-based model.

Alam et al. [11] explored different transformer models for corresponding Bangla text classification tasks. The authors showed that transformer models' fine-tuning can perform better than traditional methods and deep learning models like CNN and LSTM. They used six benchmark datasets from different domains. Rather than using the monolingual Bangla model, they fine-tune multi-lingual transformer models trained on a substantial corpus. For adapting the transformer model, they used unique tokens at the beginning and the end. Also, they used the Cross entropy loss criterion and Adam optimization algorithm [12] to train the models by splitting 80% of the dataset for training and 20% for validation. M. M. Rahman et al. [13] implemented a transformer-based Bangla document classification. For their classification task, they used BERT and ELECTRA on three open sources of data: BARD [14], `OSBC`[2], and `prothomalo`[3], collected by them. After removing unnecessary punctuation for pre-processing, they divided data into 8:2 ratios for training and validation purposes. The model weights are gained from the hugging face `hugg`[4] and then the fine-tuning process had been completed. Here 20 epochs and a batch size 16 are utilized to adjust the weights. After the training, they tested the models with test data and assessed them using precision, recall, accuracy, and F1 score. A study by Prottasha et al. [15] compared various deep learning and machine learning techniques for topic-based text classification. The data was collected from social media and labeled by an expert in the relevant domain. They pre-processed the data to remove spelling errors, noise, and missing values. Later, they extracted the features and reduced the dimensions. The BERT model was used to analyze the sentiment of the data. Further, the best weights were saved throughout the training phase, and the model was evaluated using precision, recall, accuracy, F1 score, ROC AUC, and Kappa. Bhattacharjee et al. [16] introduced BanglaBERT (a pre-trained BERT model for Bangla), a Natural language understanding (NLU) model in Bangla and presented a new ensuing dataset on Natural Language Inference (NLI). Around 35 Gigabytes (GB) of data was collected from selected 110 websites by Amazon Alexa. Following some preprocessing steps, the dataset was reduced in size to 27.5 GB. They fine-tuned BanglaBERT on the downstream tasks covering text classification,

[2]`https://scdnlab.com/corpus/`

[3]`https://www.kaggle.com/datasets/twintyone/prothomalo`

[4]`https://huggingface.co/`

sequence labeling, and span prediction. After performing the evaluation, they compared the performance of baselines and pre-trained models on different downstream tasks. Islam et al. [17] presented two newly created, manually labeled datasets in Bangla for sentiment analysis. Cutting-edge performance has been shown in Bangla for the 2-class (Positive, Negative) and 3-class (Positive, Negative, Neutral) SA tests. The $BERT_{BSA}$ deep learning model has been presented for SA in Bangla, which surpasses earlier models that have been built using word2vec or fastText embedding. $BERT_{BSA}$ was correlated with other models trained using Word2Vec and fastText embeddings.

8.3 METHODOLOGY

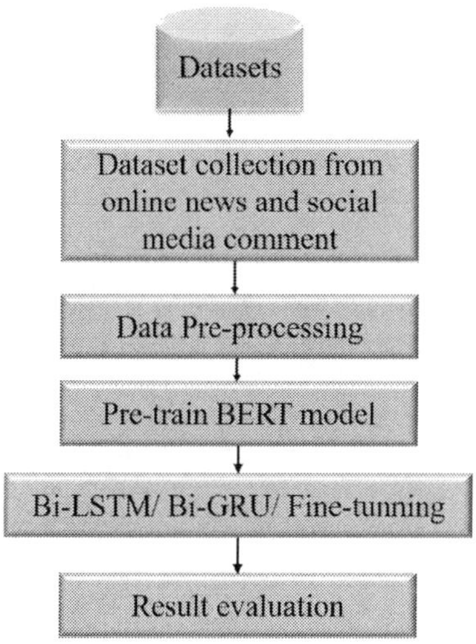

Figure 8.1: Workflow of Proposed Methodology.

Fig. 8.1 shows the workflow of the proposed methodology. There are mainly three phases in this work: (a) Collection of data from online newspapers and social media comments, (b) Pre-processing of the data, removing unnecessary fields, stopwords, punctuation, and digits from the text and (c) Training those datasets with the hybrid model and validate that model. And finally, we compare these models with three different datasets.

8.3.1 Preprocessing

Firstly, We removed the emojis and foreign characters from the text. Removing stopwords is crucial for accurately assessing public sentiment, as they are commonly used in the text but do not provide significant meaning or information. Fig. 8.2 illustrates the process of data preprocessing step by step. After all the actions are done, we get our final dataset.

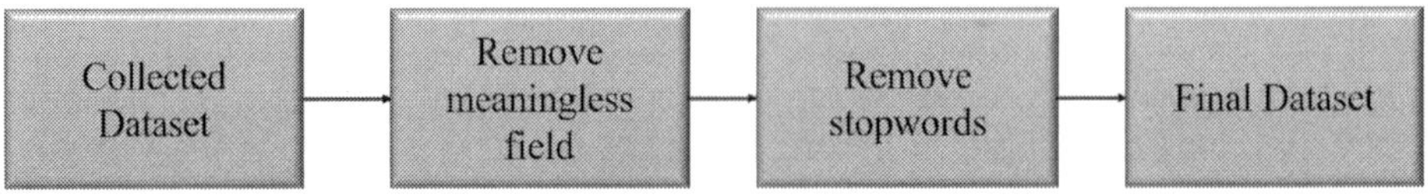

Figure 8.2: Dataset Preprocessing Steps.

8.3.2 Pre-trained Model

We utilized the "bert-base-multilingual-cased" BERT model, which is the only pre-trained model available for the Bengali language. The BERT model is pre-trained [18] using two unsupervised sub-tasks, namely masked language modeling (MLM) [19] and next sentence prediction (NSP) [20]. Through this pre-training process, the BERT model is able to learn and comprehend the language and its context, which makes it a powerful tool for NLP tasks. The steps are described below which are done by the pre-trained model.

8.3.2.1 Tokenization

When performing a classification task, it is necessary to provide a single vector that represents the entire input sentence to a classifier. BERT achieves this with the help of the hidden state of the first token of the input sequence to represent the entire sentence. So, we added an extra [CLS] token manually to the input to enable this approach. The [SEP] token is added at the end of every input sentence. If a given input sentence is shorter than the maximum length, we added padding tokens [PAD] (i.e., null tokens) to build it up to the required length.

8.3.2.2 Converting Tokens to IDs

During the training of the BERT model, we assigned a unique identification number to every token of the input. In cases where some tokens in the input data are not present in the fixed vocabulary, the model is designed to replace those tokens with a special token [UNK], which represents unknown or out-of-vocabulary tokens. Using this approach, the model can still process the input data effectively, even when some tokens are absent in the original vocabulary.

8.3.3 Hybrid Model

All the layers of our two hybrid models are the same except the 3rd layer, which is Bi-LSTM for the BERT-BiLSTM model and Bi-GRU for the BERT-BiGRU model. The working procedure of each of the layers is as follows:

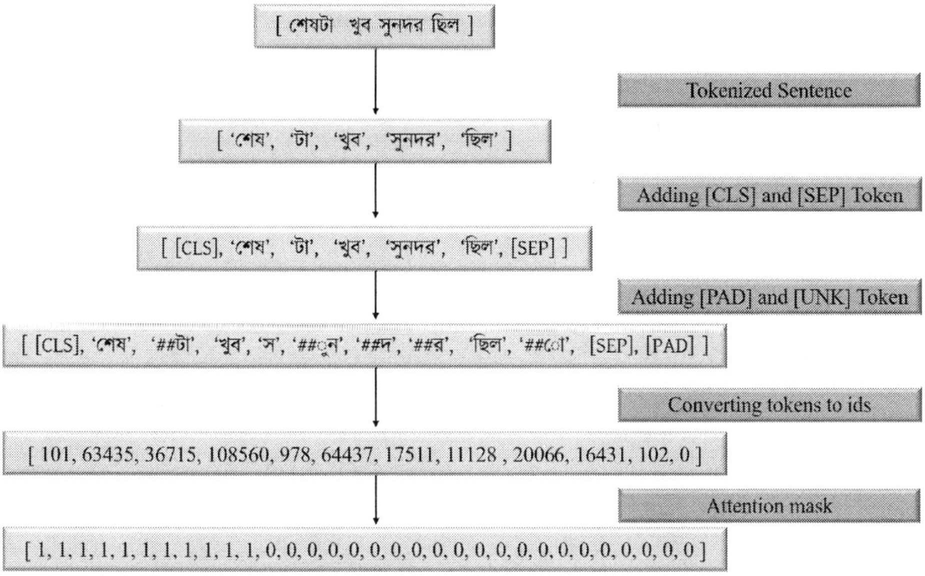

Figure 8.3: Tokenization process of the BERT model.

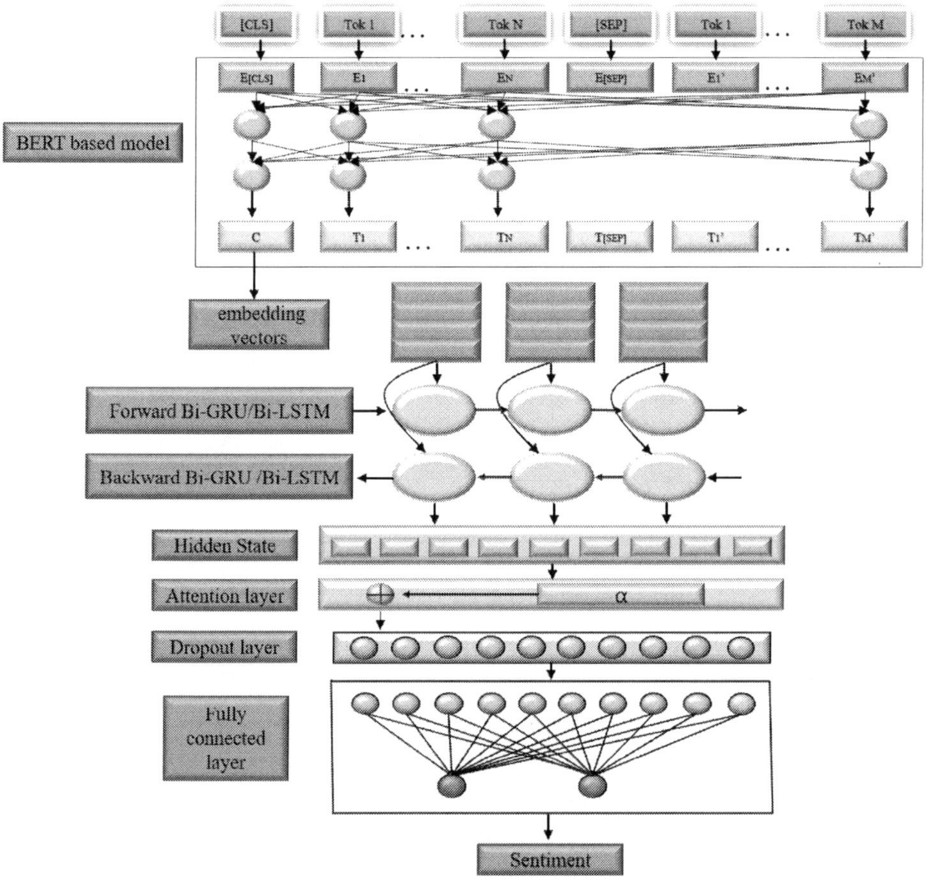

Figure 8.4: Architecture of BERT-BiLSTM and BERT-BiGRU model.

Bi-LSTM: Bi-LSTM helps the model to sense the text in both the forward and backward directions, which can get a broader range of contextual information and improve performance on tasks such as text classification and sentiment analysis. As BiLSTM has forward and backward, there are two types of LSTM. Here, the forward LSTM computes the hidden state h on the base of the previous hidden state h_{t-1} and the vector input x_t, while the backward LSTM calculates the hidden state "h", on the base of future hidden state h_{t+1} and the vector input x_t.The output sequence of hidden vectors is generated by concatenating the vectors from both the forward and backward directions of the Bi-LSTM model.

Bi-GRU: Bi-GRU can overcome the gradient vanishing problem by memorizing the info of the last nodes using a gate, making it appropriate for processing sequential data. The input value is x_t, The output of GRU is h_t, r_t is the reset gate, and the z_t is the update gate. z_t and r_t both together manage the procedure from h_{t-1} hidden state to h_t hidden state. The output value remains between 0 and 1.

Hidden State: These are the hidden vectors of the Bi-LSTM and Bi-GRU output.

Attention Layer: We added this layer to recognize essential words automatically. The word importance vector is calculated by taking the hidden states h of BiLSTM as input and applying an attention mechanism using a weight parameter W, bias b, and a hyperbolic tangent activation function.

Dropout Layer: We incorporated a dropout layer to address the overfitting issue in our model. This strategy randomly drops some training parameters during each iteration. Specifically, during training, some units in the neural layer are temporarily hidden and then optimized through the training process. During our implementation, we set the dropout rate as 0.5.

Fully-connected layer: The fully-connected layer connects each input neuron to every output neuron. This layer predicts the output, which could be positive or negative for our model. We have set the hidden dimension of this layer for BERT-BiLSTM to 200 and the output dimension to 3. For BERT- BiGRU, only the hidden dimension is different, which is 600.

8.3.4 BERT Fine-tuned Model

Here we used BertForSequenceClassification class provided by the hugging face library. Using 'BertForSequenceClassification', a single linear layer is added that takes the final hidden state outputted by BERT and maps it to the number of output classes(two) for the classification task. Here, the output labels are set to 2. Output attention and output hidden states are false. We finetuned all three datasets using this model.

8.3.5 Dataset Description

We collected our created dataset from Facebook and YouTube comments (referred to as the Online news comment dataset). Drama dataset is collected from [4] and Book review dataset is from [5]

Table 8.1: Number of training and testing data

Dataset	Split	Positive	Negative
Our Dataset	Train	2131	2431
	Test	255	270
Drama Review	Train	7238	2797
	Test	1262	510
Book Review	Train	846	854
	Test	136	164

8.4 EXPERIMENTAL RESULT ANALYSIS

The experiment was conducted on the Google Collaboratory platform using the PyTorch library with GPU acceleration and access to 12 GB of RAM and 108 GB of disk space on the Google Cloud. Three different models, BERT-BiLSTM, BERT-BiGRU, and BERT-Finetuned, were used with three distinct datasets. The dataset was divided into different training and validation sets with various splitting ratios to obtain optimal performance results. We run the model using a learning rate scheduling system. The CrossEntropyLoss function was used as the criterion and Adam as the optimizer. A dropout rate of 32 was applied to all models.

Table 8.2: Evaluation of BERT-BiGRU model

Dataset	Class	Precision	Recall	F1-Score	Support
Our Dataset	Negative	0.88	0.84	0.86	282
	Positive	0.83	0.87	0.85	243
Drama Review	Negative	0.94	0.96	0.95	1230
	Positive	0.91	0.86	0.89	542
Book Review	Negative	0.77	0.90	0.83	141
	Positive	0.90	0.76	0.83	160

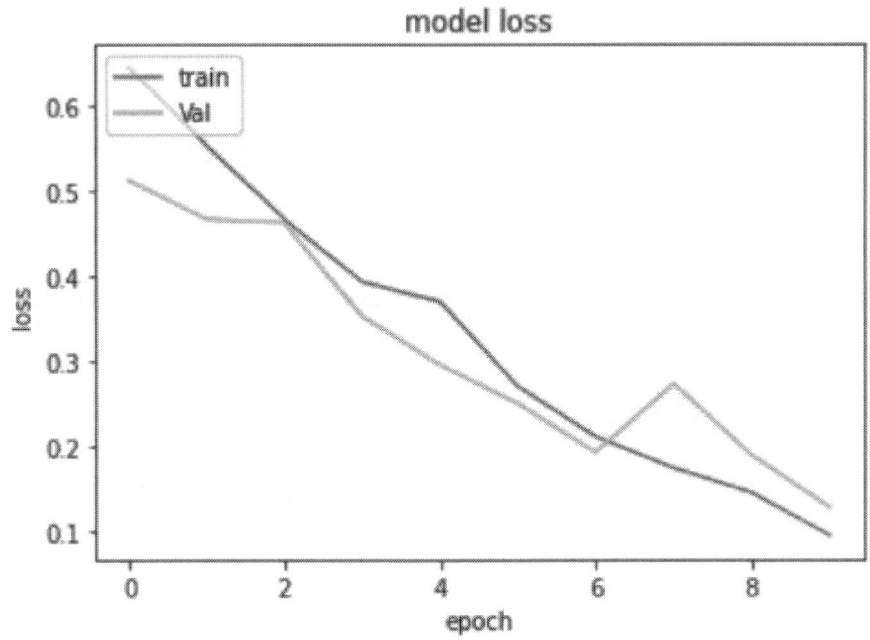

Figure 8.5: Loss vs Epoch curve for BERT-BiGRU model for Our Dataset

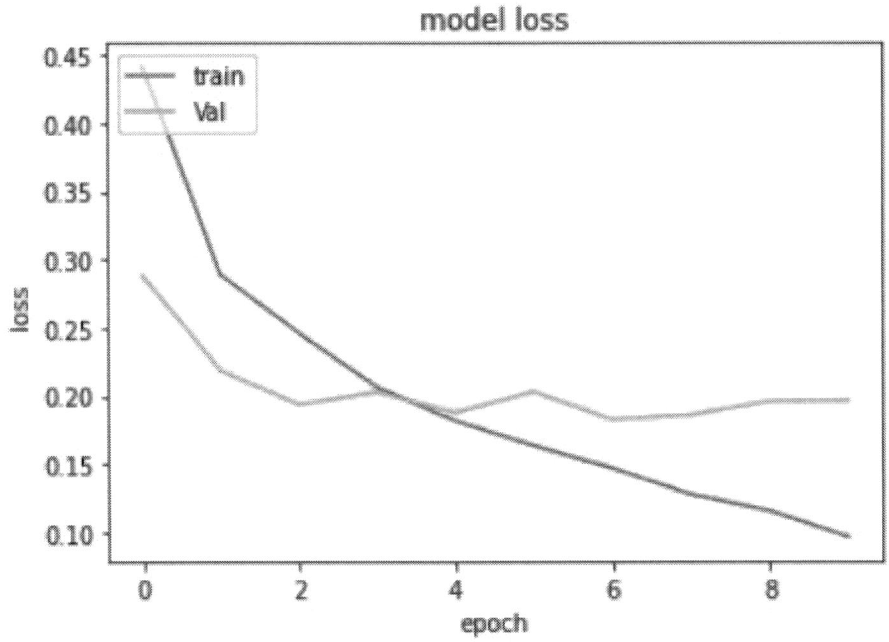

Figure 8.6: Loss vs Epoch curve for BERT-BiGRU model for Drama Dataset

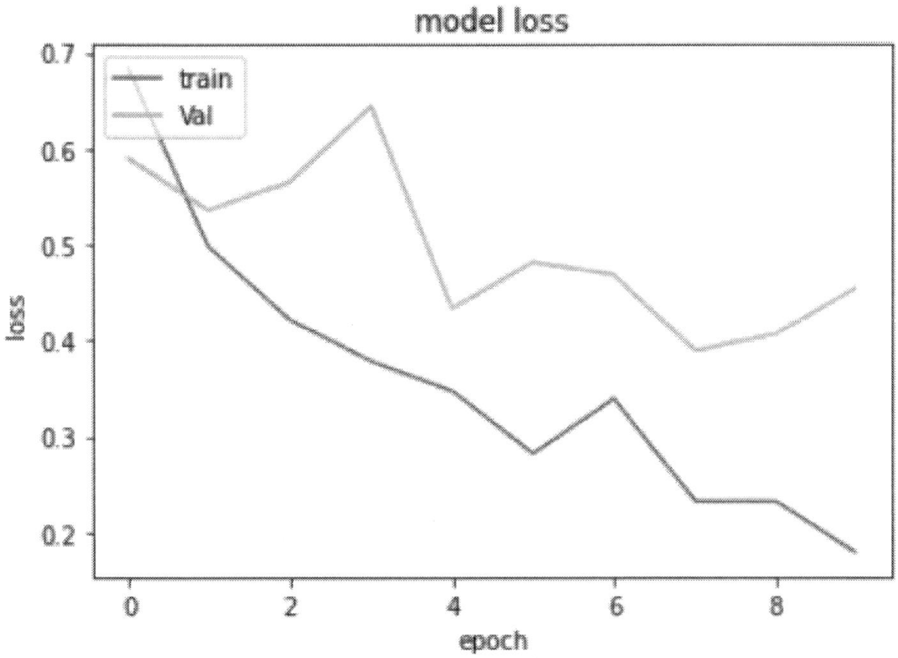

Figure 8.7: Loss vs Epoch curve for BERT-BiGRU model for book review dataset

For the evaluation of the BERT-BiLSTM model, three different datasets have been used. Then for each of the datasets, the accuracy have been calculated both for positive and negative sentiment. The result is given in Table 8.3. Here, the F1-score is higher for the drama dataset than others.

Table 8.3: Evaluation for BERT-BiLSTM model

Dataset	Class	Precision	Recall	F1-Score	Support
Our Dataset	Negative	0.89	0.78	0.83	305
	Positive	0.74	0.86	0.80	220
Drama Review	Negative	0.88	0.94	0.91	1188
	Positive	0.86	0.75	0.80	584
Book Review	Negative	0.57	0.86	0.88	109
	Positive	0.89	0.63	0.74	192

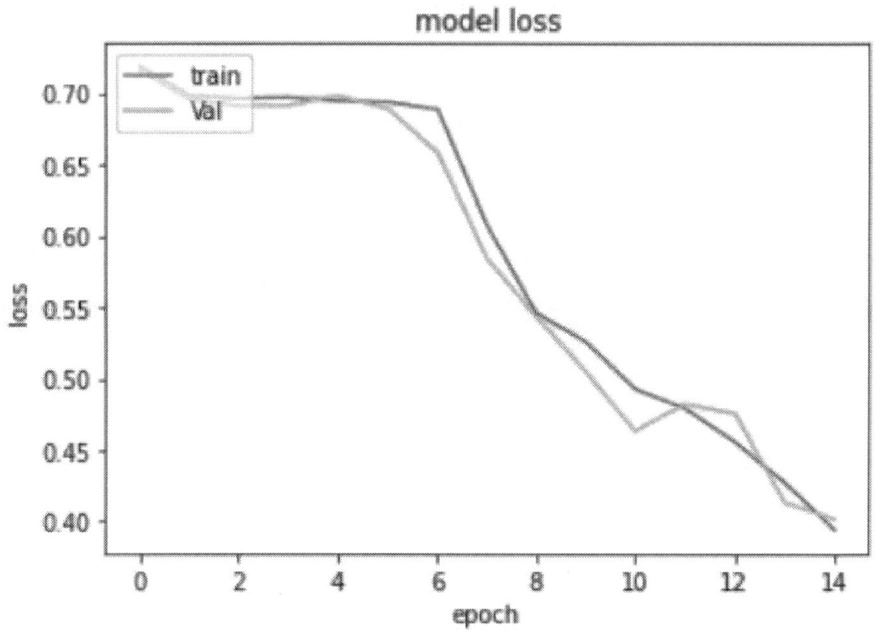

Figure 8.8: Loss vs Epoch curve for BERT-BiLSTM model for Our Dataset

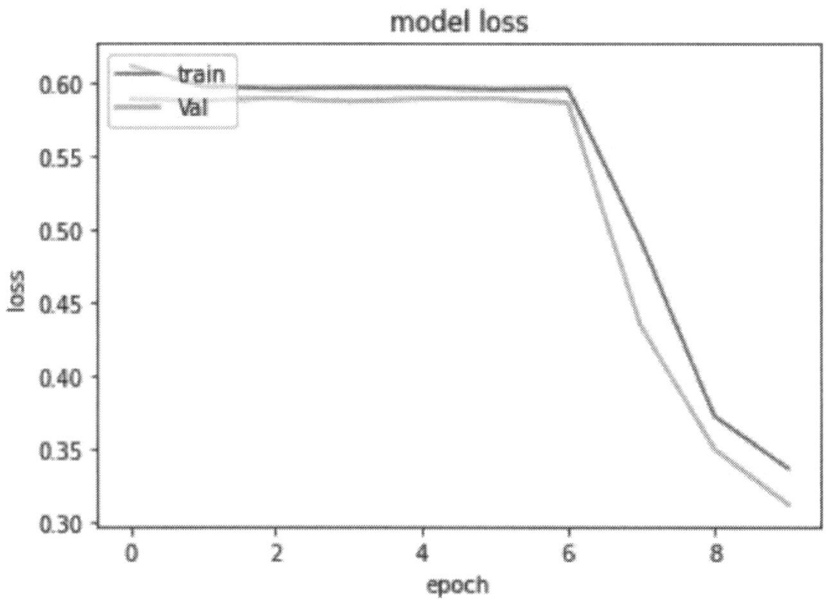

Figure 8.9: Loss vs Epoch curve for BERT-BiLSTM model for Drama Dataset

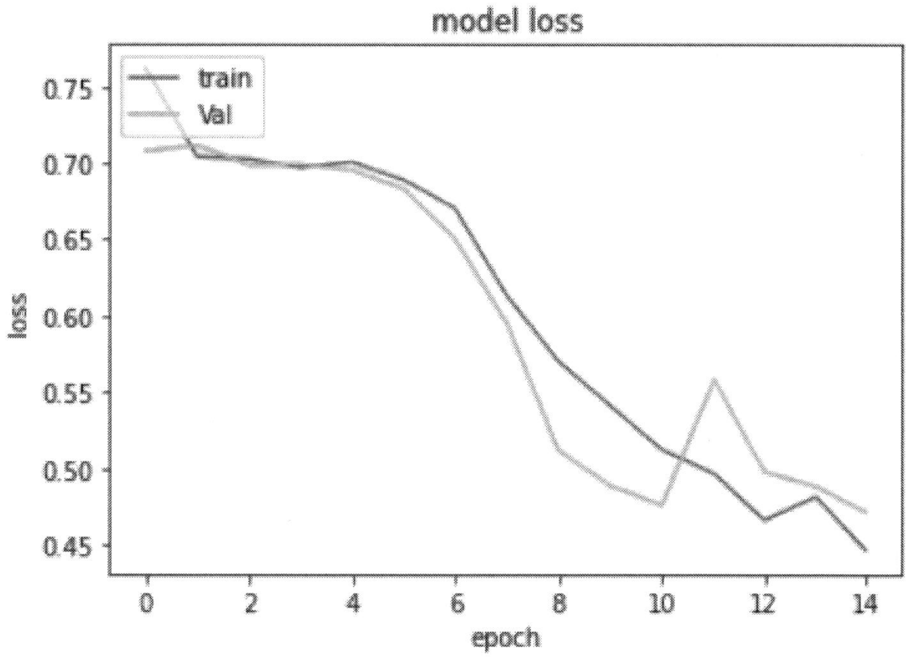

Figure 8.10: Loss vs Epoch curve for BERT-BiLSTM model for book review dataset

Table 8.4: Evaluation for BERT Fine-tuned model

Dataset	Class	Precision	Recall	F1-Score	Support
Our Dataset	Negative	0.92	0.88	0.90	279
	Positive	0.87	0.91	0.89	245
Drama Review	Negative	0.92	0.92	0.92	510
	Positive	0.96	0.96	0.96	1261
Book Review	Negative	0.86	0.94	0.90	151
	Positive	0.93	0.85	0.89	149

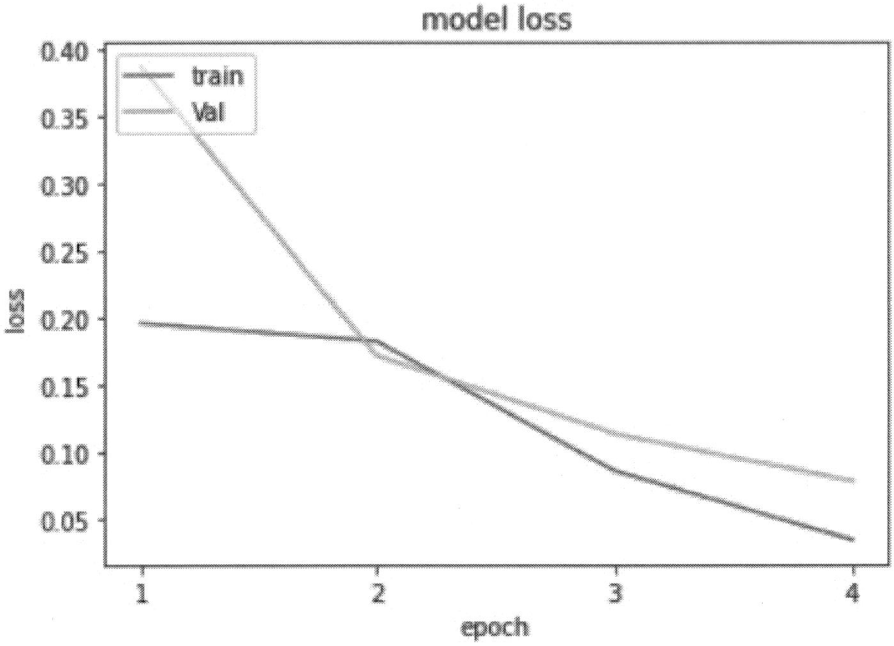

Figure 8.11: Loss vs Epoch curve for BERT Fine-tuned model for Our Dataset

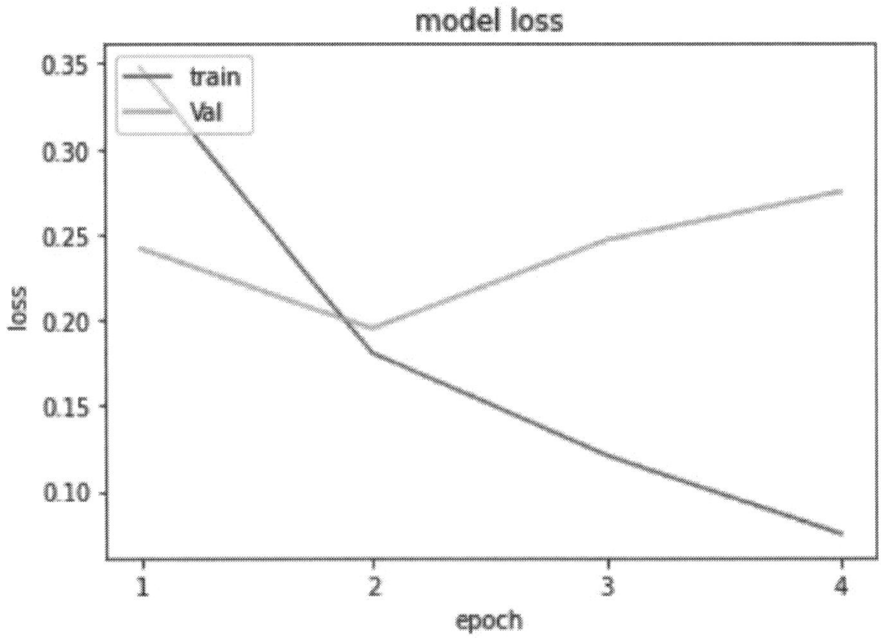

Figure 8.12: Loss vs Epoch curve for BERT Fine-tuned model for Drama Dataset

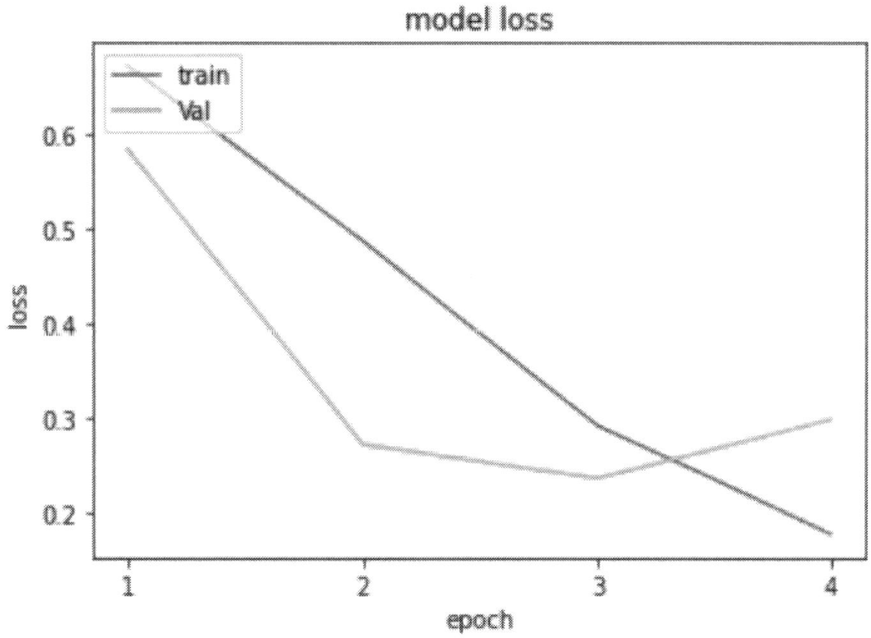

Figure 8.13: Loss vs Epoch curve for BERT Fine-tuned model for book review dataset

From the above three tables, we can see the comparison between the three models for the three datasets. Here for the performance metrics, we used precision, recall, and F1-score.

Table 8.5: Performance comparison of three models with three different datasets based on Test accuracy

Model	Our Dataset	Drama Review	Book Review
BERT-BiLSTM	82.45%	87.85%	70.84%
BERT-BiGRU	85.86%	93.45%	80.32%
BERT-Fine-tuned	**90.04%**	**95.74%**	**89.67%**

Here, the drama review dataset performs better among the three datasets and gives the highest test accuracy. Because it is a specific domain dataset, and the number of training examples here is the highest. So, it gives the maximum test accuracy of 95.74%. Our collected dataset does the second better performance. It works better than the book review dataset. Among the three models, the BERT fine-tuned model performs better for all three datasets. The BERT-BiGRU model's accuracy is better than the BERT-BiLSTM model.

8.5 CONCLUSION AND FUTURE WORK

Sentiment analysis of Bangla text is vital for many aspects of real life. In this study, we developed a benchmark dataset of approximately 5000 online news comments, which can further be used for other research work. Also, we have developed two hybrid models: BERT-BiGRU and BERT-BiLSTM. We got better accuracy in those models because of the added attention layer. The fine-tuned BERT model, where we added the extra single layer on top of BERT, gives the highest accuracy. Since only Transformer works better than Transformer with LSTM or GRU, in the future, we will build a stronger and more resilient model to classify more classes accurately. Also, the hyperparameters will be tuned well for better accuracy in the future.

Bibliography

[1] Biswas, E., Karabulut, M. E., Pollock, L., & Vijay-Shanker, K. Achieving reliable sentiment analysis in the software engineering domain using bert. *2020 IEEE International conference on software maintenance and evolution (ICSME).* pp. 162–173 (2020)

[2] Chunping, O., Wen, Z., Ying, Y., Zhiming, L., & Xiaohua, Y. Topic sentiment analysis in Chinese news. *International Journal of Multimedia and Ubiquitous Engineering.* 9(11), 385–396 (2014)

[3] Fawzy, M., Fakhr, M. W., & Rizka, M. A. Sentiment Analysis For Arabic Low Resource Data Using BERT-CNN. *2022 20th International Conference on Language Engineering (ESOLEC)* pp. 24–26 (2022)

[4] Sazzed, S. Identifying vulgarity in Bengali social media textual content. *PeerJ Computer Science* (2021)

[5] Hossain, E., Sharif, O., & Moshiul Hoque, M. Sentiment polarity detection on bengali book reviews using multinomial naive bayes. *Progress in Advanced Computing and Intelligent Engineering: Proceedings of ICACIE 2020.* pp. 281–292 (2021)

[6] Goyal, P., Pandey, S., & Jain, K. Deep learning for natural language processing. *New York: Apress.* (2018)

[7] Singhal, A., Sinha, P., & Pant, R. Use of deep learning in modern recommendation system: A summary of recent works. arXiv 2017. arXiv preprint arXiv:1712.07525, 116 (2017)

[8] Tithy, T. D., Chakraborty, S., Islam, R., & Aziz, A. A Deep Learning based Approach for Real Time Face Recognition System. *2021 International Conference on Electronics, Communications and Information Technology (ICECIT).* pp. 1–4 (2021)

[9] Fime, A. A., Ashikuzzaman, M., & Aziz, A. Audio signal based danger detection using signal processing and deep learning. *Expert Systems with Applications.* **237**, 121646 (2024).

[10] Ashikuzzaman, M., Fime, A. A., Aziz, A., & Tasnima, T. Danger Detection for Women and Child Using Audio Classification and Deep Learning. *2021 5th International Conference on Electrical Information and Communication Technology (EICT)* pp. 1–6 (2021)

[11] Alam, T., Khan, A., & Alam, F. Bangla text classification using transformers. *arXiv preprint arXiv:2011.04446.* (2020)

[12] Kingma, D. P., Ba, J., Bengio, Y., & LeCun, Y. *3rd International Conference on Learning Representations, Conference Track Proceedings arXiv preprint arXiv:1412.6980* (2015)

[13] Rahman, M. M., Pramanik, M. A., Sadik, R., Roy, M., & Chakraborty, P. Bangla documents classification using transformer based deep learning models. *2020 2nd International Conference on Sustainable Technologies for Industry 4.0 (STI).* pp. 1–5. (2020)

[14] Alam, M. T., & Islam, M. M. Bard: Bangla article classification using a new comprehensive dataset. *2018 International Conference on Bangla Speech and Language Processing (ICBSLP).* pp. 1–5. (2018)

[15] Prottasha, N. J., Sami, A. A., Kowsher, M., Murad, S. A., Bairagi, A. K., Masud, M., & Baz, M. Transfer learning for sentiment analysis using BERT based supervised fine-tuning. *Sensors 22(11), 4157* (2022)

[16] Bhattacharjee, A., Hasan, T., Ahmad, W. U., Samin, K., Islam, M. S., Iqbal, A., ... & Shahriyar, R. BanglaBERT: Language model pretraining and benchmarks for low-resource language understanding evaluation in Bangla. *arXiv preprint arXiv:2101.00204* (2021)

[17] Ittehadul Islam, K., Islam, S., & Ruhul Amin, M. Sentiment analysis in Bengali via transfer learning using multi-lingual BERT. *arXiv e-prints, arXiv-2012* (2020)

[18] Ganesh, P., Chen, Y., Lou, X., Khan, M. A., Yang, Y., Sajjad, H., ... & Winslett, M. Compressing large-scale transformer-based models: A case study on bert. *Transactions of the Association for Computational Linguistics.* 9, 1061–1080 (2021)

[19] Chung, Y. A., Zhang, Y., Han, W., Chiu, C. C., Qin, J., Pang, R., & Wu, Y. W2v-bert: Combining contrastive learning and masked language modeling for self-supervised speech pre-training. *2021 IEEE Automatic Speech Recognition and Understanding Workshop (ASRU).* pp. 244–250 (2021)

[20] Sun, Y., Zheng, Y., Hao, C., & Qiu, H. NSP-BERT: A Prompt-based Zero-Shot Learner Through an Original Pre-training Task–Next Sentence Prediction. *arXiv preprint arXiv:2109.03564* (2021)

CHAPTER 9

DDOS Detection Using Machine Learning

Diksha Sharm

Computer Science and Engineering, Chandigarh University, Mohali, Punjab, India.
Email: dikshasharma.gni@gmail.com

Kulvinder Singh

Computer Science and Engineering, Chandigarh University, Mohali, Punjab, India

CONTENTS

There is almost no place in the world today that is not connected to the internet. One of the most widely used technologies is the IOT, which allows millions of devices to be connected through the internet. As this technology grows, DoS/DDoS attacks are the most common and dangerous threats. DDoS attacks are becoming more complex and they are becoming almost impossible to detect. Distributed denial of service is a subclass of denial of service. In the order to prevent DDoS attacks, many types of research have been conducted. Machine learning and deep learning are commonly used to prevent DDoS attacks. This chapter describes different attack types, such as layer attacks. In this chapter, we carried out a comparative analysis of machine learning algorithms to discover and classify DDoS attacks. A study of the effectiveness of detecting DoS/DDoS attacks in networks has been conducted.

DOI: 10.1201/9781003605508-9

9.1 INTRODUCTION

DDoS is a server attack whose primary objective is to stop authorised users from accessing the source. In this case the security of linked devices increases. Hackers may target personal information and data that protects users from unauthorised changes [1]. Attacks and distributed denial of service attacks, hardware tempering and message suspension. DDoS attacks are a unique malware that the attacker installs to locate tools with vulnerabilities in the network. The impact of the assault in large part depends on the duration through which service is suspended and the scale of the assault. A massive amount of bots offer pc energy to increase high equipment to carry out malicious activities just as the spread of spam emails, viruses, click-on fraud and so on. It is essential to make a smart detection of this assault through the usage of machine learning techniques. Machine learning has made an incredible development in current years in detecting DDoS attacks. Machine learning techniques that can detect the maliciousness of the packets used in combating various sorts of DDoS attacks detection has been demonstrated using machine learning approaches such as K-NN, decision trees and random forest.

These machines contain infected computers and other devices that can be remotely controlled by an attacker. An individual bot is called a bot or a zombie, while a group of bots is called a botnet. An attacker can set up a botnet and then send remote instructions to each bot to direct an attack. Bots send requests to the IP address of a victim's server or network when they are targeted by the botnet, resulting in an overload of the servers or networks, resulting in an interruption in normal traffic.

9.1.1 DDOS Detection using Machine Learning

In this section we describe different DDOS detection techniques using machine learning.

9.1.1.1 Denial of Service Attacks

Denial of service is an attempt to make a website or application unusable by users, for example by flooding it with too much network traffic. Attackers interfere with legitimate user access using various techniques that consume a lot of network bandwidth or other resources. In the simplest form an attacker uses someone to perform a DOS attack on a target as shown in Figure 9.1.

9.1.1.2 Distributed Denial of Service Attacks

In a Distributed Denial of Service (DDOS) attack, an attacker uses multiple resources to launch an attack on a target. These resources can include distributed clusters of infected computers, routers IOT devices and other endpoints. Figure 9.2 shows the host network participating in the attack, creating a flood of packets or requests to bypass the target.

Figure 9.1: Denial of service attacks

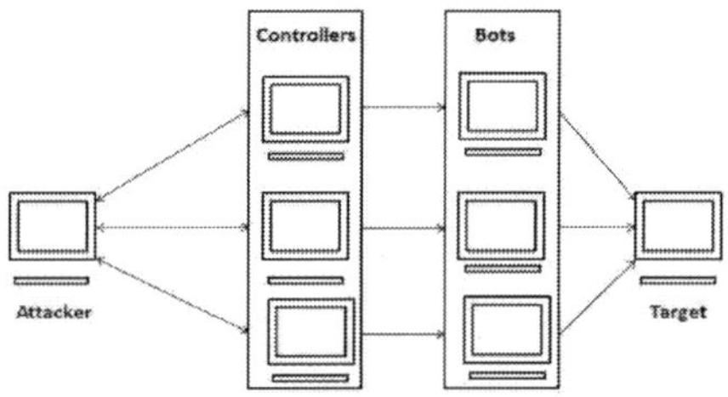

Figure 9.2: Distributed Denial of service attacks

9.2 RELATED WORK

In Jose et al. [2], it is assumed that open flow will allow sdn to collect flow data from which derived features can be obtained. Additionally, they declared that a dataset of simulated networks was used to accomplish DDoS categorization. Experimental characteristics are used for assessment to better understand and categorise DDoS. They used 7 properties from group 3 and attained a 99.999 total accuracy. They identified the top two characteristics that help us identify DDoS attacks. The attributes utilized can be used to build a simple multistage classification model. In Ahuja et al. [7], management, centralization and direction of traffic between hosts are referred to as software-defined networking (sdn). The model and mininet emulator is created using the sdn dataset. In these research works, the author combined random forest and support vector machine (SVM) to classify traffic using svc findings and random forest filtering. The class is accurate in many situations as seen by the model's 98.8% accuracy and 98.27% precision. According to the accuracy, the detection was carried out without traffic control. Sara abdalelah abbas. [10] Reduce the number of features used in training. They also use a lot of machine learning to train models. They got 99.97% accuracy using the best machine learning techniques. Pei et al. [5] carried out local attacks in their study using a DDoS attack program. The packet capture tool analyses the capture attack in comparison with regular packets to find the data attack laws and then transforms it into data attack attributes. To identify the DDoS attack, machine learning uses the random forest approach. To conduct the characteristics on a large scale, it first extracts the feature and converts the format. The random forest algorithm which recognizes the DDoS attack receives these collected features.

Table 9.1: Comparison of DDOS Detection using Machine Learning Algorithm

Author name	Topic and Algorithm
Ancy Sherin Jose, Latin R Nair, Varghese Paul [3]	Detection flooding attacks over SDN using ML techniques. (Algorithm) SVM (support vector machine).
Sara Abdalelah Abbas, Mahdi S. Almhanna [15]	DDOS attacks detection system by machine learning based on dimensionality reduction (Algorithm) PCA.
Jiangtao Pei, Yunli Chen, Wei Ji [7]	A distributed denial of service attack detection method based on ML (Algorithm) Random forest algorithm.
Nisha Ahuja, Gaurav Singal, Debajyoti Mukhopadhyay, Neeraj Kumar [12]	Automated DDOS attack detection in softwaredefined networking. (Alogrithm) Random Forest and Support Vector machine.

9.2.1 DOS and DDOS Attacks Types

DOS and DDOS attacks can be divided into three types, which are discussed in the following sections.

9.2.1.1 Application Layer Attacks

The attacker's goal in these attacks is to undermine the target server, causing a denial of service. As the server often loads multiple files and runs database queries to create a web page, a single HTTP request can be computationally cheap on the client side but costly for the server to respond to. Fig 9.3 shows an application layer attacks.

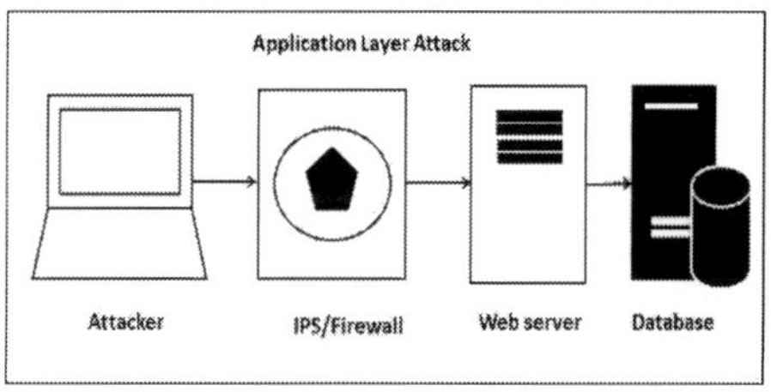

Figure 9.3: Application Layer Attacks

9.2.1.2 Protocol Attacks

Protocol attacks, also known as state exhaustion attacks, overuse server resources and network equipment like firewalls and load balancers causing service disruptions. Fig 9.4 shows protocol attaccks.

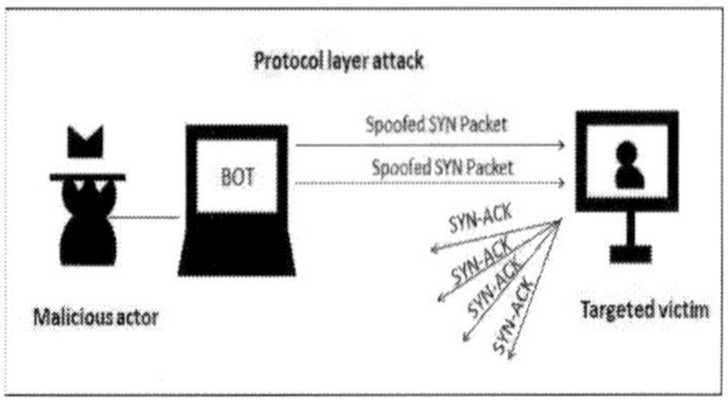

Figure 9.4: Protocol Attacks

9.2.1.3 *Volumetric Attacks*

This category of attacks attempts to create congestion by consuming all available bandwidth between the target and the larger internet. Large amounts of data are sent to a target by using a form of amplification or another means of creating massive traffic, such as requests from a botnet. Fig 9.5 shows volumetric attacks.

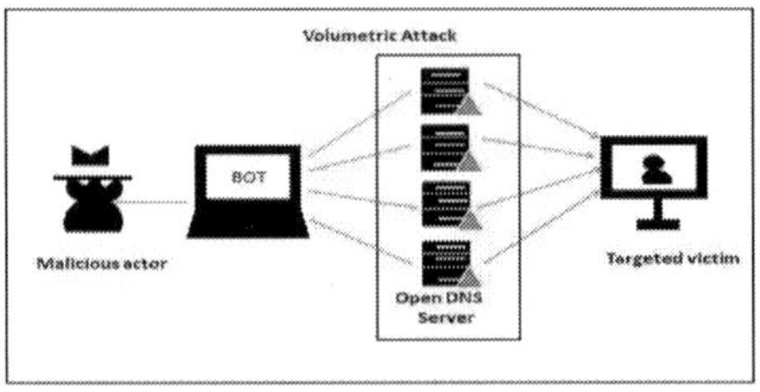

Figure 9.5: Volumetric Attacks

9.3 COMMON TYPES OF DDOS ATTACKS

In a network connection on the internet, there are many different layers or components. Each layer has its purpose, just like how a house is built from the ground up.

- TCP flood: Attacks that use TCP flooding exploit a flaw in its three way handshake to drain the target's resources.
- Syn Flood: When an attacker sends a series of syn packets to each target's port while pretending to be that target's IP address, they are considered a syn flood.
- UDP flood: The attacker attacks a target with packets using the user datagram protocol in an attempt to denial of service. As part of the assault, a random port will be overloaded on a nearby machine. This will cause the host to frequently check if there is an active programme listening on that port. If not, it will send an ICMP "destination unreachable" message.
- ICMP flood: Ping flood is a type of denial-of-service attack where the attacker floods the target with echo requests through the Internet control message protocol.

- HTTP flood: An HTTP flood occurs when an attacker sends a large number of requests to the targeted server.
- DNS flood: A DDOS attack that involves flooding the DNS servers for a domain so that the domain's resource records cannot be resolved is known as a "DNS flood".

9.4 IDENTIFICATION DDOS ATTACKS

As DDOS can cause a service to become slow suddenly, further investigation may be necessary. This is because other factors such as a legitimate spike in traffic can create similar performance problems. DDOS attacks can be detected using tools that analyse traffic.

- Malicious traffic from a single IP address or range.
- The vulnerability results from shared user behaviour such as device type, location or browser version.
- A request that rapidly advances to one page to the end of an event that causes an abnormal result, such as a butt.
- Follow these four simple steps to scan your network for IP addresses in use.
- Simple IP scanning: operating systems like windows and Linux come with simple communication protocols.
- Commands such as "ipconfig" , "arp -a" or "ping" provide easy scanning and troubleshooting.
- Ipconfig – This commond displays all network settings assigned to one or more adapters on the computer. You can view information such as IP subnets and gateways.
- Arp -a –When you issue the "arp – a" command, you get the Mac identification and IP address type for all devices on the network.
- Ping – it helps to identify the connection between two devices and find the IP address of the hostname

9.5 MITIGATING DDOS ATTACKS

In this section we describe different approaches to mitigate DDOS attacks.

9.5.1 Detection

To stop a distributed attack, a website needs to be able to distinguish an attack from a high volume of normal traffic. If a product announcement has a website swamped with legitimate new visitors, the last thing the site wants to do is stop them from viewing the content of the website.

9.5.2 Response

In this step, the DDOS protection network responds to an incoming identified threat by intelligently dropping malicious bot traffic and absorbing the rest of the traffic.

9.5.3 Routing

By intelligently routing traffic, an effective DDOS mitigation solution will break the reaming traffic into manageable chunks preventing denial of service.

9.5.4 Adaption

Good network analyses traffic for patterns such as repeating offending IP blocks, and particular protocols being used improperly.

9.6 CONCLUSION

Distributed denial of service attacks will continue to represent the greatest threat to many large and small enterprises since they harm internet users in a variety of ways. The sectors that might require human interference, complex calculations and a lack of free-to-access data. To detect a DDoS assault, however, many areas need attention. To detect DDOS attacks a variety of algorithms like liner regression, random forest, support vector machine and naïve Bayes classifier are utilised. Convolution neural networks are used in DDoS attack detection and deep learning also plays a significant role in this process. So, we have reviewed the many methods of identifying DDoS attack

Bibliography

[1] Tejaswini Ulemale, Review on Detection of DDOS Attack using Machine Learning, International Journal for Research in Applied Science & Engineering Technology (IJRASET), 10(III):764–768, 2022

[2] Dutta Sai Eswari, P.V.Lakshmi, A Survey On Detection Of DDoS Attacks Using Machine Learning Approaches, Turkish Journal of Computer and Mathematics Education, 12(11):4923–4931, 2021.

[3] Ancy Sherin Jose, Latha R Nair, Varghese Paul. Towards Detecting Flooding DDOS Attacks over Software Defined Networks Using Machine Learning Techniques REVISTA GEINTEC, 11(4): 3837–3865, 2021.

[4] Ebtihal Sameer Alghoson, Onytra Abbass Distributed Denial of Service Attacks using Machine Learning Models (2021)

[5] Parvinder Singh Saini, Sajal Bhatia, Sunny Behal. Detection of DDoS Attack using Machine Learning Algorithms, in Research Gate, March 2020.

[6] C M NalayiniI, Jeevaa Katiravan detection of ddos attcks using machine learning algorithms(2022)

[7] Jiangtao Pei, Yunli Chen, Wei Ji. A DDoS Attack Detection Method Based on Machine Learning" published in ICSP 2019.

[8] Pande, S., Khamparia, A. & Gupta, D. Feature selection and comparison of classification algorithms for wireless sensor networks. J Ambient Intel Human Compute (2021).

[9] Jeswin Johnson, Sabu George Review on DDOS Detection using Machine Learning(2022)

[10] Chin-Shiuh Shieh, Wan-Wei Lin, Thanh-Tuan Nguyen, Chi-Hong Chen, Mong-Fong Horng and Denis Miu.

[11] Detection of Unknown DDoS Attacks with Deep Learning and Gaussian Mixture Model, IEEE ICICT 2021.

[12] Nisha Ahuja, Gaurav Singal, Debajyoti Mukhopadhyay, Neeraj Kumar. Automated DDOS attack detection in software-defined networking published in Science Direct 2021.

[13] Alzahrani, R.J.; Alzahrani, A. Security Analysis of DDoS Attacks Using Machine Learning Algorithms in Networks Traffic.

[14] Pheeha Machaka, Olasupo Ajayi, Hloniphani Maluleke, Ferdinand Kahenga, Antoine Bagula, Kyandoghere Kyamakya, Modelling DDoS Attacks in IoT Networks using Machine Learning, 2021.

[15] Sara Abdalelah Abbas, Mahdi S. Almhanna, Distributed Denial of Service attacks detection System by Machine Learning Based on Dimensionality Reduction, ICMAICT 2020.

[16] Swathi Sambangi and Lakshmeeswari Gondi. A Machine Learning Approach for DDoS (Distributed Denial of Service) Attack Detection Using Multiple Linear Regression" published on 25 December 2020.

CHAPTER 10

Extracting Clinically Relevant Phrases from Patient Notes using BERT and Multi-Teacher Knowledge Distillation

Arafat Bin Hossain
Email: arafatbhossain@gmail.com

Koushik Howlader
Email: koushik.howlader@ndsu.edu

Muhammad Zubair Malik
Email: zubair.malik@ndsu.edu

CONTENTS

ONe of the key challenges in NLP tasks based on medical dataset is to identify specific clinical terms from natural texts like unstructured patient notes. Extracting clinical terms from those natural texts is a common NLP task that BERT can solve efficiently. However, using huge BERT models is a challenge because of their enormous parameter size which restricts their application to users with limited computing resources. In our study, we explore a comparable problem and adopt a

DOI: 10.1201/9781003605508-10

methodology akin to the utilization of BERT for Question Answering and Named Entity Recognition. Additionally, we compressed the large BERT model by employing multi-teacher knowledge distillation technique. For the task of clinical concept identification, we experimented with five different variants of BERT on the USMLE dataset. Our findings revealed that DeBERTa-Large, which consists of 304M parameters and 24 hidden layers, outperformed the other variants, achieving an impressive F1 score of 0.88. In our approach, we selected BERT-base and DistillBERT-SQuAD as student models. To distill knowledge from teacher models, we took the softmax of the average of logits generated by DeBERTa-Large and PubMedBERT. We then utilized the Multi-Teacher Knowledge Distillation technique to independently train both student models, ultimately creating a compressed model. Our experiment shows that the student model DistillBERT-SQuAD if it gets the distillation knowledge from its teachers, achieved an F1 of 0.81 with a parameter size of 66 M, which is almost 4.6 times smaller than DeBERTa-Large and has half the number of hidden layers.

10.1 INTRODUCTION

The precision of medical diagnosis hinges upon a doctor's ability to effectively recognize and interpret symptoms. Doctors amass considerable expertise during their training, enabling them to craft comprehensive patient notes that encapsulate a patient's complaint history, physical examination findings, possible diagnoses, and proposed treatment plans. However, mastering and assessing the skill of writing patient notes can be a laborious process that often requires feedback from other physicians. Implementing machine learning can expedite this task. The United States Medical Licensing Examination (USMLE®), for instance, includes the Step 2 Clinical Skills test. This test mandates examinees to interact with Standardized Patients who simulate specific clinical scenarios, and consequently, create a patient note [1]. Experts in the field assess these patient notes by utilizing rubrics that designate vital ideas related to each case. This could be especially beneficial in high-stress situations or when dealing with a large volume of patients, such as during a pandemic [2].

The overall concept of the problem we are trying to solve is explained below in terms of input and output –

- **Input**: *17yr M presents with palpitations. Patient reports 3-4 months of intermittent episodes of "heart beating/pounding out of my chest." Mom with "thyroid disease," dad with recent heart attack*
- **Relevant Clinical Ideas**: *Family history of **myocardial infarction**; Family history of **thyroid disorder***
- **Output**: *17yr M presents with palpitations. Patient reports 3-4 months of intermittent episodes of "heart beating/pounding out of my chest." **Mom with "thyroid disease", dad with recent heart attack***

Given the patient's note as the input, the goal is to find the key phrases in the input text which are as closely relevant to the clinical ideas which is also given as a part of the input. For example, the main goal is to identify ***"Mom"*** and ***"Dad"*** because they are related to the patient's ***"Family history"***. It is important to accurately associate ***"heart attack"*** with ***"myocardial infarction"*** and that too with a misspelled word. The nature of the input text is what makes the problem a challenging one to solve.

Natural Language Inference (NLI) is a task in Natural Language Processing(NLP) that can benefit significantly from Pre-training Language Models (PLMs) [2, 3, 4, 5]. These PLMs have shown great promise in predicting the relationships between sentences and even at the token level, leading to fine-grained outputs in NLP tasks such as Named Entity Recognition, Question Answering, and Text Summarization. The Bidirectional Encoder Representations from Transformers (BERT), a transformer-based machine learning technique for NLP, was developed by Jacob Devlin and his colleagues at Google and has become a cornerstone in the field since its introduction [6]. BERT was integrated into the Google search engine in 2019 and has consistently been one of the top choices for language-based tasks.

Two standard NLP benchmarks in the academic community are the General Language Understanding Evaluation (GLUE) [29] that contains clearly defined NLP tasks. After BERT's release, it achieved state-of-the-art performance on these tasks. However, researchers have also developed different variants of BERT to solve more robust or complex problems that might not be suitable with the BERT-based PLM. For instance, Decoding-enhanced BERT with Disentangled Attention (DeBERTa) and A Robustly Optimized BERT Pre-training Approach (RoBERTa) are two highly successful variants of BERT that address some of the existing BERT-base limitations by changing the pre-training process [7, 8]. One limitation of standard BERT is that the attention mechanism only considers the content of the word when evaluating its relative importance to other words but does not consider its position in the sentence. Therefore, DeBERTa and RoBERTa introduced changes in how the model is pre-trained to account for the position of the content. DeBERTa solved the problem and evaluated the importance of other words by considering their relative position while creating the embedding output from the attention layer. Also, the enhanced masking in DeBERTa enables the collection of complete position information just before the softmax layer, improving the masked word prediction task during pre-training. RoBERTa was produced by changing some steps in pre-training:

1. Training data is duplicated, and hence the same sentence can have multiple sets of masks;

2. Batch size is increased while pre-training;

3. In addition to Next Sentence Prediction, it tells whether the sentences are from the same document or not.

To add more meaning to the usefulness of the pretraining of models, researchers went a step ahead and extended the idea to build clinical variants of BERT by pretraining the BASE model with large medical corpora like PubMed[1], the MIMIC III dataset [32] [33], etc. Open leaderboard portals like BLURB [10] contain the list of all the lead

[1]https://pubmed.ncbi.nlm.nih.gov/

performers in clinical BERTs. Two of the most renowned variants are ClinicalBERT [9] and PubMedBERT [10], which Microsoft releases. BERT has also been heavily used with electronic health records to solve NLP tasks in the medical domain. In this study, we have tried to solve this single complex NLP problem of automating the process of translating clinical terms from exam rubrics (e.g., "diminished appetite") into the many ways that medical students have expressed these terms in their clinical patient notes (e.g., "eating less", "clothes fit looser"). We have focused on a single NLP problem because we realized it is a problem that has yet to be addressed in the research community.

PLMs are often computationally intensive and require significant resources, which may be a barrier to their adoption in clinical settings, particularly in resource-poor areas [5]. With knowledge distillation, we can create lighter models that are easier to deploy and use, making these techniques more accessible to a broader range of healthcare providers. Knowledge distillation involves transferring knowledge from a larger, more complex model (the 'Teacher') to a smaller, simpler model (the 'Student'). This process seeks to improve the performance of the student model using the learnings of the teacher model. Instead of focusing on exact labels, the teacher model generates 'soft targets,' which are probability distributions across output classes. The student model then aims to minimize the discrepancy between its output and the soft targets provided by the teacher model, rather than between its output and the actual labels.

This method effectively enables the creation of a streamlined student model from a larger teacher model, serving as a practical example of model compression and acceleration [11]. The process essentially tempers the student model's loss function by aligning it with the teacher model. The smaller model then updates its weights by capturing the knowledge differences between itself and the teacher. Recent research has indicated that, in some instances, student models can even surpass their teachers in performance [12]. Such advancements are particularly crucial in Natural Language Processing (NLP) tasks involving PLMs, where a successful model's reuse typically requires substantial computational infrastructure. However, these models can be challenging to maintain and often require a high level of coding skill due to the complex architecture of PLMs.

In our research, we applied multi-teacher knowledge distillation, utilizing two teacher models and averaging their logits to convert to softmax. A significant aspect of our study was the employment of Multi Teacher Knowledge Distillation, leveraging one or more fine-tuned PLMs to generate a lighter version. Specifically, we used DeBERTa-v3-large (a generic BERT model) and a clinical BERT model, BioMedNLP PubMedBERT, both fine-tuned on our dataset, to distill their knowledge and train the student models on the same training set used to fine-tune the teacher models.

Our findings suggest that a student model, although it may not necessarily outperform its teacher, can still deliver respectable performance in identifying medical terms while minimizing any compromise on the F1 score.

Therefore, the key contributions of this study may be summarized as follows:

1. Firstly, we tried to solve the problem of identifying key phrases inpatient notes that are relevant to particular medical terms by fine-tuning BERT, ClinicalBERT, PubMedBERT, RoBERTa, and DeBERTa with annotated patient notes

collected from the USMLE, which is a significant mix of natural and clinical text. We observed that DeBERTa, a general PLM, performs better in the identification task than the clinical variant of the BERT, showing that the choice of model depends on the dataset type.

2. We showed that such an NLP problem could be formulated as a hybrid of Named Entity Recognition (NER) and Question Answer (QA) problems. While fine-tuning, the final layer of the PLM can be given a task to classify each token (making it a NER-like task) while the model is trained with patient notes (context) and clinical terms to look for (query/question), just like how it's done while fine-tuning for QA tasks.

3. The second crucial contribution of our work is that we addressed the problem of PLMs having huge parameter sizes, which limits their use to only researchers or end users with high computing resources and massive storage. We took the two best-performing PLMs, one from both clinical and one from the available variant of BERT. We then showed that techniques like Multi Teacher Knowledge Distillation could be effectively used to distill the knowledge of larger PLMs, with parameter sizes as big as 304M, to form smaller student BERT models, with parameter sizes as small as 66M, without compromising F1 scores significantly.

The rest of the paper is structured as follows: Section 2 discusses the related works while in Section 3, we discuss – a) The Dataset, b) Data Pre-processing, c) Fine Tuning of the Pre-trained Models, and d) Multi Teacher Knowledge Distillation. In Section 4, we discuss Results & Discussion followed by Conclusion in Section 5.

10.2 RELATED WORK

BERT has been used in several Named Entity Recognition (NER) and Question Answer (QA) tasks in the clinical domain, including for multilingual purposes. Kai et al. used the multilingual variant of BERT to solve the PharmaCoNER[2] task on Spanish biomedical named entity recognition [13]. In this work, the authors employed BERT, a neural architecture based on transformers for NER. They used the multilingual BERT model trained on a Wikipedia corpus encompassing 104 languages, including Spanish, as no pre-trained Spanish BERT models were available. The initial pretraining output layers were swapped out for a Conditional Random Fields (CRF) layer, and all of the network's layers were adjusted to fine-tune the model for the NER task. For a dataset that contains clinical records but cannot be qualified as a standard Electronic Health Record (EHR), Kim et al. also used general BERT models and tackled a similar NLP task as NER with a Korean dataset collected from an online Question Answer (QA) site [14]. In both cases, authors chose general BERTs over clinical ones because the dataset they were dealing with was a mix of clinical and natural text. Qiu et al. approached a Clinical Text Structuring NLP task purely in the light of QA by passing the paragraph with clinical text and query (known clinical terms)

[2]https://github.com/chaanim/pharmaconer

to Chinese Clinical NER [31], specifically pre-trained with certain kinds of clinical notes [15]. First, a Clinical Named Entity Recognition (CNER) model is applied to the paragraph text to identify named entities and produce a one-hot CNER output tagging sequence. The contextualized representation model is then supplied with the combined query text and paragraph content to produce the contextualized representation vector. A feed-forward network is then used to calculate the start and end index of the text that is relevant to the answer using the vector and the integrated text. For each word in the calculation, a classification issue is used to identify whether it is the start or end word of the solution. This paradigm generally seeks to organize clinical text and offer solutions to clinical queries. As we can see, the choice of the PLMs highly depends on the kind of dataset we are exposed to solve the problems.

Now, for the knowledge distillation part, many different approaches have been taken in the past that explore the potential of Multi Teacher Knowledge Distillation in solving NLP tasks [23, 24]. Deep neural networks are currently seeing unheard-of success in a variety of applications [16, 17, 6]. However, applying these complicated models to embedded systems is challenging since they require a large memory footprint and computing power. The solution to this problem is Knowledge Distillation (KD), which increases the accuracy of a lightweight student model by extracting the knowledge from a pre-trained, hefty teacher model. The softmax of the teacher's logits was considered to be the only way the transferable knowledge from the teacher to the student could be formalized. In later works, it has been shown that the intermediate teacher layers can also improve the student's performance if incorporated [19, 20, 21]. Concepts like "multi-teacher distillation take advantage of multiple teachers instead of a single teacher to boost the student's performance. The most popular techniques used for this are assigning averages or weights to different teachers or calculating the weights based on cross-entropy [22, 23, 24, 25]. MultiTeacher BERT (MT-BERT) was developed in conjunction with Tsinghua University, Beijing, by Microsoft Research Asia [26]. They created a multi-teacher co-fine-tuning strategy in MT-BERT to cooperatively fine-tune numerous instructor PLMs in downstream tasks to align their output space for better collaborative teaching. This approach uses shared pooling and prediction layers. To convey the valuable information from multiple instructor PLMs to the student model in both hidden states and soft labels, they also supplied a multi-teacher hidden loss and a multi-teacher distillation loss. They tested their approach on the SST-2[3], RTE[4], and MIND[5] datasets and could show that, with fewer parameters, MT-BERT can perform very close to the teacher models and sometimes even better than the teachers. While dealing with multi-teacher knowledge distillation, Confidence Aware – Multi Teacher Knowledge Distillation(CA-MKD) seeks to lessen the impact of poor teacher predictions [27]. The basic notion is that when deciding what the pupils should learn, the teachers whose predictions are closest to the encoded ground truth are given greater weight or priority.

In order to enhance the learning experience of students, the research explores the addition of features in the intermediate layers of the CA-MKD framework. Previous

[3]https://www.kaggle.com/datasets/atulanandjha/stanford-sentiment-treebank-v2-sst2

[4]https://www.kaggle.com/datasets/nltkdata/rte-corpus

[5]https://huggingface.co/datasets/linxinyuan/mind

works in the field of computer vision have utilized pre-trained neural networks, including WRN40 [28], VGG13 [35], and ResNet32/56 [36], which share identical architectural characteristics. CA-MKD aims to generate teacher predictions that will be utilized in the distillation loss function by employing ensemble teachers and fine-tuning the weight distribution among them. The calculation of cross-entropy loss between the teacher predictions and the ground truth labels plays a crucial role in assigning unique weights that reflect sample-wise confidence. This approach seeks to protect students from being misguided by underqualified instructors. Additionally, Du and Hu et al. propose a novel information distillation technique that is specifically designed for Transformer-based models. They apply this technique to BioBERT [34], resulting in the creation of SqueezeBioBERT, a compressed version of BioBERT. SqueezeBioBERT is evaluated in the context of named entity recognition, relation extraction, and question answering tasks in healthcare text mining, demonstrating over 95% of the performance achieved by the original BioBERT despite being 4.2 times smaller in size. These findings highlight the potential of compressing deep pre-trained language models using knowledge distillation, making them more suitable for healthcare text mining applications.

10.3 METHODOLOGY

Fig. 10.1 shows our approach is unique in a way that it attempts to solve a particular NLP task using a hybrid QA/NER approach during fine-tuning and also proposes a way to compress the final model by combining information from two different kinds of BERTs, both pre-trained with varying types of data, to form a student model that is lighter than its teachers and performs decent enough. During the fine-tuning step, we decided which two models we would choose as the teacher model for our Multi-Teacher Knowledge Distillation (MT-KD) step. Also, we repeated the MT-KD step with the same teacher and student but changed the loss function.

Before diving into details, it is worth mentioning a brief overview of important steps in our methodologies –

1. Data Pre-processing includes tokenizing the patient notes and labeling the tokens, which are annotated as associated with the feature text for the patient notes.

2. In each PLMs, we feed both the patient notes and the feature text just as we would have done for the QA task. The final layer was then given the task of predicting if a token is associated with the feature, just like NER. This way, we fine-tuned all five variants (BERT-base, RoBERTa, DeBERTa, ClinicalBERT, and PubMedBERT) of BERT with our train set. The test set was also pre-processed alike and made ready for inference.

3. The five PLMs were evaluated with the test set, and the best two models were picked, one from the general BERT variants (BERT-base, RoBERTa, DeBERTa) and one from the clinical variants.

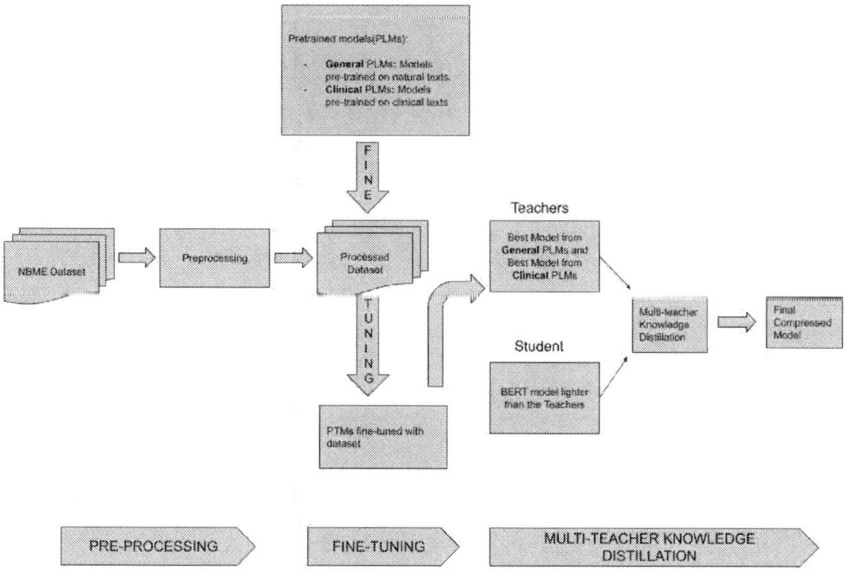

Figure 10.1: Diagram illustrating the Methodology used to Fine-tune the PLMs and then Distilled to form a Compressed Model

4. The variants picked in Step (3) is used as the Teacher Models for MTKD. For Students, BERT-base and DistillBERT-SQuAD are used.

5. Multi Teacher Knowledge Distillation is then performed using the softmax of the average of teacher logits to form a compressed version of the fine-tuned teacher models we picked in Step (4). The student is evaluated with the same test set as was done for the teachers during fine-tuning in Step (2).

10.3.1 Dataset

In our study, we have used clinical patient notes from the USMLE® Step 2 Clinical Skills patient note corpus [1]. Some important terms that are useful to understand the dataset are –

1. **Clinical Case** – The scenario (e.g., symptoms, complaints, concerns) being presented to the test taker by the standardized patients,

2. **Patient Note** – Text detailing all the details described by the patients and

3. **Feature** – Clinically relevant terms/concepts.

We received a collection of 14,300 different patient notes each corresponding to a particular clinical case with features to look for. It means it has all the texts annotated in clinical notes, which indicates a feature for a particular case. It also has character

Table 10.1: Attributes in the Clinical Patient Notes dataset.

Attributes	Description
pn_num	The unique identifier of the patient
case_num	The case a patient note represents
pn_history	The text of the encounter as recorded
feature_num	The unique identifier of the feature
case_num	The case a patient note represents
feature_text	The description of the feature
annotation	The text(s) within a patient note indicating a feature
location	Character spans indicating the location of each annotation within the note

spans indicating the location of each annotation within the note. Table 10.1 illustrates the attributes of the dataset:

These three datasets were split in as 80:20 ratio and are then pre-processed to form a single dataset which is then used for tokenization and fine tuning using the pre-trained models which is discussed in the Data Pre-processing section.

10.3.2 Data Pre-processing

We noticed some wrong annotations in the train set for the most important and time-consuming part of the pre-processing. For example, the accurate annotated text in the patient notes is "father heart attack," but the annotation says "ather heart attack." "Mother thyroid problem" was incorrectly annotated as "Mother thyroid problem." This is a human error during the recording of the annotation that has been noticed. Getting rid of this problem was crucial because training the model with false positives is not desirable. Since there was no automated way to do this, we had to do this case by case and replace the annotation in the dataset with the correctly observed annotation from the patient notes. The location of the annotated characters and the annotation texts were given in the dataset as strings, and we converted them to a list before tokenizing them. Since we are using pre-trained models like BERT, it is essential to tokenize the clinical notes. So, we have used BERT tokenizers to tokenize the clinical notes. Lastly, we added labels for each token, and an additional discussion on token-level labeling of the data is discussed in the next section because the importance of this pre-processing step heavily relies on the approach we took to fine-tune the pre-trained models with our dataset.

10.3.3 Fine Tuning Pre-trained models

We then added labels for each token where we considered the token to be labeled one if it is associated with the feature, and that way, we now have a label per token. It is an essential step for fine-tuning because our approach to solving this problem combines how BERT is used to solve QA and NER problems. In NER, the classifier tries to classify each token, and our classifier also tries to predict each token. In QA, the question is also fed along with the answer during model training, similar to how we trained the model. In our case, we also put the tokenized feature text (question) before the tokenized patient notes (answer) when we fed the data into the pre-trained model for fine-tuning.

For fine-tuning, we have added one final layer for the classification. We have used 5-fold cross-validation, recorded the F1 score for the test set, and reported the best F1 score after running the training for five epochs with Adam Optimization as the gradient descent method and 1e-5 as the learning rate. The total training time for fine-tuning each pre-trained model was around 20 hours. We repeated the fine-tuning for the three BERT models, which are pre-trained using corpora of natural texts: BERT-base-uncased, RoBERTa-large, and DeBERTa-v3-large. We also used two clinical variants of BERT - BioMedNLP, PubMedBERT-base-uncased, and Bio-Clinical BERT. We used PyTorch and the Hugging Face library to train the models and report the results. The results are shown in Table 17.2 and will be discussed in the Results and Discussion section.

10.3.4 Model Compression: Multi-Teacher Knowledge Distillation

Another vital contribution to our study is using MT-KD to resolve a few limitations posed by the finely tuned models we prepared in the previous step. As discussed in the early part of this chapter, the large pre-trained model comes with millions of parameters and huge size. If you plan to commercialize the model or make it public for use in embedded systems, this poses a problem. A model compression technique is essential to make the finely tuned model much lighter without compromising performance. In this regard, we have implemented a knowledge distillation technique by taking advantage of better-performing, larger pre-trained models from the previous step. We tried to show that a lighter PLM (student model) can take advantage of finely tuned general and clinical PLMs (teacher models) and still perform close to, if not better, than, its teachers. From the results of the fine-tuning step, we picked two models that we used as teachers for the knowledge distillation process. We decided on the best model among BERT-base-uncased, RoBERTa-large, and DeBERTa-v3-large based on the F1 score. We did the same for BioMedNLP, PubMedBERT-base-uncased, and BioClinical BERT, and in this way, we used two teacher models to perform various experiments. For students, we chose BERT-base-uncased and a variant of DistillBERT-SQuAD, which was already fine-tuned with the SQuAD dataset. We collected both models from the hugging face model repository. It is important to note that we used only one student per experiment since there has to be only one student in the knowledge distillation process, while there can be one or more teachers during the process. For our knowledge distillation, we performed two experiments, which vary only in how the loss function is defined. For the first experiment, we used the `Traditional Loss` function, which is the summation of `Student Loss` and the `Distillation Loss`. For the second experiment, we used average loss, that is - `Average` of the `Student and Distillation loss`. An explanation of how we calculated the loss functions is given in the Experiment section.

10.3.4.1 Experiment 1 (Traditional Loss)

In experiment 1, we took the softmax of the average of logits, t, produced by fine-tuned DeBERTa-v3-large and PubMedBERT-base-uncased for each training data. We then calculated the softmax, s, made by the student, BERT-base-uncased, for the same training data and calculated the loss using the traditional definition, which is the sum of student loss and distillation loss. Below are how we figured out the different loss functions in Knowledge Distillation –

$$\textbf{Student Loss} = \text{Cross Entropy(t,s)}$$
$$\textbf{Teacher Loss} = \text{Cross Entropy(t, ground truth)}$$
$$\textbf{Distillation Loss} = \text{Student Loss/ (1 + Teacher Loss))}$$
$$\textbf{Traditional Loss} = \text{Student Loss + Distillation Loss}$$

We repeated the same process but used DistillBERT-SQuAD as the student.

10.3.4.2 Experiment 2 (Average Loss)

For experiment 2, we took the softmax from teachers and students similarly to in experiment 1. Still, we defined the loss function as the average of **Student** and **Distillation Loss** to normalize the bias of the student model towards any one teacher and eventually see if we can take advantage of a clinical BERT and a general BERT to solve our problem by having a moderately performing compressed model. We ran the experiment for both students.

For each experiment, we used 5-fold cross-validation on the train set, recorded the F1 score for the test set, and reported the best F1 score after running the training for five epochs. The training time for each training was approximately 12 hours since the student model is smaller in terms of parameters than the teachers. Also, the teachers are used here only to produce the softmax, meaning only the final layer is used. The technique we used for knowledge distillation is offline knowledge distillation. Also, we tried to keep the values of temperature and alpha fixed at 0.2 and 0.5, respectively, since it's the effect of the loss function we wanted to assess. The results are shown in Table 17.3 and will be discussed in the Results and Discussion section.

10.4 RESULTS AND DISCUSSION

The reported F1 scores from fine-tuning steps gave us a few interesting insights about how BERT of different variants behaved on specific kinds of clinical notes we exposed. BERT-base-uncased achieved the worst `F1 score`; not one of the clinical variants of BERT performed best in our dataset. It was rather DeBERTa-v3-large that reached an outstanding `F1 score` of 0.88 in our test set. While none achieved the best `F1 score`, the clinical BERTs performed better than BERT-base uncased, with BiomedNLP PubMedBERT uncased performing better than Bioclinical BERT, gaining an F1 0.82. We chose F1 as the evaluation metric because the accuracy value was very high for each of the models and was not indicative of the actual performance of a model for the NLP problem we tried to solve. From the results gathered, we then picked DeBERTa-v3-large and BioMedNLP PubMedBERT uncased as teachers for the next step of our methodology, which is compressing the model using multi-teacher knowledge distillation.

We must understand that we did not chase a compressed model of the larger, better-performing, fine-tuned model for the Multi Teacher Knowledge Distillation step. Instead, we appreciated that building a student model that would generally outperform its teachers is only sometimes desirable. We tried to see if we could take advantage of an excellent performing medical BERT and models like DeBERTa-v3-large and utilize their knowledge to build a student without compromising their `F1 score` on a considerable level. As a result, taking the average of their logits does not deteriorate the F1 by a mark. For both experiments, DistillBERT-SQuAD performed better than BERT-base-uncased as a student, and the effect of taking the exact weight of the student and distillation loss in the loss function deteriorated the F1 score. In Experiment 1, DistillBERT-SQuAD achieved an `F1 score` of 0.82, while BERT-base-uncased scored 0.78. Changing the loss function in Experiment 2 reduced the `F1 score` for both the students to 0.81 and 0.74, respectively, but it hampered BERT-base-uncased more than it did for DistillBERT-SQuAD.

Interestingly, knowledge distilled using the traditional loss function yielded a fine-tuned DistillBERT-SQuAD student model that performs better than the fine-tuned BioClinical BERT and BERT-base-uncased while performing equal to the fine-tuned BioMedNLP PubMedBERT-base-uncased in terms of F1 score. BERT-base-uncased, when used as a student and using the distilled knowledge from two good teachers, performs better, with an F1 score of 0.78, compared to its finely tuned variant shown in Table 17.3 (Param Size in millions, M). However, none of the students could outperform RoBERTa-large and DeBERTa-v3-large fine-tuned with the same dataset. Therefore, we could see that the student, DistillBERT-SQuAD, with only six hidden layers and 66M parameters, could achieve a decent F1 of 0.82 compared to 0.88 conducted by one of its teachers, DeBERTa-v3-large, which contains 304M parameters with 24 hidden layers. Also, the BERT distilled using our approach performed better than the fine-tuned BERT base, F1 score of 0.74, with 12 hidden layers and 110M parameters.

Table 10.2: Performance of PLMs on Test set after Fine-Tuning

BERT Variant	**Param Size (in millions, M)**	**Hidden Layers**	**F1 Score**
BERT-Base-Uncased	110M	12	0.74
BioClinical BERT	110M	12	0.80
PubMedBERT	110M	12	0.82
RoBERTa Large	354M	12	0.83
DeBERTa Large	304M	24	0.88

Table 10.3: Performance of the Student PLMs built with Multi-Teacher Knowledge Distillation

Student BERT	Param Size	Hidden Layers	Loss Function	F1 Score
BERT Base	110M	12	Traditional	0.78
BERT Base	110M	12	Average	0.74
DistillBERT-SQuAD	66M	12	Traditional	0.82
DistillBERT-SQuAD	66M	12	Average	0.81

10.5 CONCLUSION

Our study addresses several key challenges that researchers often face when working with clinical notes for Natural Language Processing (NLP) tasks. Notably, the content of these notes often straddles between clinical and non-clinical language, complicating the task of identifying specific clinical terms. This uneven distribution of clinical content also impacts the performance of specialized models like clinical BERT variants, as our experiments with the NBME dataset illustrated. Interestingly, our results showed that DeBERTa, a model not specifically designed for the medical domain, performed best in extracting key terms, possibly due to its better handling of non-clinical natural language in the notes.

We further developed a hybrid QA/NER method, mirroring the methodologies used in utilizing BERT, and observed favorable results. However, a prominent concern is the considerable computational cost of large pre-trained language models (PLMs), which might limit their usability for researchers or users with restricted computational resources. To address this, we selected the best performing models on our dataset and used them as "teachers" to train lighter, smaller "student" models, leveraging the multi-teacher knowledge distillation technique. Notably, our "student" model, DistillBERT-SQuAD, demonstrated commendable performance even with a smaller size – it's nearly 4.6 times smaller than the top-performing DeBERTa, with half the hidden layers, yet still achieved an F1 score of 0.81.

In conclusion, our findings suggest that a carefully trained, lighter model can feasibly approximate the performance of its larger counterparts in challenging NLP tasks, even when dealing with datasets that contain a significant mix of clinical and non-clinical language. This opens a promising path toward developing effective and resource-efficient models for processing clinical texts.

Bibliography

[1] The USMLE® Step 2 Clinical Skills Patient Note Corpus. Available at: `https://aclanthology.org/2022.naacl-main.208/`. Accessed: 10-04-2022.

[2] Dai, A. M., and Le, Q. V. Semi-supervised sequence learning. *Advances in Neural Information Processing Systems*, 28, (2015).

[3] Howard, J., & Ruder, S. Universal language model fine-tuning for text classification. *arXiv preprint arXiv:1801.06146*, (2018).

[4] Peters, M. E., Neumann, M., Iyyer, M., Gardner, M., Clark, C., Lee, K., et al. Deep contextualized word representations. *arXiv preprint arXiv:1802.05365*, 441, (2018).

[5] Radford, A., Narasimhan, K., Salimans, T., & Sutskever, I. Improving language understanding with unsupervised learning. *Citado*, 17, 1–12, (2018).

[6] Rogers, A., Kovaleva, O., & Rumshisky, A. A primer in BERTology: What we know about how BERT works. *Transactions of the Association for Computational Linguistics*, 8, 842–866, (2021).

[7] He, P., Liu, X., Gao, J., & Chen, W. DeBERTa: Decoding-enhanced BERT with disentangled attention. *arXiv preprint arXiv:2006.03654*, (2020).

[8] Liu, Y., Ott, M., Goyal, N., Du, J., Joshi, M., Chen, D., et al. RoBERTa: A robustly optimized BERT pretraining approach. *arXiv preprint arXiv:1907.11692*, (2019).

[9] Alsentzer, E., Murphy, J. R., Boag, W., Weng, W.-H., Jin, D., Naumann, T., & McDermott, M. B. A. Publicly Available Clinical BERT Embeddings. *CoRR*, abs/1904.03323, (2019). Available at: `http://arxiv.org/abs/1904.03323`.

[10] Gu, Y., Tinn, R., Cheng, H., Lucas, M., Usuyama, N., Liu, X., et al. Domain-specific language model pretraining for biomedical natural language processing. *ACM Transactions on Computing for Healthcare (HEALTH)*, 3(1), 1–23, (2021).

[11] Hinton, G., Vinyals, O., & Dean, J. Distilling the knowledge in a neural network. *arXiv preprint arXiv:1503.02531*, (2015).

[12] Deng, X., & Zhang, Z. Can students outperform teachers in knowledge distillation-based model compression? *Unpublished Manuscript.*

[13] Hakala, K., & Pyysalo, S. Biomedical named entity recognition with multilingual BERT. In *Proceedings of the 5th Workshop on BioNLP Open Shared Tasks*, 56–61, (2019).

[14] Kim, Y.-M., & Lee, T.-H. Korean clinical entity recognition from diagnosis text using BERT. *BMC Medical Informatics and Decision Making*, 20(7), 1–9, (2020).

[15] Qiu, J., Zhou, Y., Wang, Q., Ruan, T., & Gao, J. Chinese clinical named entity recognition using residual dilated convolutional neural network with conditional random field. *IEEE Transactions on Nanobioscience*, 18(3), 306–315, (2019).

[16] He, K., Zhang, X., Ren, S., & Sun, J. Deep residual learning for image recognition. In *Proceedings of the IEEE Conference on Computer Vision and Pattern Recognition*, 770–778, (2016).

[17] Silver, D., Schrittwieser, J., Simonyan, K., Antonoglou, I., Huang, A., Guez, A., et al. Mastering the game of Go without human knowledge. *Nature*, 550(7676), 354–359, (2017).

[18] Devlin, J., Chang, M.-W., Lee, K., & Toutanova, K. BERT: Pre-training of deep bidirectional transformers for language understanding. *arXiv preprint arXiv:1810.04805*, (2018).

[19] Romero, A., Ballas, N., Kahou, S. E., Chassang, A., Gatta, C., & Bengio, Y. FitNets: Hints for thin deep nets. *arXiv preprint arXiv:1412.6550*, (2014).

[20] Zagoruyko, S., & Komodakis, N. Paying more attention to attention: Improving the performance of convolutional neural networks via attention transfer. *arXiv preprint arXiv:1612.03928*, (2016).

[21] Chen, D., Mei, J. P., Zhang, Y., Wang, C., Wang, Z., Feng, Y., et al. Cross-layer distillation with semantic calibration. In *Proceedings of the AAAI Conference on Artificial Intelligence*, Vol. 35, 7028–7036, (2021).

[22] You, S., Xu, C., Xu, C., & Tao, D. Learning from multiple teacher networks. In *Proceedings of the 23rd ACM SIGKDD International Conference on Knowledge Discovery and Data Mining*, 1285–1294, (2017).

[23] Fukuda, T., Suzuki, M., Kurata, G., Thomas, S., Cui, J., & Ramabhadran, B. Efficient knowledge distillation from an ensemble of teachers. In *Proceedings of Interspeech*, 3697–3701, (2017).

[24] Wu, M.-C., Chiu, C.-T., & Wu, K.-H. Multi-teacher knowledge distillation for compressed video action recognition on deep neural networks. In *ICASSP 2019 IEEE International Conference on Acoustics, Speech and Signal Processing (ICASSP)*, 2202–2206, (2019).

[25] Kwon, K., Na, H., Lee, H., & Kim, N. S. Adaptive knowledge distillation based on entropy. In *ICASSP 2020 IEEE International Conference on Acoustics, Speech and Signal Processing (ICASSP)*, 7409–7413, (2020).

[26] Wu, C., Wu, F., & Huang, Y. One teacher is enough? Pre-trained language model distillation from multiple teachers. *arXiv preprint arXiv:2106.01023*, (2021).

[27] Zhang, H., Chen, D., & Wang, C. Confidence-aware multi-teacher knowledge distillation. In *ICASSP 2022 IEEE International Conference on Acoustics, Speech and Signal Processing (ICASSP)*, 4498–4502, (2022).

[28] Zagoruyko, S., & Komodakis, N. Wide residual networks. *arXiv preprint arXiv:1605.07146*, (2016).

[29] Wang, A., Singh, A., Michael, J., Hill, F., Levy, O., & Bowman, S. R. GLUE: A multi-task benchmark and analysis platform for natural language understanding. *arXiv preprint arXiv:1804.07461*, (2018).

[30] Gu, Y., Tinn, R., Cheng, H., Lucas, M., Usuyama, N., Liu, X., Naumann, T., Gao, J., & Poon, H. Domain-specific language model pretraining for biomedical natural language processing. *arXiv preprint arXiv:2007.15779*, (2020).

[31] Li, X., Zhang, H., & Zhou, X.-H. Chinese clinical named entity recognition with variant neural structures based on BERT methods. *Journal of Biomedical Informatics*, Vol. 107, 103422, (2020).

[32] Pollard, T. J., & Johnson, A. E. W. The MIMIC III clinical database, version 1.4. *The MIMIC III Clinical Database. PhysioNet*, (2016).

[33] Johnson, A. E. W., Pollard, T. J., Shen, L., Lehman, L. H., Feng, M., Ghassemi, M., Moody, B., Szolovits, P., Celi, L. A., & Mark, R. G. MIMIC-III, a freely accessible critical care database. *Scientific Data*, Vol. 3, No. 1, 1–9, (2016).

[34] Lee, J., Yoon, W., Kim, S., Kim, D., Kim, S., So, C. H., & Kang, J. BioBERT: A pre-trained biomedical language representation model for biomedical text mining. *Bioinformatics*, Vol. 36, No. 4, 1234–1240, (2020).

[35] Simonyan, K., & Zisserman, A. Very deep convolutional networks for large-scale image recognition. *arXiv preprint arXiv:1409.1556*, (2015).

[36] He, K., Zhang, X., Ren, S., & Sun, J. Deep residual learning for image recognition. *arXiv preprint arXiv:1512.03385*, (2015).

CHAPTER 11

An Approach to Ensure Public Safety Using Masked Face Recognition

Jafrin Iqbal Chowdhury

International Islamic University Chittagong, Dept. of CSE, Chittagong, Bangladesh.

Shahanaz Ahmed Nishi

International Islamic University Chittagong, Dept. of CSE, Chittagong, Bangladesh.

Umma Aymon

International Islamic University Chittagong, Dept. of CSE, Chittagong, Bangladesh.

Tanveer Ahsan

International Islamic University Chittagong, Dept. of CSE, Chittagong, Bangladesh.

Shayhan Ameen Chowdhury

International Islamic University Chittagong, Dept. of CSE, Chittagong, Bangladesh.

M. Jamshed Alam Patwary*

Rangamati Science and Technology University, Dept. of CSE, Rangamati, Bangladesh.

CONTENTS

DOI: 10.1201/9781003605508-11

THe masked face recognition system is currently one of the most sought-after technologies in the security business, with applications in access control, automated checkout, and other identity verification scenarios. In order to focus on facial recognition, researchers are increasingly using deep convolutional neural networks and other cutting-edge techniques. When it comes to feature extraction and recognition, neural networks have demonstrated incredibly outstanding performance. In this study, a deep learning-based model is portrayed that can identify a person based on their face, obstructed face, and whether or not they are wearing a mask. The proposed approach employs two models, one based on a convolutional neural network architecture for masked face recognition and the other on a MobileNetV2 architecture for mask detection. For these two tasks, two separate datasets have been employed. We have used 10,000 images to train and validate our model for masked and unmasked faces for specific individuals for masked face recognition. The mask detection model has also been developed using around 12,000 images. In comparison to other cutting-edge technologies, both of our methods for mask detection and masked face recognition performed admirably.

11.1 INTRODUCTION

Masked face recognition system has now become the most hyped system for security purposes. Face recognition (FR) systems typically show non-occluded faces with three major facial features: mouth, nose, and eyes. Pandemics, laboratories, medical operations, and major pollution require people to wear masks that completely or partially conceal their features. Due to security and identification concerns, it is critical to recognize people wearing masks without having to uncover them. Much research has been conducted in recent years on masked face recognition in general.

Ai et al. [1] developed a facial mask identification system that utilizes transfer learning along with a neural network to recognize masks. While the MobileNetV2 model was employed to distinguish faces, the ResNet-50V2 model was utilized to find and categorize the existence of masks in images. Li et al. [7] created a methodology for face recognition that trains data using a CNN approach utilizing a deep learning approach. TensorFlow is employed in the system classifier section.

There are several frequently employed strategies in the branches of pattern recognition and image analysis. These techniques encompass Support Vector Machines (SVM) [21], Multi-task Cascaded Convolutional Networks (MTCNN) [20], and Local Binary Patterns (LBP) [19]. By comparing pixel values with those of nearby pixels, the LBP algorithm detects local patterns in images. It has been widely used in many applications, proving its efficacy and efficiency. MTCNN, on the other hand, is a face identification method based on deep learning. To find faces in an image, it uses a cascaded framework of convolutional neural networks. MTCNN has demonstrated exceptional accuracy in accurately identifying faces. For classification tasks, supervised machine learning techniques like SVM are typically utilized. It creates a

[1*] Corresponding author.

hyperplane by optimizing the margin for splitting various classes. Pattern recognition and visual analysis are two areas where SVM has received an immense amount of study and application. Numerous research studies and real-world applications have been used to evaluate and validate the effectiveness and efficiency of these strategies. They have considerably aided the development of computer vision and recognition of patterns.

Qi et al. [11] developed a system for masked face recognition that makes use of LBP for feature extraction, an enhanced version of MTCNN, and SVM to identify faces. Face masks have become an accepted norm in the midst of the epidemic because they help prevent virus flare-ups. Many countries have begun to mandate that people wear masks when leaving their homes or entering public places. Aside from wearing face masks, the government advocates for other safety and hygiene precautions. As a result, recognizing and identifying occluded faces is becoming an increasingly important requirement for every nation in order to meet health and safety obligations, monitoring, access restrictions, attendance, and other security purposes. Furthermore, evidence suggests that criminals from all walks of life frequently conceal their individuality. This poses a significant threat to society. However, maintaining touchless access control in buildings while identifying faces hidden from view on any monitoring systems will be difficult. Face recognition algorithms and performance have been hampered by masked faces. In this study, we attempted to recognize faces even when people's faces were obscured by masks, as well as determine whether or not they were wearing masks.

Our research is divided into two main sections: The first section is concerned with identifying individuals (masked-unmasked), while the second is concerned with determining whether or not a person is wearing a mask. For the two tasks, we used the traditional multilayer CNN models and MobileNetv2 respectively. Our first model recognizes individuals by automatically extracting facial features from images. The second model recognizes a masked or unmasked face in images. Our method identifies whether or not an identified person is wearing a mask. The key contributions of this paper are as follows:

- For masked face recognition and mask detection, we recommend a Convolution neural hybrid model. We used the perks of CNN to shrink the size of processed data while also enabling parallel implementation and trimming the number of layers to reduce the use of calculation blocks.

- Our proposed model shows promising performance even when the mask color varies.

The structure of this paper is as follows. The second section provides context for feature extraction and classification algorithms. The model we proposed, which is based on CNN for feature extraction and classification and mobilentv2, is described in the third section. The fourth part demonstrates and explains the experimental results. A summary of the paper is presented in the last section.

11.2 RELATED WORK

Face recognition and mask detection are two well-known studies in the area of information processing that have seen a sharp increase in popularity in recent years. Due to their crucial applications in numerous fields, these disciplines have attracted wide interest and prominence. Tackling the challenges and demands provided by functioning such apps across different platforms is crucial as information processing technologies continue to develop.

Face recognition technology has made significant advancements in recent years, making it practicable to identify and authenticate people based on their facial traits. Security systems, access control, surveillance, and personal device authentication are just a few of the fields where the ability to recognize faces with accuracy has found uses. Researchers have concentrated on creating cutting-edge algorithms and methodologies that can handle alterations in lighting conditions, position, and occlusions induced by masks due to a rising need for trustworthy face recognition systems.

Corresponding to this, mask detection has become more popular since the global COVID-19 outbreak. The need for automated systems capable of recognizing and identifying people wearing masks has been brought to light by the widespread use of face masks as a form of defense against the virus's spreading. Mask detection is essential for guaranteeing adherence to safety regulations, especially in public areas, busy transportation hubs, and medical facilities.

The importance of these technologies in addressing practical problems and enhancing general safety and security is reflected in the rising importance of facial recognition and mask detection as scientific disciplines. The potential applications of facial recognition and mask detection are projected to grow further, having a favorable effect on numerous businesses and sectors, as scientists and technologists strive to pursue new directions and build innovative solutions.

In the arena of face recognition, a widespread range of conventional approaches and other tactics have been expansively explored and documented in the related work, as demonstrated by orientations [5, 6, 8], among others. Meanwhile, as shown by the references [2, 4, 6, 5, 7, 8, 11, 15, 18], numerous research studies have concentrated in particular on the arduous challenge of masked face identification. Support Vector Machines (SVMs), which are referenced in [11], have developed into unique learning machines with a variety of benefits, such as their speedy problem-solving and strong generalization abilities. SVMs have so been extensively used in the field of face identification. It's crucial to recognize that such techniques have certain inherent drawbacks. Particularly when faced with obstructed faces or oscillations in postures, aging, and facial emotions, they frequently struggle to obtain high identification rates. Research and development efforts are still being put into dealing with these issues and enhancing how well face recognition systems operate in these circumstances.

A number of studies have also been studied for mask identification, including [1, 3, 9, 11, 13, 14, 16, 17, 24]. These studies attempted to create reliable technology and algorithms for completely recognizing and distinguishing between people wearing masks. The results of these studies have proven excellent performance and increased mask identification precision in circumstances with high illumination conditions. The obstacles provided by brightly lit settings have been successfully addressed by the algorithms and methods suggested in these studies, enabling accurate detection and recognition of masked faces. It is crucial to remember that this research' performance in low-intensity illumination settings has been determined to be insufficient. Under these circumstances, it can be difficult to identify faces because poor lighting may interfere with visibility and make it difficult to discern masks meticulously. The shortcomings seen in low-intensity lighting circumstances indicate the necessity for additional study and the creation of advanced algorithms that are capable of handling difficult lighting scenarios.

We are active in both problems in this manuscript, which are how to extract features and the process of designing a classifier for masked face recognition and mask detection while reducing running memory and processing time requirements. As a result, our study includes two technical components. The first is used to extract features. We discovered PCA to be one of the most widely used techniques for dimension reduction through facial feature extraction. According to [2], the utilization of PCA as a great extractor of facial features has been validated, and it could be used for face recognition. Furthermore, the use of a large number of random vectors necessitates a large amount of storage space as well as a high computational cost. Concerning the second specialized portion, namely the advancement of novel classifiers for face recognition, many alluring methods for classifiers have been developed, including SVMs, nearest neighbor classifiers, and minimum distance classifiers. SVM, PCA, and other machine learning approaches [12, 16] perform admirably in straightforward recognizing pattern operations, but they might have difficulty with the complexity of facial features and looks, leading to unsatisfactory face identification accuracy. Recently, CNN-based face recognition models have outperformed others. Multiple applications, including RetinaFace [16], Vgg-16 [5], have accomplished critical findings in this frame of reference. Matter of fact, massive amounts of data and deep networks are vital factors in the success of these kinds of applications.

As an aspect of our contribution to this work, we attempt to resolve this problem by integrating masked face recognition and mask detection models to produce a result that is useful in real-world contexts and also explores memory and processing time concerns. Furthermore, we concentrate on developing a CNN-based model for masked face recognition and a MobileNetv2 model for mask detection in order to extract features and optimize their use for classification. Our model is deliberately developed to decrease both its running time and memory requirements, allowing us to achieve the ideal balance of speed and accuracy while also tracking down a meaningful application, which is essential.

11.3 OVERVIEW OF THE FRAMEWORK

Our proposed solution employs two models that were trained on mask detection datasets from Kaggle and individual photos from GitHub that were adjusted due to the low GPU issue. We aim to establish whether an identified individual is wearing a mask or not based on the hybrid output of these models, which is significant in security or health-related contexts. The mask detection model determines if a person is wearing a mask or not. The masked face recognition model then recognizes persons by extracting facial traits from images automatically, and the final outcome fits our expectations. In this scenario, our models are based on MobileNetv2 and sequential CNN. Datasets have been preprocessed as relevant. The proposed framework's process diagram is depicted in Fig. 11.1. Images were preprocessed using grayscale conversion, resizing, augmentation, and scaling. Label encoding is used to train classifier models.

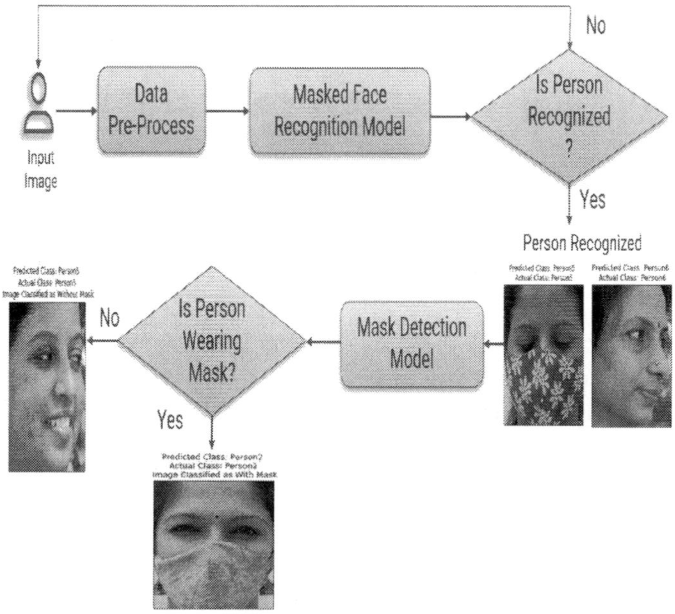

Figure 11.1: Workflow diagram of the proposed framework

11.3.1 Masked Face Recognition Model

A CNN model with four 2D convolutional layers was utilized to identify masked and unmasked faces for the individuals' masked face recognition inspection. As a matter of fact, model size and complexity are reduced. Depth-wise convolution is used to fetch features, while pointwise convolution is used to concatenate them to create more features. The model is made up of 5 by 5 and 3 by 3 filters. Max pooling layers with a 2-by-2 filter are utilized in line with the Relu activation function to minimize the input size. The model includes a dropout layer [10] and one fully connected layer to avoid overfitting. The Softmax classifier is used for classification at the classification layer. To reduce input size and avoid overfitting, max pooling is used.

11.3.2 Mask Detection Model

The proposed mask detection method makes use of MobileNetv2. We loaded the MobileNetV2 model and used ImageNet to initialize the weights. We made the entire model untrainable to prevent weight updates in the layers. The model needs (224, 224, 3) image size as input. Next, we added a flattened layer to reduce the dimensionality of the MobileNetV2 output. Finally, the output layer consists of two receptors that employ the softmax activation function to determine whether an input image contains a face mask.

11.3.3 Classification Loss Function

Loss function evaluates the effectiveness of a classification model or maps decisions to their costs. Since our job is multi-class classification, we used Categorical cross-entropy for both frameworks. It is the predicate of the distinction between two probabilistic. The loss is figured in Equation 11.1 by adding the following sums:

$$Loss = -\sum_{i=1}^{outputsize} y_i.log(\hat{y_i}). \tag{11.1}$$

As it rescales the simulation results, the categorical cross-entropy loss method performs optimally when used with Softmax.

11.3.4 Optimization Algorithm

Optimizers [22] in a neural network endeavor to minimize loss by adjusting features like weights and learning rates. We utilized the ADAM optimizer to create the masked face recognition model and the SGD optimizer to find masks because these optimizers performed well with the specific challenge. Adam is a versatile gradient descent method that combines the greatest characteristics of AdaGrad and RMSProp while remaining simple to use. It estimates the learning rate and surpasses other optimization approaches, although it is more expensive. The SGD optimizer is a fast and memory-efficient deep learning optimization technique. It boosts model generalization and adapts to various neural network designs and hyperparameters.

11.4 EXPERIMENTAL ANALYSIS

This section describes the performance assessment of the proposed strategy on the GPU. Keras, an open-source deep learning framework, was utilized to train all of the methods. We initially present the findings of evaluating the proposed strategy in terms of time complexity, memory consumption, and recognition and detection performance. What's more, we compare the proposed method with other cutting-edge methods.

11.4.1 Datasets

In this study, we introduced a technique that makes use of two datasets. One dataset is for masked face recognition, which authenticates individuals whether they are masked or not, and the other is for mask detection, which establishes whether an individual is hidden or exposed.

- **Dataset I:** `Dataset I`[1] was obtained from Kaggle for the mask detection task, and it contains nearly 12k images divided into two classes (WithMask and WithoutMask). Fig. 11.2 shows some evidence from this database.

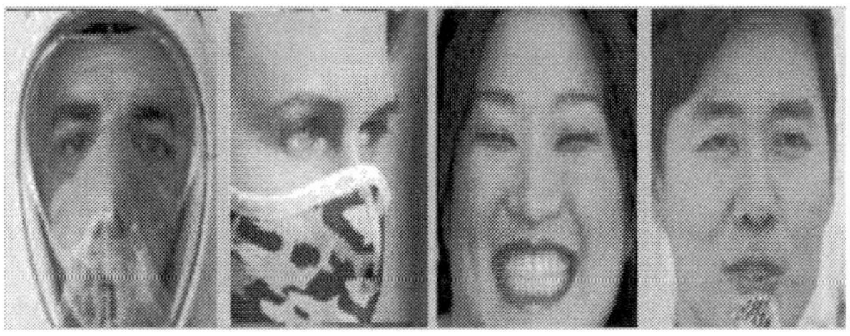

Figure 11.2: Some instances of Dataset I

- **Dataset II:** `Dataset II`[2] was obtained from GitHub. There are 10000 images for 10 classes. Fig. 11.3 illustrates some examples from this database. We customized the masked face recognition dataset to enhance the performance and accuracy of our models due to a lack of relevant and specific datasets and a low GPU issue.

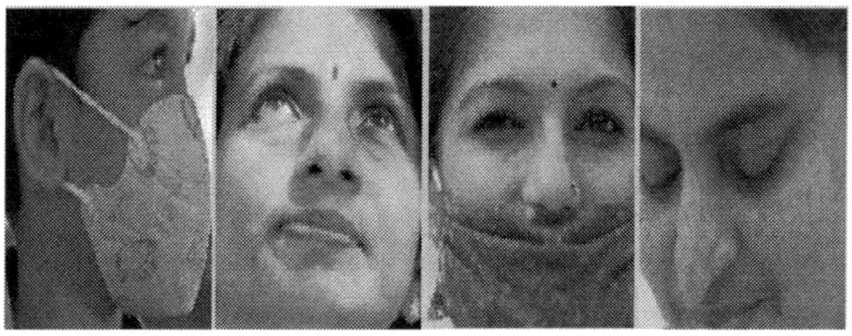

Figure 11.3: Some instances of Dataset II

To evaluate the impact of various components of the two architectures on their performance, we carried out an ablation study for our entire work. This suggested assessment can be carried out by investigating the effects of changing parameters, such as the quantity and size of convolutional layers, filters, pooling layers, and batch normalization. The results enabled us to understand how each element contributes to our model's overall quality and choose the best parameters for it.

11.4.2 Ablation study for Masked Face Recognition model

Through an ablation study, we determined the best settings for the layer, dropout, and Optimizer in the Masked Face Recognition model.

- The accuracy of the model improves as the number of layers increases, but it starts to lose accuracy once it becomes overly complex and starts reexamining the training data rather than projecting to new data. As shown in Fig. 11.4, the model performs better with four layers.

---

[1] `www.kaggle.com/datasets/ashishjangra27/face-mask-12k-images-dataset`

[2] `https://drive.google.com/drive/folders/1fb3-vfDmfuvI7RhqhyIYIJjHbh5tbNA0w`

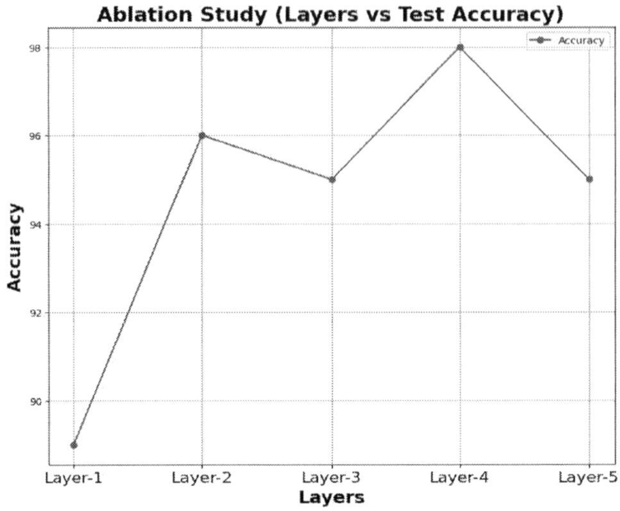

Figure 11.4: Layers vs Test Accuracy

- Fig. 11.5 shows that the optimized dropout rate for the developed framework is 0.5, as accuracy improves as the dropout rate up to 0.5 and then lessens after 0.6.

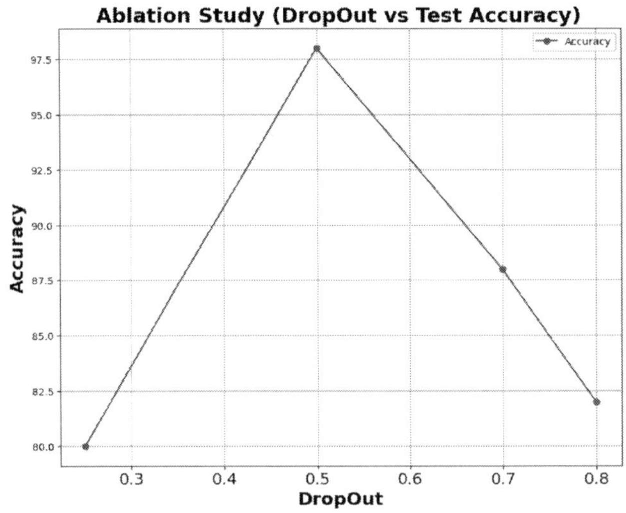

Figure 11.5: Dropout vs Test Accuracy

- As shown in Fig. 11.6, the Adam optimizer is the strongest optimizer for the proposed masked face recognition model and outperforms other optimizers.

11.4.3 Ablation Study for Mask Detection Model

Another ablation study was also conducted to determine the ideal settings for the mask detection model's optimizer, activation function, and number of epochs.

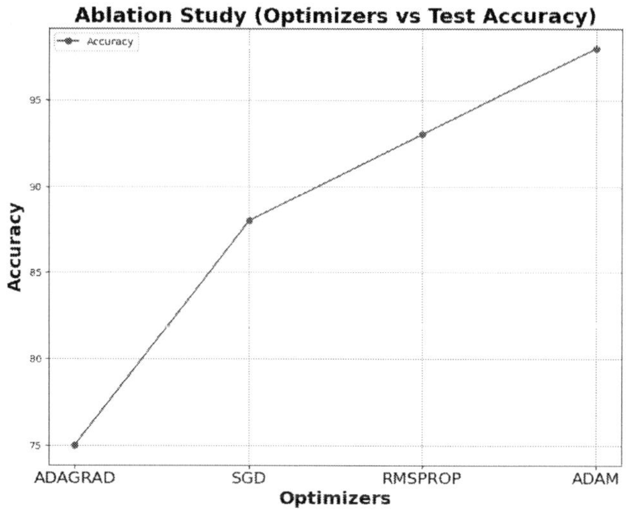

Figure 11.6: Optimizers vs Test Accuracy

- As shown in Fig. 11.7, the SGD optimizer is the finest optimization technique for the developed framework.
- The provided mask detection model works best with softmax as the activation function, as shown in Fig. 11.8, because accuracy peaks at this function and begins to fall off when sigmoid is utilized.
- The accuracy of the model increases with an increasing number of epochs but starts to decline after reaching a certain point, where the model becomes too complex and begins memorizing the training data rather than generalizing it to new data. The best performance of the model was observed with 15 epochs as seen in Fig. 11.9.

11.4.4 Analysis of Experimental Results

Based on classic CNN, the suggested masked facial recognition model shines on two datasets, achieving 95% accuracy on Dataset II and 89% accuracy on the CISIA Web facial dataset [7], in which they employed a mixed dataset that includes CISIA Web Face. However, in simple facial recognition, the attention process [7] does not work well. It has been demonstrated that our model performs well for basic face recognition as well as accepting modifications in mask and position while being effective in terms of execution time and memory use. Finally, our suggested framework has shown that a simpler design can produce results that are similar to or better than sophisticated models while being more effective and simpler to implement. Table 11.1 summarizes the Masked Face Recognition Model's performance on several datasets.

The proposed mask detection model, which is based on MobileNetv2, excels over the competition, achieving 99.80% accuracy on the Dataset I and 99.75% accuracy using the approach of Taneja et al. [14]. While dealing with differences in lighting,

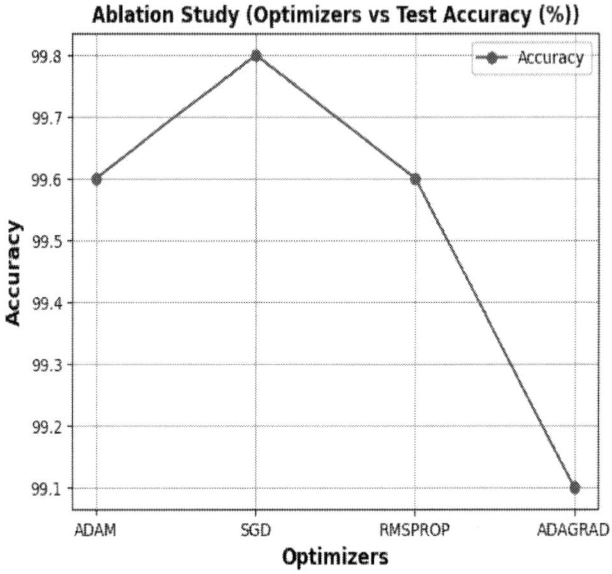

Figure 11.7: Optimizers vs Test Accuracy

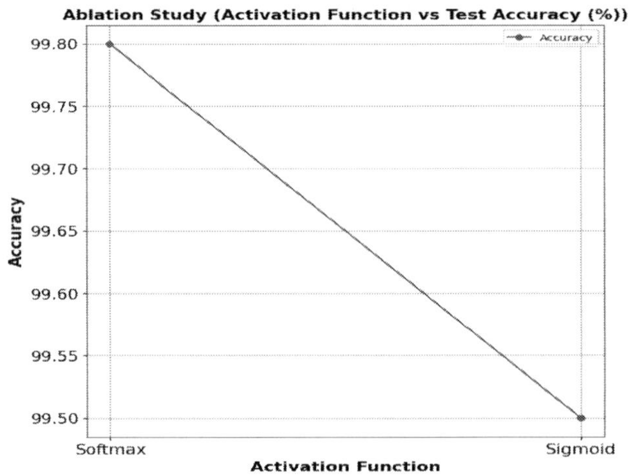

Figure 11.8: Activation Function vs Test Accuracy

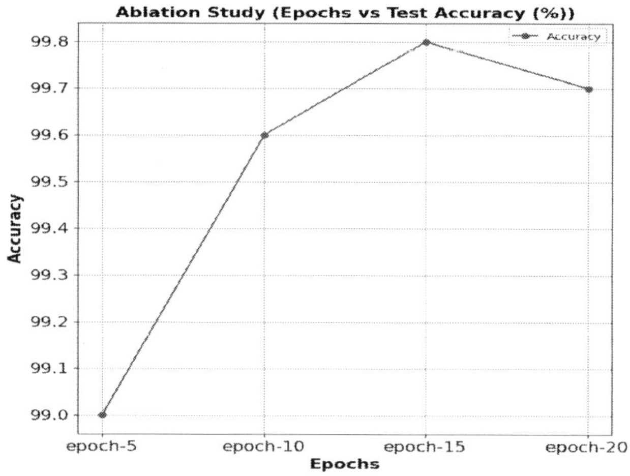

Figure 11.9: Epochs vs Accuracy

Table 11.1: Masked Face Recognition Model's performance on several datasets.

Model	Dataset	Accuracy
Masked Face Recognition	Dataset-II	95%
	CISIA Web Face [7]	89%
Attention Approach [7]	CISIA Web and others	91.5%

face posture, and expressions, our proposed model has proven to be fast and memory-efficient. Finally, our proposed model indicates that a basic design can give results that are equivalent to, if not superior to, those of more advanced models while being more computationally efficient and easier to implement. Table 11.2 shows a summary of the performance of mask detection methods on the same datasets, comparing our proposed model to other cutting-edge models.

Table 11.2: Performance of Mask Detection Method on the same Dataset

Model	Dataset	Accuracy
Mask Detection	Dataset-I	99.80%
Deep learning approach [14]	Dataset-I	99.75%

In the end, the proposed hybrid architecture outscored traditional models in both masked face recognition and mask detection experiments, demonstrating exceptionally high accuracy rates. The models demonstrated excellent computational efficiency, ease of deployment, and accurate recognition of faces and masks in images.

11.5 CONCLUSION

In this study, we recommend a convolutional neural hybrid model for mask detection and masked face recognition. As a result, we utilized the benefits of CNN to reduce the size of processed data while simultaneously enabling concurrent implementation. In addition, we shortened the number of layers to reduce the utilization of computation blocks. Our research made use of the well-known Face Mask Detection Dataset and Face Recognition Dataset from Kaggle and GitHub, respectively. Furthermore, the proposed optimization procedure is simple yet powerful in terms of completion time and memory usage on the GPU integration for the model set MobileNetv2-CNN, as well as performing well in different mask variations. In the future, a more sophisticated CNN classifier will be used to handle a much larger dataset and achieve a good balance between accuracy, processing speed, and memory utilization.

Bibliography

[1] Ai, M. A., Shanmugam, A., Muthusamy, S., Viswanathan, C., Panchal, H., Krishnamoorthy, M., & Orban, R. Real-time facemask detection for preventing COVID-19 spread using transfer learning based deep neural network.*Electronics.*, **11(14)**, p. 2250 (2022)

[2] Ejaz, M. S., Islam, M. R., Sifatullah, M., & Sarker. A. Implementation of principal component analysis on masked and non-masked face recognition. In *2019 1st international conference on advances in science, engineering and robotics technology (ICASERT)*, pp. 1–5, IEEE (2019)

[3] Fan, X., & Jiang, M., RetinaFaceMask: A single stage face mask detector for assisting control of the COVID-19 pandemic. In *2021 IEEE international conference on systems, man, and cybernetics (SMC)*, pp. 832–837, IEEE (2021)

[4] Golwalkar, R., & Mehendale, N. Masked-face recognition using deep metric learning and FaceMaskNet-21. *Applied Intelligence*, **52(11)**, 13268–13279 (2022)

[5] Hariri, W. Efficient masked face recognition method during the covid-19 pandemic. *Signal, image and video processing*, **16(3)**, 605–612 (2022)

[6] Cao, W., Wu, Y., Huang, C., Patwary, M. J., & Wang, X., MFF: Multi-modal feature fusion for zero-shot learning. *Neurocomputing*, **510**, 172–180 (2022)

[7] Li, Y., Guo, K., Lu, Y., & Liu, L. Cropping and attention based approach for masked face recognition. *Applied Intelligence*, **51**, 3012–3025 (2021)

[8] Patwary, M. J., Cao, W., Wang, X. Z., & Haque, M. A. . Fuzziness based semi-supervised multimodal learning for patient's activity recognition using RGBDT videos. *Applied Soft Computing*, **120**, 108655 (2022)

[9] Patwary, M. J. A., Parvin, S., & Akter, S. Significant HOG-histogram of oriented gradient feature selection for human detection. *International Journal of Computer Applications*, **132**, no. 17 (2015)

[10] Piotrowski, A. P., Napiorkowski, J. J., & Piotrowska, A. E. Impact of deep learning-based dropout on shallow neural networks applied to stream temperature modelling. *Earth-Science Reviews*, **201**, 103076 (2020)

[11] Qi, C., & Yang, L. Face recognition in the scene of wearing a mask. In *2020 International Conference on Advance in Ambient Computing and Intelligence (ICAACI)*, pp. 77–80, IEEE (2020)

[12] Karim, S., Akter, N., & Patwary, M. J. Predicting autism spectrum disorder (ASD) meltdown using fuzzy semi-supervised learning with NNRW. In *2022 International Conference on Innovations in Science, Engineering and Technology (ICISET)*, pp. 367–372, IEEE (2022)

[13] Sethi, S., Kathuria, M., & Kaushik, T. Face mask detection using deep learning: An approach to reduce risk of Coronavirus spread. *Journal of biomedical informatics*, **120**, 103848 (2021).

[14] Taneja, S., Nayyar, A., & Nagrath, P. Face mask detection using deep learning during COVID-19. In *Proceedings of Second International Conference on Computing, Communications, and Cyber-Security*, pp. 39–51, Springer, Singapore. (2021)

[15] Vu, H. N., Nguyen, M. H., & Pham, C. Masked face recognition with convolutional neural networks and local binary patterns.*Applied Intelligence*, **52(5)**, 5497–5512 (2022)

[16] Karim, S., Akter, N., & Patwary, M. J. Predicting autism spectrum disorder (ASD) meltdown using fuzzy semi-supervised learning with NNRW. In *2022 International Conference on Innovations in Science, Engineering and Technology (ICISET)*, pp. 367–372, IEEE (2022)

[17] Yu, J., & Zhang, W. Face mask wearing detection algorithm based on improved YOLO-v4.*Sensors*, **21(9)**, 3263 (2021).

[18] Ashraf, M., & Hussain, M. Z. Eigenfaces Technique, an Improved Face Recognition Approach Using Neural Network. *Asian Journal of Computer Science and Technology*, **8(2)**, 98–104 (2019).

[19] Karanwal, S., & Diwakar, M. OD-LBP: Orthogonal difference-local binary pattern for Face Recognition.*Digital Signal Processing*, **110**, 102948 (2021)

[20] Zhang, N., Luo, J., & Gao, W. Research on face detection technology based on MTCNN. In *2020 international conference on computer network, electronic and automation (ICCNEA)*, pp. 154–158, IEEE (2020).

[21] Geetha, M., Latha, R. S., Nivetha, S. K., Hariprasath, S., Gowtham, S., & Deepak, C. S. Design of face detection and recognition system to monitor students during online examinations using Machine Learning algorithms. In *2021 international conference on computer communication and informatics (ICCCI)*, pp. 1–4, IEEE (2021)

[22] Ning, C., & You, F. Optimization under uncertainty in the era of big data and deep learning: When machine learning meets mathematical programming. *Computers & Chemical Engineering*, **125**, 434–448 (2019)

[23] Asma, A., Mostafa, D. N., Akter, K., Mahmud, M., & Patwary, M. J. Fuzziness Based Semi-supervised Deep Learning for Multimodal Image Classification. In *International Conference on Machine Intelligence and Emerging Technologies*, pp. 91–105, Cham: Springer Nature Switzerland (2022)

[24] Sen, S., & Sawant, K. Face mask detection for covid 19 pandemic using pytorch in deep learning. In *IOP Conference Series: Materials Science and Engineering*, Vol. 1070, No. 1, p. 012061, IOP Publishing (2021)

CHAPTER 12

Improving the Efficiency of Waste Management with a Residual Network-Based Framework

Md. Asfakul Islam

Department of Computer Science and Engineering, East West University, Dhaka, Bangladesh

Nafiz Faiyaz

Department of Computer Science and Engineering, East West University, Dhaka, Bangladesh

Mehedi Hasan Adnan

Department of Computer Science and Engineering, East West University, Dhaka, Bangladesh

A. M. Feroz Ehses Shishir

Department of Computer Science and Engineering, East West University, Dhaka, Bangladesh

Faysal Ahmed

Department of Computer Science and Engineering, East West University, Dhaka, Bangladesh

Farhana Alam

Department of Computer Science and Engineering, East West University, Dhaka, Bangladesh

Ahmed Wasif Reza

Department of Computer Science and Engineering, East West University, Dhaka, Bangladesh

CONTENTS

DOI: 10.1201/9781003605508-12

WAste management is a significant challenge for cities and communities around the world. It not only has environmental impacts but also has economic and social consequences. Machine learning techniques can help waste management by analyzing massive volumes of data and identifying behaviors and patterns that can assist decision-making. Our main goal is to use a machine learning algorithm in waste management for detecting different types of waste materials from Images. We presented a machine learning framework PyTorch for categorizing waste into various groups such as metal, paper, glass, trash, plastic, and cardboard. The ResNet model had already been pre-trained on the ImageNet dataset, which comprises 14 million images and 1000 image categories. We collected 2527 images of different classes and achieved a validation accuracy of 94% and a testing accuracy of 93.14%. Machine learning in waste management can enhance waste collection and disposal efficiency and effectiveness, resulting in lower costs and environmental impacts.

12.1 INTRODUCTION

Waste management involves the movement, collection, treatment, reuse, and disposal of various types of waste, such as solid, liquid, dangerous, electronic waste, and more. Effective waste management is crucial for protecting the environment and public health, conserving natural resources, and enabling the efficient usage of resources by recycling and composting.

Machine learning is like artificial intelligence with training algorithms to recognize patterns and make decisions based on data. Machine learning is able to be applied to waste management to evaluate data collected from sensors, cameras, and other sources in order to improve waste management efficiency and efficacy [1]. Machine learning algorithms, for example, may be employed to forecast the quantity of waste produced in a given place by considering parameters such as population density, weather conditions, and economic activity. This can assist waste management agencies in optimizing the allocation of their resources and more accurately forecasting the requirement for additional capacity.

Other potential applications of machine learning algorithms in waste management include detecting the sorts of items in a trash stream, projecting the likely composition of future waste streams, and improving waste pickup vehicle routes [2]. Overall, implementing machine learning algorithms in the management of waste can improve resource efficiency, lower costs, and reduce the environmental effect of waste disposal activities.

The chapter is organized as follows: The section 12.2 provides comprehensive literature review and what is already known about the topic. In section 12.3, the proposed methodology is discussed, explaining the methods and models used for the study. Section 12.4 describes the system architecture and design. In section 12.5, the implementation methods and the obtained results are presented. Section 12.6 critically analyzes and interprets the results and discusses what they mean. Finally, in section 12.7, the paper concludes by summarizing the main points of the study.

12.2 RELATED WORK

Proper waste management is crucial for maintaining cleanliness and health in a city. Various techniques for managing waste utilize learning, which has helped promote sustainability by reducing trash, pollution, and costs while improving the quality of recovered waste. Developed economies should consider how waste collection methods have affected resource recovery.

Conventional object recognition methods depend on manual feature construction and extraction, as well as the development of classifiers. However, multi-layer convolutional neural networks can automatically extract and filter features at different layers, resulting in more accurate detection and substantial generalization. There are two types of deep learning-based object identification methods: two-stage and one-stage [3][4].

The YOLOv1 algorithm, which Redmon presented, is a type of one-stage method [4]. YOLOv1 has limitations Because it directly matches the position dimensions and confidence level. In order to identify and classify six distinct types of waste, Abeywickrama employed SVM and CNN, and the resulting recognition accuracy was 83% [5]. Chen used the Faster RCNN method for recognizing 199 garbage items in a pipeline, achieving a system-failed detection accuracy of 3% and an error identification rate of 9% [6].

Because of the growing need for waste identification and categorization on handheld gadgets, much of the equipment used in these cases is edge hardware with limited processing capabilities, making more widespread detection networks difficult to implement. Matching approaches have been used in the literature to establish task offloading plans between consumers and edge devices. The topic of task offloading in complex edge device implementations has been studied in the literature by Zhihong et al. [7] and Shen et al. [8]. A cooperative alliance has been formed between various edge devices using alliance game theory to perform the end user's computing tasks collectively.

Chen and Xu [9] use a distributed game method to implement task offloading between edge devices to reduce the total execution wait time of the assignment. Managing waste in the future will require a delicate balancing act of diplomacy, technology, information accessibility for the general public, and, most importantly, goodwill to improve things [10].

The decomposition of biodegradable waste, food, and yard waste in trash produces methane gas, 28 times more harmful than carbon dioxide [11]. These organic materials in household waste also attract pests and can cause health problems. It is important to have a system in place to manage waste and prevent the release of harmful gases. This could include alerts for when trash cans are full and need to be cleaned and utilizing the methane gas produced for energy production.

The government of India categorizes waste into biodegradable, non-biodegradable, and domestic hazardous categories. To make it easier for citizens to understand, they have introduced a system for separating wet and dry waste. Certain items, such as plastic and thermal coal, are banned and must be treated separately, according to the Bureau of Indian Standard Specification. It is vital for citizens to be aware of these regulations and for there to be monitoring of waste segregation practices. The main objective is to identify the contents of dry and wet waste based on images of the collected garbage.

To ensure compliance with waste segregation regulations, a system is proposed to use image processing on garbage bin pictures to determine whether the waste is dry, wet, or mixed. These data will be analyzed to generate reports for specific areas, which will be used to identify areas not adhering to the government's waste segregation norms. This information can then be used to conduct awareness campaigns or other efforts to educate and motivate citizens in those areas to follow the regulations.

12.3 PROPOSED MODEL

Figure 12.1 shows our proposed methodology interface; it illustrates our proposed model's different components and connections, the inputs and outputs it receives and gives, and how it interacts with other external dependencies or interfaces to achieve the desired results.

We made a model of waste detection using machine learning. In our model, it has different types of materials such as:

1. Camera.
2. A system for detecting waste (By using Machine Learning)
3. Two types of Trucks (recyclable waste & non-recyclable waste)
4. Roller Machine, which includes Motors, Roller Pads, and Path changer in three ways, centers of the roller machine.

To correctly detect the waste, manually put it in the roller machine one by one, so the camera-captured images are first stored in the cloud, where they can be easily accessed and shared. After that, they are imported into a machine learning model, where they are analyzed and processed using advanced algorithms. These algorithms are then used to detect and classify the different types of images in the dataset, allowing us to accurately identify and categorize each image based on its content and characteristics. When the camera detects a waste, a signal will go to the waste detection system. According to the instructions in the system, the roller path changer

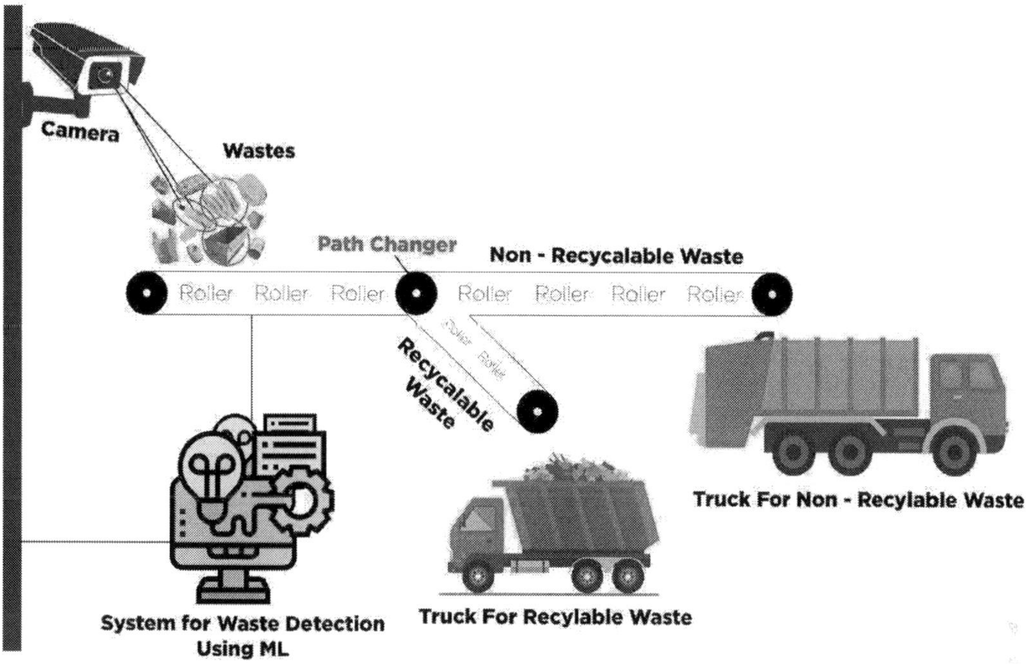

Figure 12.1: Proposed Methodology Interface

will give directions on whether it is reusable or non-reusable. If the waste is recyclable, then it goes to one side. If non-recyclable, then it goes to the other side, and the recyclable waste goes to the green color truck, and on the other hand, the non-recyclable waste goes to the red color truck. This is how this process will continue. A flow chart, such as the one in Fig. 12.2, is a diagram that represents a process or a workflow.

After the process starts of waste detection, the camera will be prepared for acquisition. Then the system takes the input from the captured image by camera and compares the image with the pre-trained model using machine learning. Then the system recognizes whether it is recyclable and separates them into a container. If recyclable, send a command to the corresponding motor to put the waste into a recyclable waste truck, otherwise in a non-recyclable waste truck.

12.4 SYSTEM ARCHITECTURE AND DESIGN

This section describes our proposed waste image classifying system architecture and design in full detail. This approach are divided into some sections which are described below.

12.4.1 Dataset Description

In our dataset, we gathered 6 different types of waste images. The Waste images used in our dataset are Cardboard, Glass, Meta, Paper, Plastic, and Trash (mixed of all kinds of waste images). Figure 12.3 shows examples of images classified into

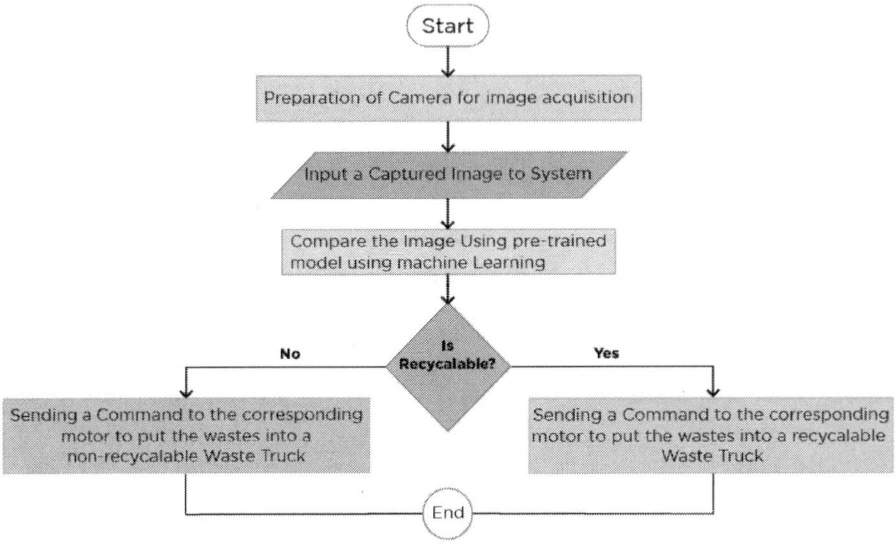

Figure 12.2: Model Flowchart

different classes by a machine learning model. The figure shows different images, such as Cardboard, Glass, Metal, Paper, Plastic, and Trash.

Our dataset has six classes of images, each representing a different category or label. These classes can include objects, scenes, or actions and will be used to train

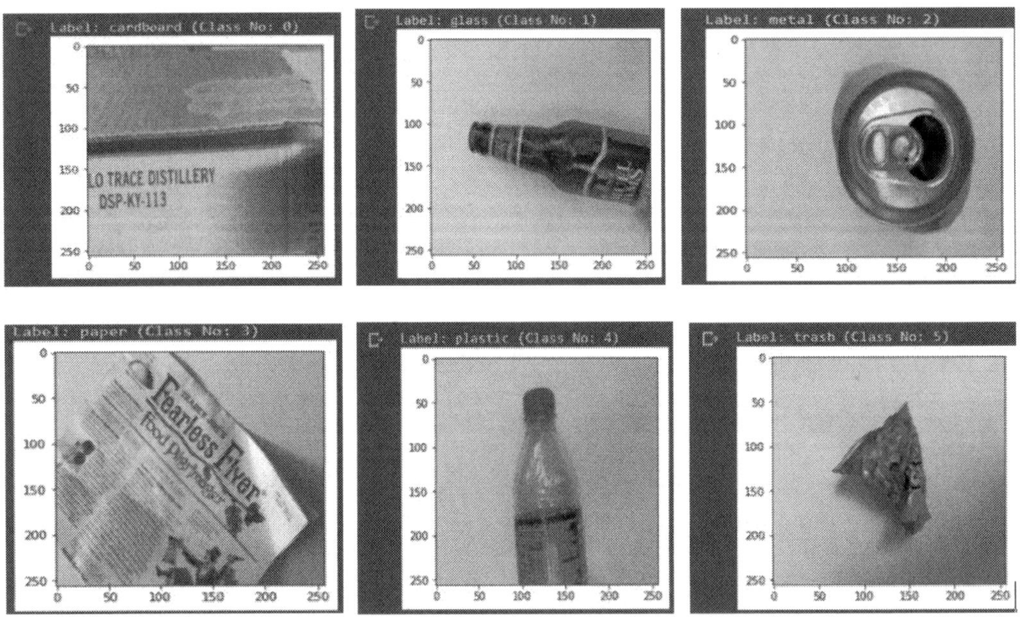

Figure 12.3: Different Classes of Images.

and test our machine-learning model. Figure 12.4 shows the count of different types of images in a dataset that is used to train the model.

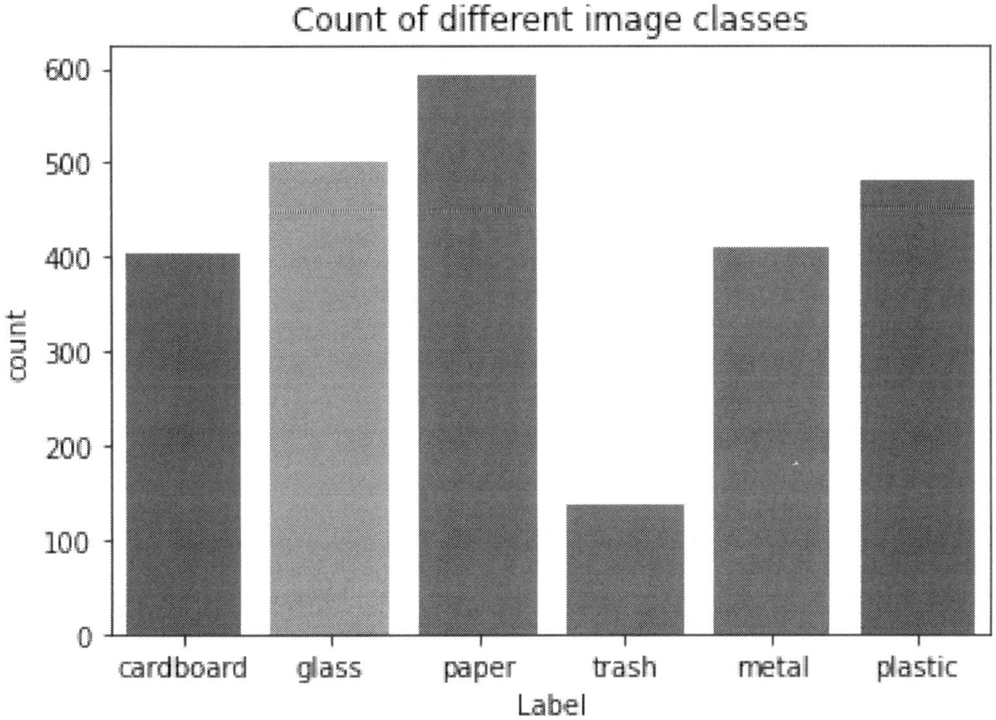

Figure 12.4: Different Image Count in the Dataset.

There are 2527 images in our dataset, including 137 trash, 482 plastics, 594 pieces of paper, 410 pieces of metal, 501 pieces of glass, and 403 pieces of cardboard.

When handling class imbalance in machine learning, one approach is to use different evaluation metrics that consider the imbalance in the data. Some commonly used metrics include:

- **Precision:** This measures the proportion of true positive predictions among all positive predictions.
- **Recall:** This measures the proportion of true positive predictions among all actual positive instances.
- **F1-score:** This is the harmonic mean of precision and recall, and is a good overall measure of the model's performance.

12.4.2 Pre-Proposing of Data

We performed pre-processing steps to enhance the model's accuracy and performance. Initially, we labeled the image classes, resized every image to 256x256 pixels, and converted them into tensors using the PyTorch library. We also divided the data into training, validation, and testing sets to evaluate the model's accuracy. The batch size controls how many samples are passed on via the network. Therefore, we defined the train dataset batch size as 32, the validation batch size as 64, and the test dataset as 32.

Figure 12.5 shows a sample of images used in a batch during the training process of a machine-learning model. These images update the model's parameters to improve its performance.

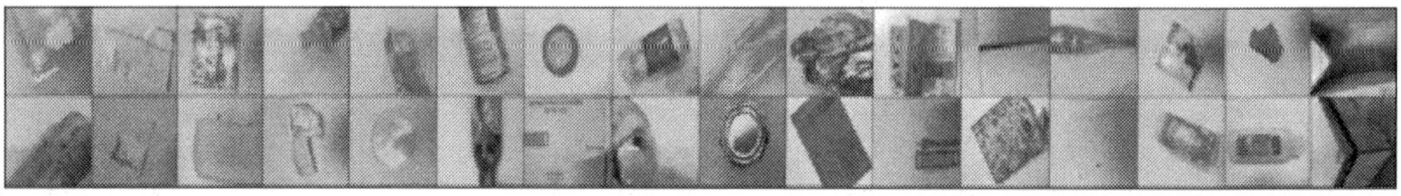

Figure 12.5: Train-dataset Batch Images.

12.5 IMPLEMENTATION AND RESULT

In this section, the model description and experimental results are discussed.

12.5.1 Experimental Set-up

The systems used for this project comprised an Intel(R) Core(TM) i7-6700k CPU, 16 GB of RAM, Nvidia GTX 1060 3GB GPU, Samsung 970 EVO Plus 500GB NVMe, and 1 TB hard drive. The system was operated on a Windows 10 64-bit operating system.

12.5.2 Why we used ResNet

The key feature of ResNet (Residual Neural Network) is the use of residual connections, which allow for the training of much deeper networks than traditional architectures by addressing the problem of vanishing gradients.

The residual block is the fundamental building element of a ResNet model. The residual block is made up of layers such as one or more convolutional layers, batch normalization layers, and a shortcut connection. The convolutional layers are used to extract features from the input image. The batch normalization layers, in contrast, stabilize the learning procedure by normalizing all inputs for every layer. The shortcut connection skips one or more tiers and connects the input to the output of the block immediately.

The shortcut connection allows the network to bypass the layers in the residual block and learn the identity function, which is the function that maps the input to itself. By learning the identity function, ResNet allows the network to learn the residuals, or the differences between the output and the input, instead of the actual output. This makes it easier for the network to learn and improve performance.

During training, the layers' weights in the residual block are learned by minimizing the difference between the output and the input using a backpropagation algorithm. The architecture of ResNet allows for an easy flow of gradients during the backpropagation phase, which enables the model to learn features and improve performance more efficiently.

Figure 12.6 shows a diagram of the working flow of a ResNet model, a type of deep neural network architecture. It is a convolutional neural network architecture that uses residual connections to improve the training of deep networks. A ResNet model can have several layers composed of convolutional, pooling, and activation layers.

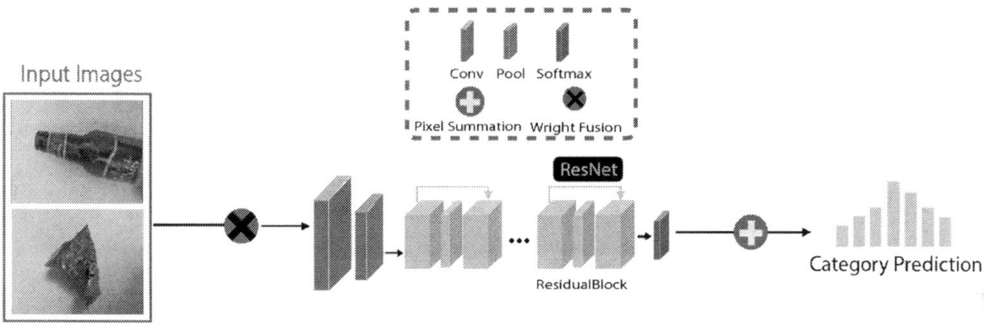

Figure 12.6: Workflow of ResNet.

12.5.3 Experimental Result

The present study suggests a garbage detection and classification approach based on a pre-trained ResNet utilizing CNN. Our dataset has six waste classes: cardboard, glass, plastic, paper, trash, and metal. We collected 2527 images, trained 1593 images, tested 758, and validated 176 images. In our training phase, we generate predictions and calculate accuracies. In our multiclass testing, there are 758 samples with 93.14% accuracy. Our average accuracy is 93%, which is an excellent result.

Figure 12.7 shows a confusion matrix, which shows the number of correct and incorrect predictions made by the model for each class.

Table 12.1 presents, cardboard, which predicts 106/113, glass predicts 136/144, metal predicts 125/129, paper predicts 171/175, plastic predicts 132/149, and trash predicts 36/48 correctly.

Figure 12.8 shows an image that is taken from a test dataset and the predicted class assigned to it by a trained machine learning model.

Figure 12.9 shows an image taken from an outside test dataset that is not used in the model's training and the predicted class assigned to it by a trained machine learning model.

In Figure 12.10, Accuracy vs. Epoch, we see peek accuracy at 94% at Epoch 4 and in Figure 12.11 Loss vs. Epoch. There is a linearity when the epoch is increasing. The loss of training and validation is also decreasing.

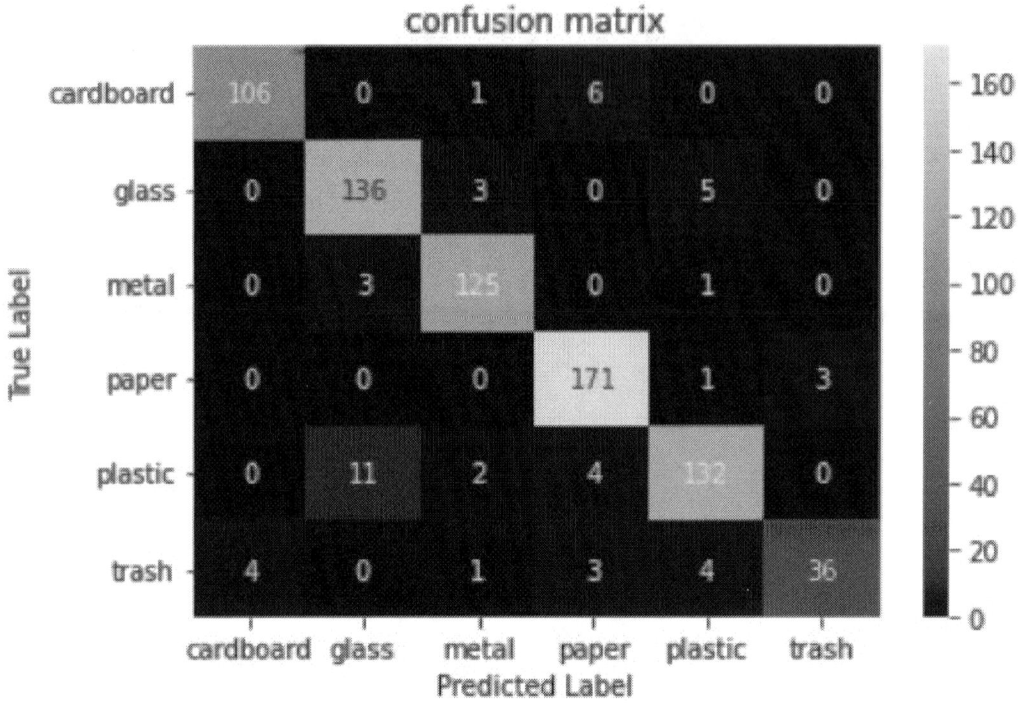

Figure 12.7: Confusion Matrix.

Table 12.1: Prediction Results of All Classes.

Material	Correctly Predicted	Total
Cardboard	106	113
Glass	136	114
Metal	125	129
Paper	171	175
Plastic	132	149
Trash	36	48

12.6 RESULT ANALYSIS

Table 12.2 presents the results of testing using images that were not utilized in the training or testing phase of the project. These images were employed to evaluate the performance of the model's predictions.

ResNet is a deep learning model known for effectively training deeper neural networks without the problem of vanishing gradients. Residual connections (shortcut connections) allow information to flow more easily between layers, which helps to reduce the vanishing gradient problem. It can be trained on very deep neural networks (up to 152 layers) without the need for additional techniques, such as unsupervised pre-training.

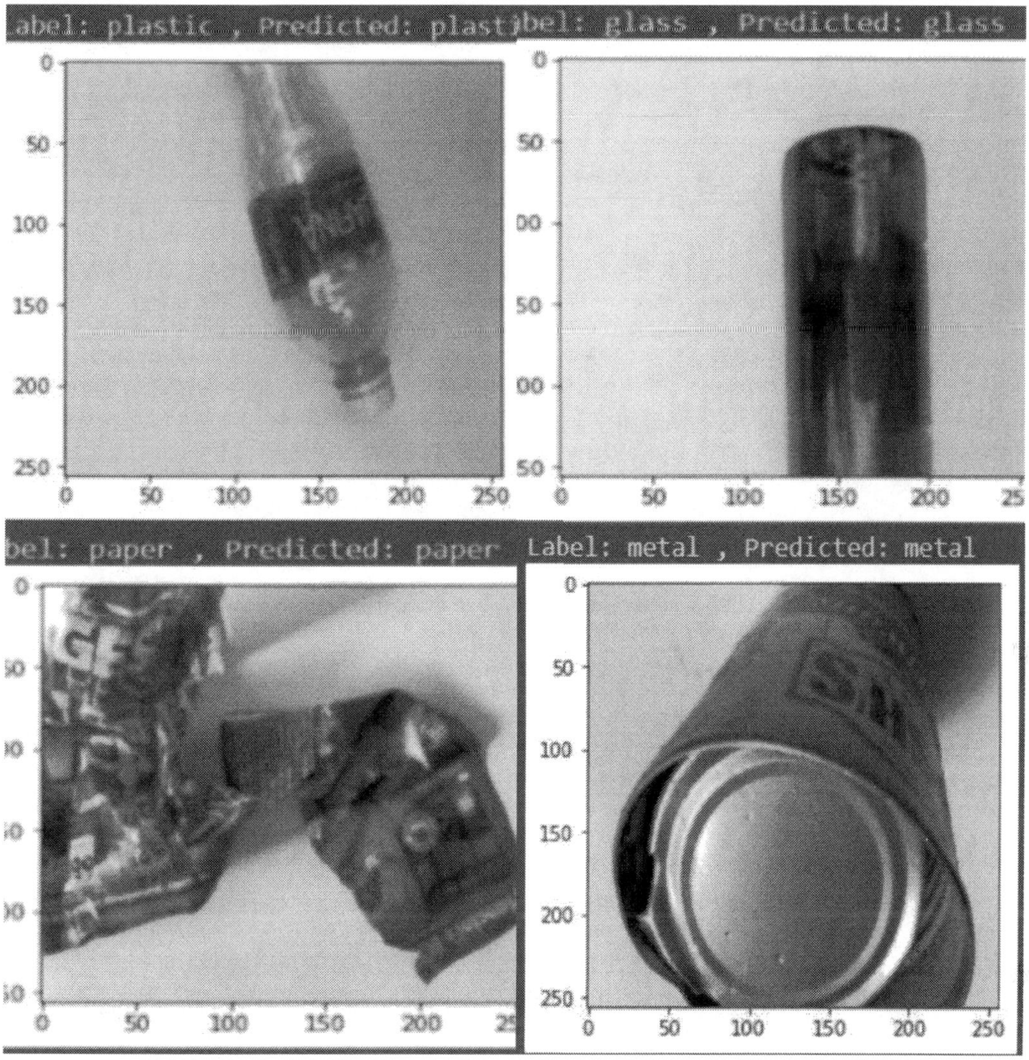

Figure 12.8: Prediction from the Test Dataset.

Table 12.2: Classification Report of ResNet.

Image Classes	Precision	Recall	F1-Score
Cardboard	0.96	0.94	0.95
Glass	0.91	0.94	0.93
Metal	0.95	0.97	0.96
Paper	0.93	0.98	0.95
Plastic	0.92	0.89	0.90
Trash	0.92	0.75	0.83

On the other hand, VGG16 (Visual Geometry Group 16) is a deep learning model known for its use of small convolutional filters and very deep architectures. VGG16

Figure 12.9: Prediction from an External Test Dataset (Cloudinary).

demands an extensive amount of memory and computing resources to train, making it difficult to use on some systems. It can be prone to overfitting, primarily when used on small datasets.

ResNet is suitable for training deeper neural networks and reducing the vanishing gradient problem. At the same time, VGG16 is known for using small convolutional filters and very deep architectures. However, it necessitates a significant amount of memory and processing resources and is prone to overfitting. Table 12.3 shows the comparison between the ResNet model we used and the VGG16 model.

In summary, ResNet and VGG16 are both popular architectures for image classification, with ResNet known for its very deep architecture and effective use of residual connections and VGG16 known for its deep architecture and use of smaller filters to recognize fine-grained features.

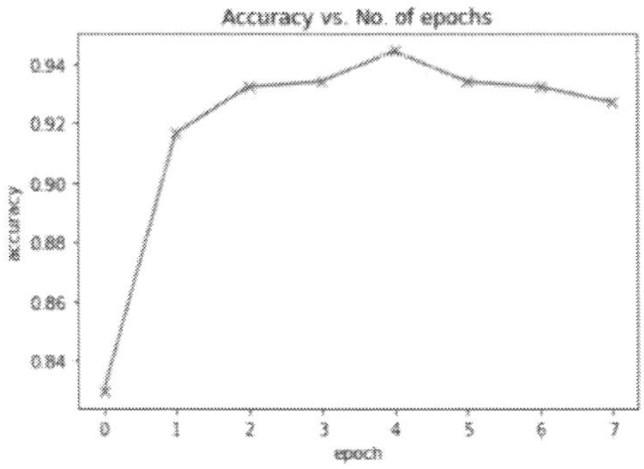

Figure 12.10: Accuracy vs. No. of Epochs Graph.

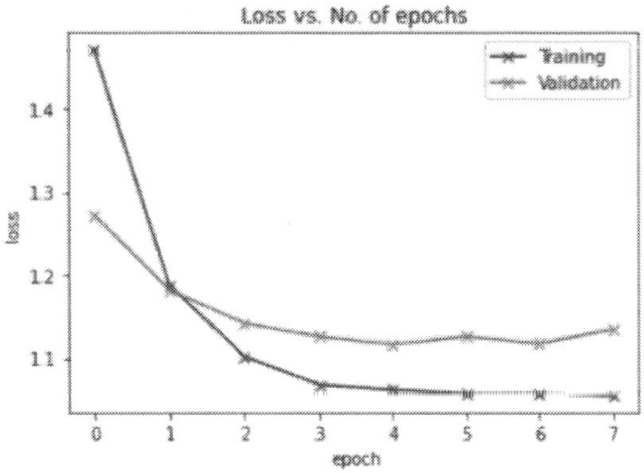

Figure 12.11: Loss vs. No. of Epochs Graph.

Table 12.3: Comparison between ResNet and VGG16.

Image Classes	Precision	Recall	F1-Score
Paper	0.83(VGG16), 0.93(ResNet)	0.84(VGG16), 0.98(ResNet)	0.88(VGG16), 0.95(ResNet)
Plastic	0.76(VGG16), 0.92(ResNet)	0.78(VGG16), 0.89(ResNet)	0.77(VGG16), 0.90(ResNet)
Glass	0.84(VGG16), 0.92(ResNet)	0.68(VGG16), 0.75(ResNet)	0.76(VGG16), 0.83(ResNet)

12.7 CONCLUSION

In conclusion, waste management is a difficult task that affects the environment, the economy, and society. Machine learning can enhance waste management by enabling the analysis of massive volumes of data and the detection of patterns and trends. Machine learning is utilized in trash management in two ways: optimizing collection routes and forecasting rubbish creation rates. The use of machine learning could improve the efficiency and efficacy of garbage collection and disposal, resulting in cheaper costs and fewer negative environmental effects. However, it is crucial to address the ethical implications of using machine learning in waste management, such as bias and privacy problems.

Bibliography

[1] Gutberlet J, Uddin SM (2017) Household waste and health risks affecting waste pickers and the environment in low- and middle-income countries. International Journal of Occupational and Environmental Health 23:299–310. doi: 10.1080/10773525.2018.1484996

[2] Xiaoling Xia, Cui Xu, Bing Nan (2017) Inception-V3 for Flower Classification. 2017 2nd International Conference on Image, Vision and Computing (ICIVC). doi: 10.1109/icivc.2017.7984661

[3] Zhang Y, Song C, Zhang D (2020) Deep learning-based object detection improvement for tomato disease. IEEE Access 8:56607–56614. doi: 10.1109/access.2020.2982456

[4] Algabri M, Mathkour H, Bencherif MA, et al (2020) Towards deep object detection techniques for phoneme recognition. IEEE Access 8:54663–54680. doi: 10.1109/access.2020.2980452

[5] Gutberlet J, Uddin SM (2017) Household waste and health risks affecting waste pickers and the environment in low- and middle-income countries. International Journal of Occupational and Environmental Health 23:299–310. doi: 10.1080/10773525.2018.1484996

[6] Abeywickrama T, Cheema MA, Taniar D (2016) K-nearest neighbors on road networks: A journey in experimentation and in-memory implementation. In: arXiv.org. https://arxiv.org/abs/1601.01549. Accessed 14 Jan 2023

[7] Zhihong C, Hebin Z, Yanbo W, et al (2017) A vision-based robotic grasping system using deep learning for Garbage Sorting. 2017 36th Chinese Control Conference (CCC). doi: 10.23919/chicc.2017.8029147

[8] Shen X, Wu Y, Chen S, Luo X (2021) An intelligent garbage sorting system based on edge computing and visual understanding of social internet of vehicles. Mobile Information Systems 2021:1–12. doi: 10.1155/2021/5231092

[9] Chen L, Xu J (2017) Socially trusted collaborative edge computing in Ultra Dense Networks. Proceedings of the Second ACM/IEEE Symposium on Edge Computing. doi: 10.1145/3132211.3134451

[10] Chen L, Xu J, Zhou S (2017) Computation peer offloading in Mobile edge computing with energy budgets. GLOBECOM 2017 - 2017 IEEE Global Communications Conference. doi: 10.1109/glocom.2017.8255052

[11] Tulebayeva, Nazym & Yergobek, Dimash & Pestunova, Galina & Mottaeva, Angela & Sapakova, Zhansaya. (2020). Green economy: waste management and recycling methods. E3S Web of Conferences. 159. 01012. 10.1051/e3sconf/202015901012.

CHAPTER 13

Designing the Most Eco-Friendly Spatial Landscape and Natural Environment Using A* Path Finding Algorithms

Izazul Hoq Imran

East West University, Dept. of CSE, Dhaka, Bangladesh.

Tropa Das

East West University, Dept. of CSE, Dhaka, Bangladesh.

Sharmin Akther Rima

East West University, Dept. of CSE, Dhaka, Bangladesh.

Meftahul Zannat

East West University, Dept. of CSE, Dhaka, Bangladesh.

Ahmed Wasif Reza

East West University, Dept. of CSE, Dhaka, Bangladesh.

Paresh Dhakan

Intelligent Systems Research Centre University of Ulster, Londonderry, UK.

CONTENTS

DOI: 10.1201/9781003605508-13

Conventional architectural landscapes often suffer from shortcomings in visual appeal and resident satisfaction, which can be addressed through green architectural landscapes. However, incorporating landscaping into architectural designs can be difficult due to space constraints. To tackle this challenge, the authors of a recent study propose using various path-finding algorithms to optimize green building landscapes and environmental design. These algorithms aim to optimize multiple objectives, including construction costs and the overall environment of the space. According to the findings, the suggested path-finding algorithms help optimize the architectural environment and increase the enjoyment of the building's residents. Results show that the numerous proposed path-finding systems enhance the architectural landscape setting, leading to happier residents.

13.1 INTRODUCTION

The construction industry is beginning to see green buildings as a promising new area for growth. Environmental protection, resource conservation, energy efficiency, and cost reduction are just a few of the many factors that must be considered throughout a green building's complete life cycle and in the development of the building's operations. The layout should consider the beneficial impacts on the environment and the economy. The building facade is the first point of contact for visitors interested in learning more about environmentally friendly growth and green construction. The structure is significant since it is the primary means of access for the residents. Drawing on garden art and engineering ideas, garden planning and design include strategically planting and cultivating an area's flora to enhance the space's natural, liveable, and recreational qualities. This method might be seen as an expansion of landscape design. New plant life may be cultivated and planted to this end. Sustainable and environmentally friendly building methods heavily influence garden landscape design. The research and development of eco-friendly architecture have yet to start [1]. Incorporating eco-friendly building practices into a garden's plan may create specific difficulties, discussed below, along with possible solutions. The standard method is arranging for temporary buildings and on-site machinery to be used in concert based on previous projects' successes and failures. It has to be altered to fulfill the demands of contemporary structures in light of the increasing stringency of the criteria for site layout. As a result, it's not uncommon for the site plan to fall short of the standards set by the building process and fail to foster an environment flexible enough to accommodate any alterations. For architects, the rules for creating energy-efficient structures may be simpler to understand and more arduous to put into practice. This holds more accuracy for more prominent buildings than smaller ones. The practice of enriching a space via the use of aesthetic and engineering gardening principles is known as garden planning and design. To achieve this goal, garden art, and engineering principles are used to construct a natural, living, and recreational place. This may be achieved, for example, by cultivating plants whose primary purpose is to improve the surrounding environment [2]. Take, for instance, the residential complexes found in Wuhan. Alterations can be applied to the construction's beginning period, glass-to-wall ratio, siding framework, rooftop framework, thermostat set point, circulating temporal air fix the mark, conjunction factor of darkening variable, and the glazing type. These alterations are designed to improve the building's energy efficiency. Because of this, we can reduce our energy use and costs. Both of these processes must take place at the exact moment [3]. The results of the optimization are then contrasted with the current guidelines. Alterations may be applied concerning the construction's beginning period, pane-to-wall proportions, external wall framework, roof organization, thermostat set point, air conditioner heat setup particular, combined factor determining screening component, and the glazing species. Alterations may also be made to the planning to build's heated heat setting and air-cooling temperature set point. The A* method is often used for pathfinding due to its rapidity and precision in determining the shortest route between sites. When figuring out the quickest path from one place to another is crucial, as it often is in games and other applications, this is the tool of choice. There are several reasons why it could be a good idea to use A* to design the placement of green spaces in a building or other environment. A* locates the quickest route rapidly between two places. Therefore, it can assist in designing the park's most efficient route. It works effectively in complex environments with barriers and other limitations, making it suitable for creating green areas with many features and constraints. A* makes the green space more accessible to disabled people by optimizing its layout. It reduced paved areas and increased green zone flora to improve sustainability. To enhance vegetation availability. To improve the visual appeal of the organization of green spaces, A* might be used to consider the visual appeal of the various pathways and routes available in the area. This strategy can enhance the design of parks and other green spaces for jogging, cycling, and walking, among other activities. Roads and paths conducive to these activities are built to achieve this goal. Parks and other outdoor spaces may be utilized to maximize their pedagogical potential. A* may be used to carefully arrange trails that guide tourists on a tour of a park or other green area while pointing out significant features or landmarks. To make the most of green areas for social interaction, this method was also employed to construct walkways that encourage people to converse with one another and explore the surroundings. Plants were put at critical intersections to improve pedestrian safety and traffic movement.

13.2 METHODOLOGY

The issue: To create an eco-friendly landscape, explain the project's goals and constraints. Consider biodiversity, natural resources, and carbon footprint reduction. Data collection: Gather topography, land cover, and infrastructural data for the area under consideration. Map ecosystems, species distributions, and environmental sensitivity ratings.

Preprocessing: Preparing data for analysis and integration. Transform data into an A*-compatible graph or grid.

Environmental Goals: Ecological factors determined; define environmental constraints and goals. Avoid fragile ecosystems and create minimal land disturbance while pathfinding.

A*-Adapted Algorithm: Change the A* algorithm to include environmental limitations and objectives. Weight and heuristics should prioritize ecologically friendly routes. Create cost functions with environmental impact, resource conservation, and carbon emissions.

Pathfinding: Successfully implementing the customized A* technique requires a programming language or pathfinding library. Create algorithms or functions to handle environmental constraints and objectives. Make sure the algorithm considers both speed and environmental impact.

Validation and Assessment: Evaluate the strategy for finding the greenest path. Compare the results to conventional wayfinding. Assess the landscape's long-term sustainability and ecological balance.

Case Study Use: Apply the generated strategy to the specified study topic. Run eco-friendly pathfinding for the most efficient spatial environment. Examine the route, land utilization, and environmental benefits.

Assessing Performance: This stage evaluates the algorithm's computational efficiency. Consider execution time, memory use, and scalability.

Final Thoughts: Examine and understand the case study's results. Discuss the technique's benefits, drawbacks, and improvements. Highlight eco-friendly spatial landscape design and A* pathfinding algorithm applications.

13.3 RELATED WORK

Research suggests that a classical garden takes inspiration from its surrounding landscape and includes significant regional and individual characteristics in its layout. The traditional garden landscape may have a simplistic look, but it retains some features regarding cultural inheritance and assimilation [4]. Many of China's beautifully preserved classical landscapes showcase national cultural characteristics from various times and serve as historical records of the development of civilization over thousands of years. In terms of furthering the cause of social progress and the design of modern garden landscapes, this one-of-a-kind garden landscape provides a helpful proposal. Several problems with conventional landscape design have only become apparent with time. As people's understanding of society and their ideas about it evolve continuously, the traditional garden landscape no longer meets their varied needs [5].

For instance, garden landscapes must now adhere to people's expectations of what a natural environment should seem like to be a suitable complement to contemporary structures. Therefore, the concept of historically planned garden landscapes needs to be simultaneously updated. When navigating a static environment, robots need to be wise and capable of employing local path-planning algorithms in situations where they have only a partial picture of their surroundings. This planning has many real-world applications, including warehouse and supply chain management. A* may determine the shortest path in a static environment by including information about the obstacles along the way [6]. There are two layers to the challenge of determining the shortest path in a fixed setting. The work may be split into two phases: identifying viable node pairs and evaluating the most straightforward approach given the identified node pairs [7]. The algorithm is useless and impractical because none of those conditions hold in a dynamic setting. The method for active route planning is not applicable in this case since it is not suited to a static environment. The A* method was chosen since it is typical of the techniques employed by static-environment real-time route planning systems nowadays. New research is expanding on this algorithm to achieve greater efficiency and effectiveness. Research suggests that a classical garden takes inspiration from its surrounding landscape and includes significant regional and individual characteristics in its layout. Although traditional garden landscapes are minimalist in design, they retain certain aspects regarding cultural inheritance and incorporation [8]. Most of China's well-maintained ancient gardens serve as historical records of the advancement of civilization over thousands of years and representations of national cultural traits from various periods. In terms of furthering the cause of social progress and the design of modern garden landscapes, this one-of-a-kind garden landscape provides a helpful proposal [9]. Several problems with conventional landscape design have only become apparent with time. This research emphasizes the importance of using an effective way-finding algorithm in designing landscape spaces in green buildings. This is due to people's increasing awareness and consideration of the environment.

13.4 PROBLEMS IN PLANNING GREEN BUILDINGS, GARDENS, AND LANDSCAPES

This section describes the planning problems.

13.4.1 Ignore Economic and Environmental Risks

Ignoring pricing and ecological setting When it comes to the planning and design of gardens and landscapes for environmentally friendly buildings, safety may be a challenge since it can lead to the use of techniques and materials that are not environmentally friendly or acceptable [10]. To minimize the negative impact of a garden or landscape on the environment, it's imperative to properly handle materials with a significant carbon footprint, which should be acquired locally. Additionally, it's crucial to factor in the maintenance cost as it can pose a financial burden for both the building owner and the residents.

13.4.2 Lack of regional identity and style

Garden and landscape design for green buildings suffers when they fail to reflect local aesthetics and traditions since doing so is essential to creating a sense of place and adding value to the structure itself. Native plant species, for instance, don't need as much water or other maintenance since they're already adapted to the local environment. The garden's or landscape's beauty will be enhanced by including cultural elements like historical garden designs or plant species.

13.4.3 Concerns About the Feasibility of Green Construction

When planning and developing gardens and landscapes inside green buildings, it is essential to solid comprehend the concept of environmentally friendly construction. The term" green building" refers to designing, constructing, and operating facilities in a way that is responsible for the surrounding environment and effectively uses available resources.

13.4.4 Integration of Complex Ecological Factors

Even if this technique considers ecological issues, such as biodiversity and resource conservation, more study is still needed on integrating more complicated environmental elements. This may include taking into consideration the individual habitat needs of various species, taking into account the services provided by ecosystems, and taking into account the interactions between various ecological components [11].

13.4.5 Dynamic Environmental Changes

Changes in the Dynamic Environment The present technique operates on the assumption of a stable environment. However, settings in the actual world are ever-changing due to various variables, including land use, climate, and other aspects. Developing algorithms capable of adapting to changing environmental circumstances and enabling the ongoing optimization of environmentally friendly spatial landscapes should be the primary focus of study in the future.

13.5 A BREAKDOWN OF THE STEPS INVOLVED IN THE PATH-FINDER ALGORITHM

It is possible to accomplish this if many lines of inquiry are wanted to be pursued. It's over, and no matter what happens, a solution that can be implemented will be discovered. If there is not currently one, it will generate a new one. By using a heuristic, one may significantly cut down on the amount of time needed to complete the operation. It is conceivable that the expenses involved in traveling from one node to another might differ. This may make it more difficult to travel between specific nodes or along certain paths, such as when a gaming character comes across rough terrain and is forced to walk more slowly or when an aircraft must take more time to go from one spot to another.

If there is a solution, the A* algorithm will locate it. This is true irrespective of the kind of the issue being addressed. Using a heuristic can cut the time needed to finish the procedure down by a significant amount [12]. Some paths may be more challenging due to the varied expenses associated with traveling from one node to another. In video games, this may imply that it takes an explorer longer to cross rugged terrain or an aircraft to go from one city to another. These examples illustrate how the costs of moving between nodes might make it impossible to access specific nodes or paths. It can conduct several different searches in response to the input provided by the user

A node's g-score, or its baseline, reflects the distance between that node and the tree's root. Generalized growth may be calculated using the following formula:

$$g(n) = g(n.parent) + \text{cost}(n.parent, n)$$

The cost of moving from node $n1$ to node $n2$ is denoted by the expression $\text{cost}(n1, n2)$ in this context. It is possible to accurately estimate the distance between each node and the destination by using heuristic evaluation scores. One example of this is the H-score.

It is essential that the calculation uses as few processing resources as possible, given that the H-score is computed at least once for each node along the road to success. The H-score may be calculated in several different methods, each determined by the properties of the graph being searched. The following are some of the most frequent approaches:

$$\begin{aligned}
\text{Manhattan distance:}\quad h(n) &= |n.x - \text{goal.x}| + |n.y - \text{goal.y}| \\
\text{Diagonal distance (uniform cost):}\quad h(n) &= c \cdot \max(|n.x - \text{goal.x}|, |n.y - \text{goal.y}|) \\
\text{Diagonal distance:}\quad d_{\text{max}} &= \max(|n.x - \text{goal.x}|, |n.y - \text{goal.y}|) \\
d_{\text{min}} &= \min(|n.x - \text{goal.x}|, |n.y - \text{goal.y}|) \\
h(n) &= c_{\text{dd}} \cdot d_{\text{min}} + c_{\text{n}} \cdot (d_{\text{max}} - d_{\text{min}}) \\
\text{Euclidean distance:}\quad h(n) &= \sqrt{(n.x - \text{goal.x})^2 + (n.y - \text{goal.y})^2}
\end{aligned}$$

The f-score is the sum of the g-score and h-score, representing the total path cost via the current node:

$$f(n) = g(n) + h(n)$$

13.5.1 Path finding Algorithm

My absolute focus is on the A* algorithm, which has become the preferred method for pathfinding due to its versatility. A* aims to find the shortest possible routes, like the Dijkstra algorithm. In optimization problems, A* uses the Greedy Best-First-Search heuristic to its advantage. Utilizing Greedy's Best-First Search is straightforward yet highly effective due to its ability to combine different approaches. One effective technique for finding a path is called Dijkstra's method. This method uses a greedy

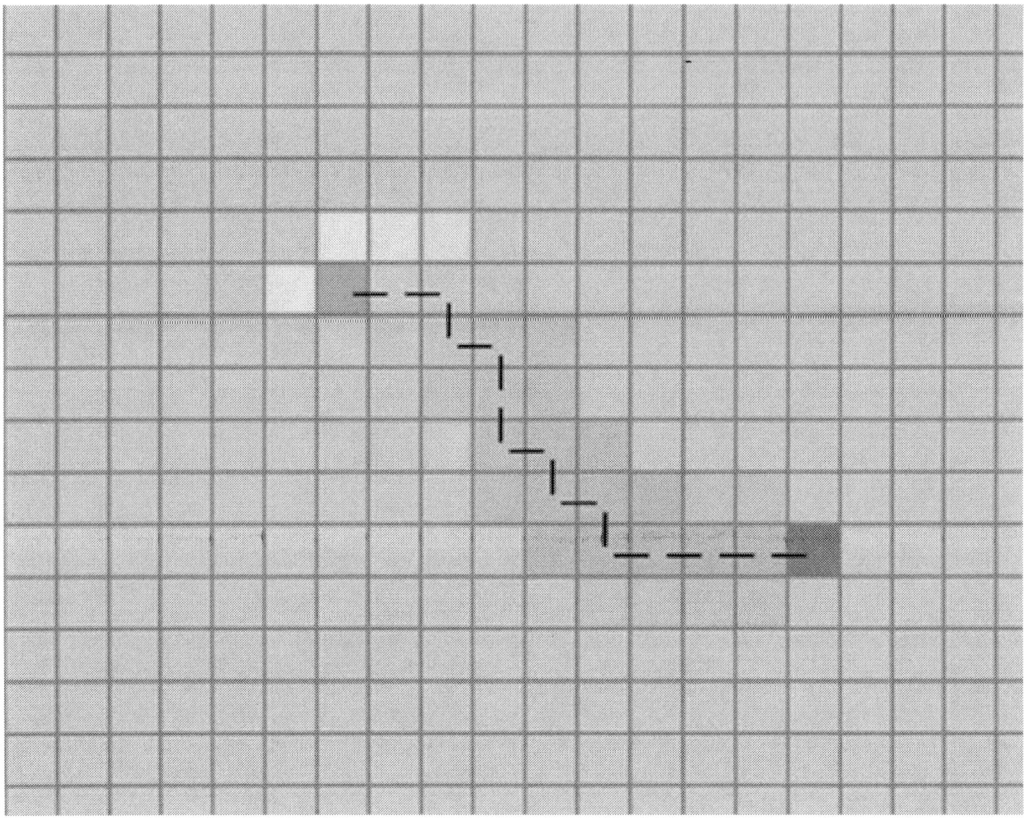

Figure 13.1: Path Finding.

best-first search to prioritize vertices closest to the target. We use the symbol h(n) to represent the cost of moving from vertex n to the objective and g(n) to represent the cost of traveling from the starting point to vertex n. The yellow edges on the figures show the furthest edges (h), while the teal origins indicate the most distant starting points (g). During the main loop (n), the algorithm maintains a balance between these two factors, looking at the lowest $f(n) = g(n) + h(n)$ vertex [13]. This section covers different aspects of pathfinding heuristics, such as how to create them, how to use them, and how to define maps. While some parts of the discussion are finished, others are still being developed.

13.5.2 The Process of Preprocessing for Pathfinding Algorithms

The labyrinth can be navigated in all four cardinal directions, but some areas are restricted due to the presence of red tiles. When starting off, you can only move downwards due to walls above and to the left, and limited progress to the right due to red square bars. To solve this maze, I used code from two blogs and referenced Patrick Lester's A* article for background and discussion. Both of these resources are available online. As a starting point, create the "Node" class and method, which constructs an object for each node and considers the node's parent, position, and

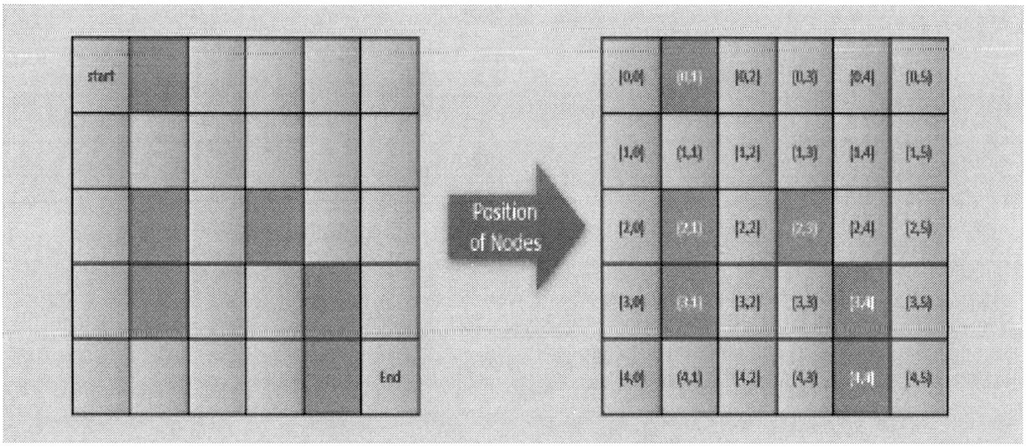

Figure 13.2: Optimize site planning

cost values (g, h, and f). Define a route function that returns the path from the starting point to the end when called. The search function drives the code's logic. Initialize variables (3.1), include the starting node in the "to visit" phrase (3.2), and define a stop condition to avoid infinite loops in the code. Move to the next position until all stop prerequisites are met (3.3). Find the square with the lowest "yet-to-visit" cost and ensure we have not exceeded the maximum number of iterations (3.4). Compare the current square with the required square to determine if we have found the correct route (3.5). Make changes to the child node by utilizing the current court and the four adjacent squares Figure 13.2. If you come across any items that cannot be moved or have been visited before, simply disregard them. In such a scenario, designate the previous node as the parent of the new node and adjust its location accordingly. It is important to ensure that all the child nodes that have been created are thoroughly examined. Kindly note that the cost is associated with squares f, g, and h. To determine the most suitable route, we will utilize the g cost, regardless of whether the destination is on the "to visit" list. This approach guarantees a lower overall price. To recalculate the square's g and f scores, we need to switch the square's parent to the presently active court. Let me provide an in-depth explanation of the primary program - we will commence with a labyrinth and identify its starting and ending points. Afterward, we will utilize the search function, and if a route is already accessible, we will leverage the path function. Now, let me guide you through the code in greater detail (refer to the numbers in brackets). We will start by creating a node class that encompasses its parent, location, and three costs (g, h, and f). We will then establish an asymmetry checker and proceed to create the node.

13.5.3 Algorithms for Navigating Paths and Energy-saving Devices for Environmentally Friendly Gardens

Cost-effective gardening practices are crucial to help spread the scientific idea regarding growth and encourage innovation in creating eco-friendly, low-maintenance landscapes. Energy-efficient landscaping is excellent at doing this task. Energy-efficient

Algorithm 2 A* Path Finding Algorithm

```
function A*(start, goal, heuristic)
    frontier ← PriorityQueue()
    frontier.put(start, 0)
    came_from ← {}
    cost_so_far ← {}
    came_from[start] ← None
    cost_so_far[start] ← 0
    while not frontier.empty() do
        current ← frontier.get()
        if current == goal then
            break
        end if
        for neighbor in current.neighbors() do
            new_cost ← cost_so_far[current] + current.cost(neighbor)
            if neighbor not in cost_so_far or new_cost < cost_so_far[neighbor] then
                cost_so_far[neighbor] ← new_cost
                priority ← new_cost + heuristic(neighbor, goal)
                frontier.put(neighbor, priority)
                came_from[neighbor] ← current
            end if
        end for
    end while
    path ← [goal]
    while path[−1] ≠ start do
        path.append(came_from[path[−1]])
    end while
    path.reverse()
    return path
end function
```

gardening is a crucial part of building a culture that makes good use of energy [14]. A thorough examination of the factors influencing the local weather will be conducted, and those findings will be firmly incorporated into the planning and layout of the homes to produce a pleasant atmosphere to live in and a micro-climate beneficial to energy usage. Primary content considerations include site location, zoning, building and roadway orientation, building body architecture, building distance, the prevailing directions of winter and summer rainy seasons, ultraviolet rays, along additional environmental elements. Construction and construction of icy residential environments are only getting underway. Thus chaotic expansion, cultural appropriation, and a race to the top are to be anticipated. Planting seedlings out of season, growing in unfavorable conditions, bringing in a wide variety of alien plants, and draining lakes to make way for gardens are all examples of such strategies. A few research projects, if any, have focused on ecological landscape design in cold urban populations outside of private homes [15]. Ecological design from a holistic perspective should be included in the planning of freezing urban environments to make these areas more habitable and encourage people to value and conserve their natural surroundings. The planet's resources will be better protected and preserved due to this.

13.5.4 Experiment and Analysis

Here, we use not just the algorithm developed by Dijkstra but also the BFS algorithm and the algorithm called A*. By comparing the times required by various algorithms, we can observe that some take more time than others. Using the provided data, we'll compare and contrast these three methods.

The following comparison of path-finding algorithms, Dijkstra, BFS, and A*, is offered for your review in the table. The analyzed metrics are the length and distance traveled, time spent between nodes, and the total number of nodes visited. Let's go further into the specifics of each statistic and the repercussions of the discoveries, shall we?

Length Distance: Each method's case study route length is the distance. A* always creates shorter circuits than Dijkstra and BFS. This is significant because shorter travels require fewer resources, do less environmental damage, and emit less carbon dioxide. The A* algorithm's ability to prioritize efficient paths while incorporating ecological characteristics may assist in designing environmentally friendly spatial landscapes with optimal connectivity.

Time Duration: The best route algorithms' computation efficiency is shown by the time duration. The A* algorithm outperforms the Dijkstra and BFS algorithms in time. Practical implementation requires faster computation to reduce landscape generation time and allow apps to operate in real-time or near-real-time. The A* algorithm optimizes landscape planning and management decision-making by finding optimal paths.

Visited Node: The number of visited nodes shows how extensively the algorithm searches. The algorithm's exploration efficiency. A* visits fewer nodes than Dijkstra and BFS. This drop suggests that A* intelligently explores search space by investigating fewer nodes and finding the optimum paths. The A* technique reduces path-finding algorithms' computational burden by assessing nodes as few times as feasible. Evaluate nodes as few times as possible.

In conclusion, the table shows that the A* technique computes the shortest path faster and longer. It provides shorter courses with fewer node evaluations than Dijkstra and BFS. The A* algorithm's ability to optimize route selection while considering ecological factors may reduce environmental impact and resource consumption, which are significant implications for building environmentally friendly spatial landscapes. A*'s computing efficiency allows real-time decision-making and application implementation. However, additional research and validation are needed to assess the algorithm's efficacy across more case studies and resolve any limits or unique conditions Figure 13.3.

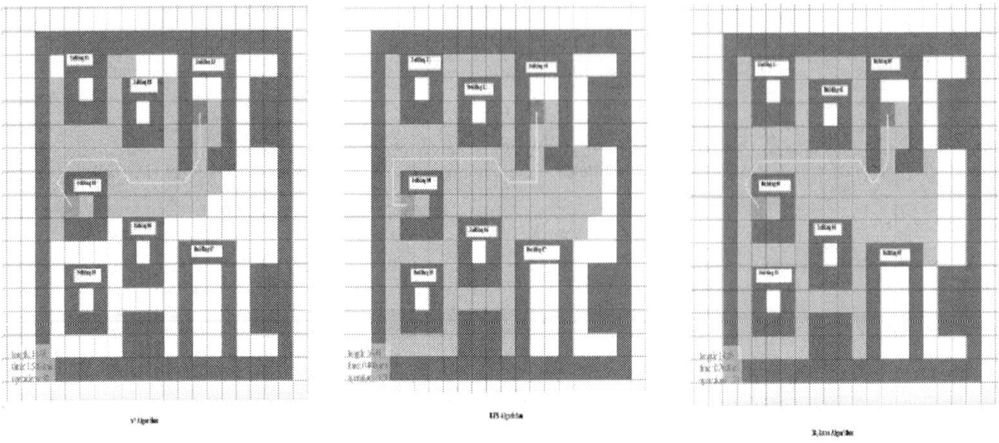

Figure 13.3: Time and count routed area for case 01.

Table 13.1: Case Study Results 01

Case Study	Dijkstra	BFS	A*
Length Distance	19	19	15.5
Time Duration (ms)	1.01	0.42	0.31
Visited Nodes	136	131	63

In the first case study, the A* algorithm achieved the shortest path distance of 15.5, while Dijkstra and BFS had distances of 19 each. The time durations for A*, Dijkstra, and BFS were 0.31 ms, 1.01 ms, and 0.42 ms, respectively. The number of visited nodes for A*, Dijkstra, and BFS were 63, 136, and 131, Table 13.1.

In the second case study, A* achieved a length distance of 22.07, while Dijkstra and BFS had distances of 25 each. The time durations for A*, Dijkstra, and BFS were 0.51 ms, 0.51 ms, and 0.32 ms, respectively. The number of visited nodes for A*, Dijkstra, and BFS were 68, 221, and 224, Table 13.2.

A* achieved a length distance of 19.9 in the third case study, whereas Dijkstra and BFS both had distances of 24.2. A*, Dijkstra, and BFS had time durations of 0.42 ms, 0.66 ms, and 0.51 ms, respectively. For A*, Dijkstra, and BFS, the corresponding numbers of visited nodes were 228, 237, and 111, Table 13.3.

Table 13.2: Case Study Results 02

Case Study	Dijkstra	BFS	A*
Length Distance	25	25	22.07
Time Duration (ms)	0.51	0.32	0.51
Visited Nodes	221	224	68

Table 13.3: Case Study Results 03

Case Study	Dijkstra	BFS	A*
Length Distance	24.2	24.2	19.9
Time Duration (ms)	0.66	0.51	0.42
Visited Nodes	237	111	228

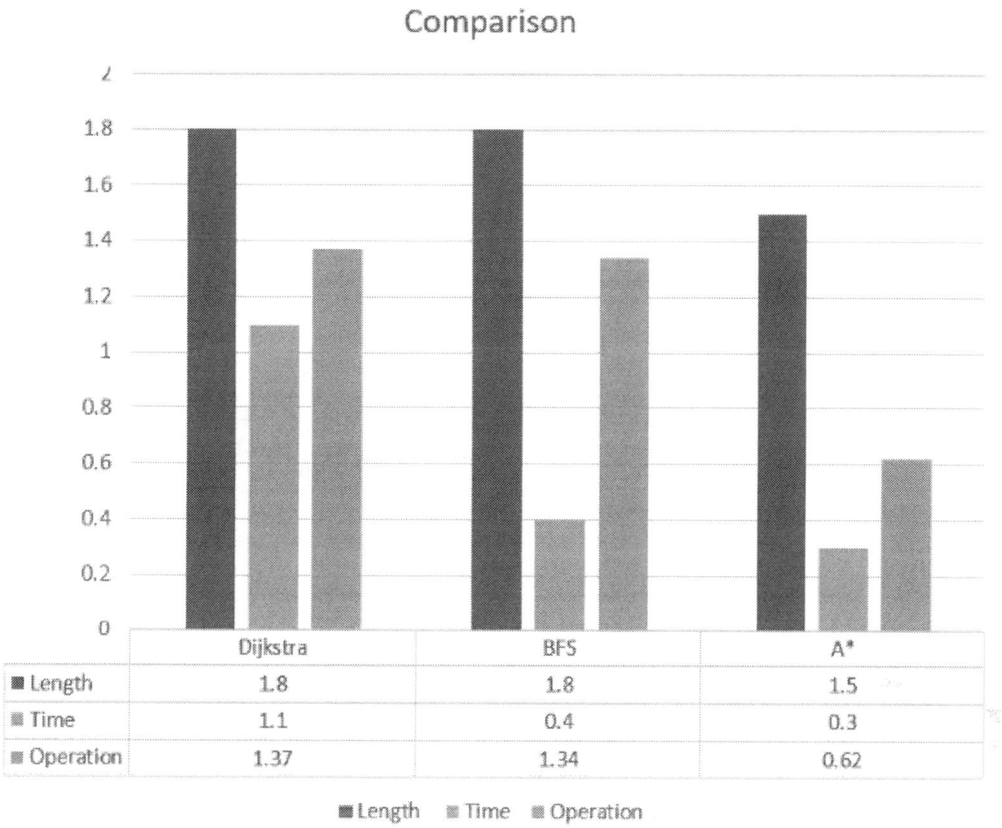

Figure 13.4: Case number 01's time and count traveled area.

In these case studies, the A* algorithm performs best overall. It can establish a balance between real path costs and heuristic assistance that yields the best pathfinding outcomes. A* is a useful tool for a variety of real-world applications, such as robotics, game development, logistics, and more, because of its shorter path distances and lower processing demands (Figure 13.4).

13.6 CONCLUSION

The A* Pathfinding algorithm and green landscape design may appropriately portray the area's traditions. This not only marks a change in the path of development, but it also contributes to ensuring that humankind will continue to evolve in an environmentally friendly and responsible way. An ecologically sustainable landscape that is thoughtfully planned can effectively promote the concept of green buildings and prioritize the well-being of individuals in development. Additionally, it can contribute to environmentally friendly growth while ensuring the preservation of biodiversity and ecosystems during the construction process. This is possible if the preparation and construction of the vegetated landscaping are done well. These objectives are attainable if careful preparation and the creation of environmentally friendly landscapes are carried out successfully. This exemplifies the core concept of development: it is possible for humans and the created surroundings within which people inhabit to thrive happily and cooperatively with their fellow creatures. This constitutes a few of the essential ideas behind the improvement.

Bibliography

[1] Hart, P. E., Nilsson, N. J., and Raphael, B. (1968). A formal basis for the heuristic determination of minimum cost paths. IEEE Transactions on Systems Science and Cybernetics, 4(2), 100–107. doi:10.1109/TSSC.1968.300136

[2] Elshafei, G., Vilčeková, S., Zelenčáková, M., and Negm, A. M. (2021). An Extensive Study for a Wide Utilization of Green Architecture Parameters in Built Environment Based on Genetic Schemes. Buildings, 11(11), 507.

[3] Alwaer, H., and Croome, D. J. C. (2010). Key performance indicators (KPIs) and priority setting use the multi-attribute approach for assessing sustainable, intelligent buildings. Building and Environment, 45(4), 799–807.

[4] Wang, D. D., Wang, X. D., Li, Y., and Xie, J. (2020). Landscape environment design of urban public space based on visual enhancement technology. Fresenius Environmental Bulletin, 29(10), 9005–9013.

[5] Mao, D., Zhang, Y. C., Kong, D. Z., He, S. L., and Li, X. Y. (2018). Evaluation of urban multi-scale landscape ecological pattern based on open space classification: a case study in xinxiang, China. Applied Ecology and Environmental Research, 16(5), 6787–6799.

[6] Wan, J., adn Zhao, Z. B. (2020). Analysis of landscape pattern change and ecosystem service value of urban green space - taking beidaihe district as an example. Fresenius Environmental Bulletin, 29(7), 5382–5389.

[7] Sadeghi, A. R., and Ahmadi, F. (2017). Using Sustainable Landscape Design Criteria in Order to Regeneration of Urban Seasonal Rivers' Bed. European Journal of Sustainable Development, 6(2).

[8] Karur, K., Sharma, N., Dharmatti, C. and Siegel, J. E. (2021). A Survey of Path Planning Algorithms for Mobile Robots. Vehicles, 3(3), 448–468.

[9] Dudi, T., Singhal, R. and Kumar, R. (2020). Shortest Path Evaluation with Enhanced Linear Graph and Dijkstra Algorithm. In: 2020 59th Annual Conference

of the Society of Instrument and Control Engineers of Japan (SICE). IEEE; 2020. p. 451–456.

[10] Liu, Q., Zhao, L., Tan, Z. and Chen, W. (2017). Global path planning for autonomous vehicles in off-road environment via an A-star algorithm. International Journal of Vehicle Autonomous Systems, 13(4), 330–339.

[11] Min, W., Hui, H. X. and Wang, J. Q. (2017). Rural Landscape Planning under the Aesthetic Dual Objective System: Conceptual Framework and Practical Approach. Landscape Architecture, 2017(6), 95–104.

[12] Mingqing, D. (2019). Research on environmental protection and energy-saving design method based on modern urban public landscape. Environmental Science and Management, 44(12), 5.

[13] Heymans, A., Breadsell, J., Morrison, G., Byrne, J. and Eon, C. (2019). Ecological urban planning and design: A systematic literature review. Sustainability, 11(13), 3723.

[14] Rhoads, W. J., Pruden, A. and Edwards, M. A. (2016). Survey of green building water systems reveals elevated water age and water quality concerns. Environmental Science: Water Research and Technology, 2(1), 164–173.

[15] Wang, M. (2021). Investigation of remote sensing image and big data analytic for urban garden landscape design and environmental planning. Arabian Journal of Geosciences, 14(6), 473. doi: 10.1007/s12517-021-09174-w.

CHAPTER 14

Rainwater Harvesting and Reducing Biological Threat of Utilizing

Md. Tahmid Zoayed

East West University, Dept. of CSE, Dhaka, Bangladesh.

Sayma Haque Arshe

East West University, Dept. of CSE, Dhaka, Bangladesh.

Taohid Hossain Ayon

East West University, Dept. of CSE, Dhaka, Bangladesh.

Ayesha Akter

East West University, Dept. of CSE, Dhaka, Bangladesh.

Ahmed Wasif Reza

East West University, Dept. of CSE, Dhaka, Bangladesh.

CONTENTS

RAinwater harvesting is an old way to preserve water and reuse it for multipurpose but still, the number of its uses is less than the expectations. People usually assume it to be harder than regular technique. They prefer the regular way of water use. As collecting the water is not the only thing but it also needed purification. In this project, we have figured out how to collect rainwater and filter it for daily usage.

DOI: 10.1201/9781003605508-14

We have worked on the perspective of East West University to show how the cost of water reduces as we use rainwater for use. Again, excessive rainwater causes floods and natural disasters. Reusing rainwater in such a way is also a method to reduce the possibilities of natural calamities like floods and earthquakes.

14.1 INTRODUCTION

Water is widespread on Earth. It is one of the most important elements for human survival and necessary for every life on Earth to live. According to the availability, this element is considered a human resource. From the very beginning, this human resource has been exploited and at present, we come to that point where we feel a shortage of water. Depending to the World Water Council, 785 million people in the whole world do not have access to safe water at this time. They have estimated that by the year 2050, almost 40% people of the world's population will reside in such locations where water resources are limited [1]. Between 165 million people, 68 million (41% of the population) do not have access to a dependable, clean supply of water, and 100 million (61%) do not have access to safe home sanitary facilities. And 29% of the population lives in poverty on less than $3.10 a day [2]. So now is the time to choose a different alternative water supply. Three different types of water sources include surface water, groundwater, and Rainwater. The megacity of Bangladesh is Dhaka City. According to the report for 2022, the population density of this city is 23,234 per square kilometer. Water is in great demand because of Dhaka's fast expansion. To meet this demand an authority that has been working since 1963 is Dhaka Water Supply and Sewerage Authority (WASA) [3]. Their current supply of water is around 2.11 million cubic meters per day, which does not meet our 2.25 million cubic meters per day demand [4]. As a result, people from Low-Income communities (LIC) are suffering. Most importantly an alarming pace of groundwater depletion is taking place. Hussain and Ziauddin [5] conducted research that took into account the overall state of the water sources, storage procedures, water quality management systems, and the organizations participating in enhancing the area's drinking-water supply. Dacope upazila residents mostly utilize water from ponds, canals or ditches, rivers, tubewells, the DPHE-UNICEF sand-filter facilities, and precipitation [5]. WASA authorities are alleged to have afterward launched an action plan to recharge subsurface aquifers. Similarly, the RAJUK was allegedly considering amending the Construction Construction Rules-2008 to include an obligatory provision for rainwater collecting for all forthcoming construction projects in the city. Experts believe that with an average annual rainfall of 2,000 mm in the country, the government may implement a program to educate the public about the use of rainwater in the aftermath of the potable water crisis [6]. This cost and power-effective system need to be implemented in every aspect of human life. Community leaders and social workers identified the lack of clean water sources as the primary development issue facing their community some two years ago. Following a series of climate adaptation assessments, a Performance Based Climate Resilience Grant of US$ 11,428 was approved to fund the rainwater harvesting project and to save lives in Manikkhali [7]. Many works are done in this field, for instance, Haslenda Hashim et al. [8] discussed a new technique for designing a Large Scale RWH for a community of 200 houses with an average of 4 persons per house and an average total daily water demand of 160 m^3. It was found that the optimal size storage tank for a 20,000 m^2 roof area is 160 m^3 with 60% reliability. The application of this model to the case study revealed a significant water saving of up to 58%. The total cost for this system is 443,861 USD over a life span of 25 years. Based on a questionnaire survey of 35 households in a residential complex in Bucaramanga, Colombia, Edgar Ricardo Oviedo-Ocaa et al. [9] used a bottom-up design approach, this article considers the initial financial feasibility of three RWH and GWR system configurations proposed for high water-using households (equivalent to 203 L per capita per day). The predicted potable water savings for the chosen configuration ('Alt 2') were 44% (equal to 131 m^3/year), with a return on investment of 6.5% and a payback period of 23 years. In this paper, we aim to do Rainwater Harvesting and Analyse the Biological Threat of Reusing it at East West University for Green infrastructure. The research objectives of our paper are:

- We want to implement a system of Rainwater Harvesting.
- We want to reduce the cost of water usage.
- We want to omit health risk issues using sensors and chemicals.
- We want to make sure of the utilization of rainwater in different aspects.

14.2 RELATED WORK

Water shortages exist everywhere. To relieve this problem, many actions are being taken. RWH will be the subject of an extensive worldwide effort. Some works are listed below:

- A. M. Ashrafull and M. M. Islam et al. [10] (2015) have researched Comparative Cost Analysis and Rain Water Harvesting Underground Water Withdrawal Near Gazipur City. In this paper, the applicability of rainwater gathering systems for home water supply was the study's main focus. Systems that were poorly planned, in terms of storage space and area, might only provide water for a portion of the dry season after the rainy season.

- Sadia Rahman, M. T. R. Khan, et al. [11] (2014) have discussed the Water quality sustainability of rainwater harvesting systems. This report uses a case study to examine the acceptability and appropriateness of rainwater for use as domestic water. Here, assessing the water's quality in comparison to industry standards, evaluating the financial advantages in terms of cost, taking into account water and energy saving, and finally proposing the system as a suitable supply of water in both regular and emergency circumstances.

- Md. Mafizur Rahman et al. [12] (2011) work on the potential for collecting rainwater in Dhaka city. Here, an examination of the variation in roof size across the different wards of the search has begun in Dhaka City for a suitable catchment region for RWH. It calculated that RWH may provide up to 10% of the water demand for community-based RWH systems and up to 33% of the water demand for individual harvesting systems.

- Van Meter, K. J., Steiff,et. al. [13] have researched recognizing the cycles of water release and storage between scales ranging. First the analysis of measurements of tank water levels and explore the changes in tank water depth during the rainy season. Secondly, about restrictions on the water balances in the tank and catchment area as well as ways to collect rainwater in tanks for irrigation demand.
- Johannes G. Paul, et. al. (2004) [14] discussed the utilization of rainwater. There are many sizes at which rainwater collecting systems may be put up. Water harvesting is, generally speaking, a comparatively straightforward method of water supply and placement as a system of distributed potable water.

Table 14.1: Comparative analysis of the related works

Works	Area of Study	cost	successful aspect
Ashraful et.al. [10] (2015)	A Study on Rain Water Harvesting and Comparative Cost Analysis around Gazipur City.	calculated	2.56 TK will be saved every day
Rahman et al. [11] (2014)	Sustainability of Rainwater Harvesting System in Terms of Quality	calculated	analysis of the power consumption successfully.
Rahman et al. [12] (2011)	Potentiality of rainwater harvesting in Dhaka.	not calculated	10% of the water demand may be satisfied
Meter et. al. [13] (2016)	Understanding water storage and release dynamics across spatial scales	not calculated	Implement the RWH tank and balance of catchment waters.
Paul et. al. [14] (2004)	A key measure to improve water resource management	not calculated	utilization of freshwater resources in various aspects.

Here in Table 14.1, all the works are done in RWH but all papers did not include cost calculation. which is very important to implement a certain system. System implementation is common but without having any cost calculation users cannot judge whether it is efficient or not. Here every discussed paper includes a certain topic some are implementing the RWH system, some are on the utilization of rainwater, and some analysis on the factors of rainwater. On the other hand, our prototype system includes all together. This is the reason we implement a new model with better, more efficient, and more sustainable results.

14.3 MATERIALS AND METHODS

In this section, our objective is to specify the materials and methods we are going to use for this entire project. Our goal is to propose a new model of rainwater harvesting from the perspective of East West University.

14.3.1 Materials

In this project, we need two types of materials, some are software, and some are hardware. Though the hardware part is mostly needed for the project. The list of all the materials given below will be needed to implement the system.

14.3.1.1 Software

- Surface Water
- Groundwater and
- Rain Water

14.3.1.2 Hardware

- Surface Water
- Groundwater and
- Rain Water

14.3.2 Datasets

As the project is being done from the perspective of East West University, the datasets for this project are mostly taken from East West University authority [15]. Some other data related to the rain is collected from the weather office websites of Bangladesh [16]. Firstly, we have collected the monthly cost of water at East West University for a whole year. From the cost, we have come up with the approximate water usage in liters by some calculations. The dataset is in Table 14.2

The mathematical Equation 14.1 is used for calculating water consumption.

$$(PPU \times Total\ water(in\ Liter))/1000 = Total\ Cost(in\ Taka) \tag{14.1}$$

Table 14.2: Monthly Cost of Water

Month Name	Cost (In Taka)
January	39000
February	36000
March	38000
April	44000
May	48000
June	47000
July	48000
August	50000
September	46000
October	45000
November	43000
December	37000

Where PPU = Price Per Unit. From the recent government economic year, the price per 1000-liter water is 42 Taka. So, PPU = 42.

$$(Total\ Cost(in\ Taka) \times 1000)/PPU = Total\ water(in\ Liter) \tag{14.2}$$

Using the equation 14.2, we get a new Table of data that has the total amount of water usage at East West University per month.

This Table 14.2 indicates that the use of water is maximum in August and Minimum in February. It also gives an insight that the use of water is comparatively less in winter than in summer. We have also collected the annual rate of rain in Dhaka city from the weather office websites. From the weather report, we get this data about the rainfall across Bangladesh shown in Table 14.2. From this data, we will only extract the data of Dhaka City as East-West University is in Dhaka City as shown in Figure 14.1.

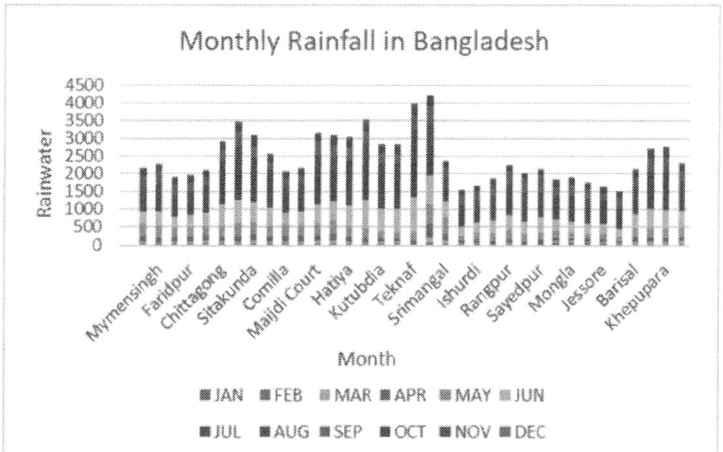

Figure 14.1: Monthly Average Rain Fall in Dhaka (in mm)

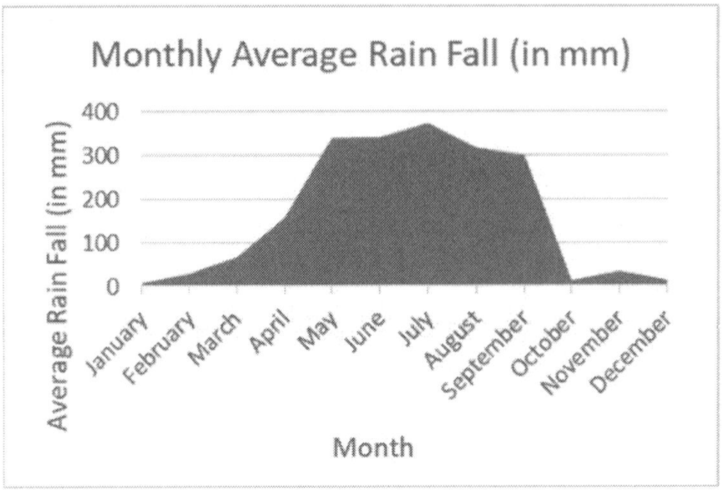

Figure 14.2: Monthly Average Rain Fall (in mm)

From the data table, we have extracted the data of Dhaka city only. Let's visualize the data in a plot shown in Figure 14.2. We can see in Figure 14.2 that the maximum number of rainfall occurs in May, June, July, and August than in the rest of the months. So, we can assume from now that these months are our peak point to harvest rainwater. We have also collected the area information of East West University to detect the approximate rainfall at East West University. The area of East West University is 4,58,957.04 square feet. These are the data we have collected for this research.

14.3.3 Mode/Process flow

The goal of the model is to use the rainwater again safely and direct the excess water into the ground for ecological balance. The whole process can be divided into sub-process.

1. **Collecting the water:** At first, we need to store the water by the drainage system of the East West University area and pass the water by a pipe into a large space. The challenge of collecting the water can be easily overcome. We need to make sure that the water drainage system from the terrace and ground should connect to a single pipe at last. The pipe will have all the water and the water will flow to a large chamber under the university area.

2. **Passing water into 1st chamber:** At the end of the pipe, we'll place a filter system that removes the big particles and cleans the water at a primary level. We'll use the basic sand and rock technique to find a large portion for the primary filtration of the water.

3. **Mechanism into 1st chamber:** In the first chamber, we are going to implement a PH sensor and two buckets of Sulfuric Acid and Sodium Hydroxide. As soon as the water arrives the sensor will send a notification to the system

about the water. The system will check if the PH level is more than 7.5. If it is more than 7.5 then it means it has base properties. As it is acidic, it will send a notification to the bucket of Sulfuric acid. As soon as the sulfuric is released the water gets neutralized. The same system will repeat until the PH of the water is 6.5 to 7.5. As soon as the PH is in range, it will release the water towards the second chamber. Cisco networking is shown in Figure 14.3 and the flowchart is given in Figure 14.4.

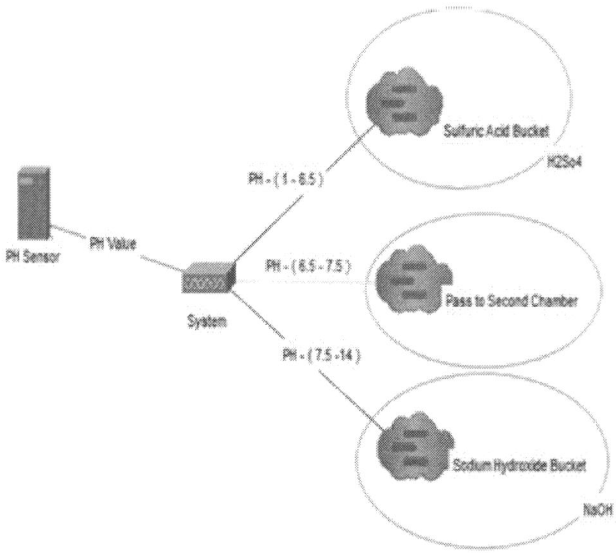

Figure 14.3: Cisco Design of PH System

The total flow of the system is described in Figure 14.5. through a diagram.

4. **Mechanism into 2nd chamber:** This chamber is basically like a tank system. It'll store water and supply it when the water is needed.

5. **What if the water exceeds storage:** Excessive water can't be stored and neither we can just throw it and create a mess. For this reason, we'll make another pipe and link it to the underground for at least 5 meters deep. It'll help to maintain ground-level water. The authority's ability to provide an uninterrupted clean water supply in Dhaka is becoming increasingly difficult as ground and surface water is decreased over time. In this context, even at the individual plot level, good rainwater management has the ability to make a substantial impact on the current problem. The overall physiochemical quality evaluation of Dhaka city's harvested rainwater is reported in this study and compared to the World Health Organization (WHO) standard [17]. Due to the lack of ground-level water, natural disasters like earthquake occurs, so it is a way to reduce the chances of an earthquake [18].

Figure 14.4: Diagram of PH sensor system

14.3.4 Mathematical Formulation and Equation

To calculate the amount of water getting used at East West University, we have used some mathematical equations. In the analysis part, we would also need some other equations for calculating rainfall in the East West University area and show some comparisons. The equations 14.3-14.5 are used or will be used in the future.

$$(PPU \times Total\ water\ (in\ Liter))/1000 = Total\ Cost\ (in\ Taka) \tag{14.3}$$

Where, PPU = Price Per Unit = 42 Taka (according to the 2021 economic year in Bangladesh) Formula to convert square feet to square meters:

$$(Area\ (Square\ feet))/10.764 = Area\ (Square\ meter) \tag{14.4}$$

For East West University we use this Equation 14.4, Area (Square meter) = 42,638.14 Formula to calculate total rainfall at East West University on a particular

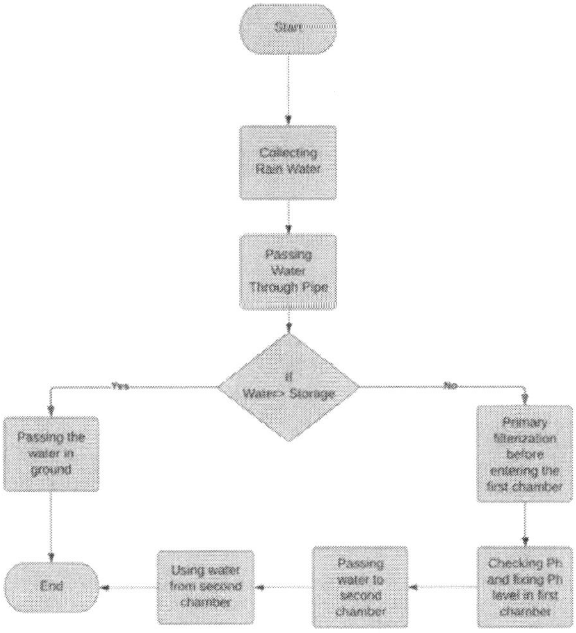

Figure 14.5: Flowchart of the full system

month:

$$Total\ rainfall\ (in\ Liter) = Averagerainfall\ (inmm) \times Area\ (Squaremeter) \tag{14.5}$$

Where, Area (Square meter) = 42,638.14 (Constant for East West University)

14.4 RESULT AND ANALYSIS

As we have proposed the system, will collect the water from rain and clean it for reuse purposes. From Equation 14.5 we get the amount of rain East West University can harvest in every month.

Table 14.3 states the monthly rainfall at East West University. Now if we can utilize 100% of all the water collected from the rain, we can decrease the cost of water by a large amount. But it is impossible due to the storage barrier. But still, if we can use even 10%, it will reduce water needs at a large quantity. The probable water need from WASA by 50%, 25%, and 10% is shown in Figure 14.6.

In Figure 14.6, the blue line represents the need for water for East West University without rainwater harvesting and the red line represents the water consumption reduction. As we can see, it's quite impossible to use rainwater to the fullest extent so we are assuming that to be 50%. Here, we can see the downfall of needed water from the very beginning of the year. After 50% harvest,

- In January, the need for water will be below 800000 L where it was 1000000 L approx. before harvesting

Table 14.3: Monthly rainfall at East West University (Round Figure)

Month Name	Average Rainfall (in mm)	Total rainfall at East West University	Total rainfall at East West University after 10% waste	Total rainfall at East West University (Round Figure)
January	7.7	328313.678	295482.3102	295482
February	28.9	1232242.246	1109018.021	1109018
March	65.8	2805589.612	2525030.651	2525030
April	156.3	6664341.282	5997907.154	5997907
May	339.4	14471384.72	13024246.24	13062620
June	340.4	14514022.86	13062620.57	13062620
July	373.1	15908290.03	14317461.03	14317461
August	316.5	13494971.31	12145474.18	12145474
September	300.4	12808497.26	11527647.53	11527647
October	12.3	524449.122	472004.2098	472004
November	34.4	1466752.016	1320076.814	1320076
December	12.8	545768.192	491191.3728	491191

- In February, the need for water will be below 200000–300000 L where it was 800000 L approx. before harvesting
- In (March-September) the need for water will be 0 where it was (900000–1200000) L approx. before harvesting
- In (October-December) the need for water will be (600000–800000) L approx. where it was (800000–1000000) L approx. before harvesting.

After the analysis, the need for water from WASA will be ZERO from (March to September) and for the other months, it will be much lesser than before. Cost analysis of water after rainwater harvesting is shown in the Table 14.4. Negative values are zero (0) to be considered.

In Figure 14.7, the blue line represents the cost of water at east west university without rainwater harvesting and the red line represents the cost reduction. After 50% harvest,

- In January, the cost for water will be (30000–35000) tk where it was 40000 tk approx. before harvesting
- In February, the cost for water will be (10000–15000) tk where it was (35000–40000) tk approx. before harvesting
- In (March-September) the cost for water will be 0 where it was (35000–50000) tk approx. before harvesting

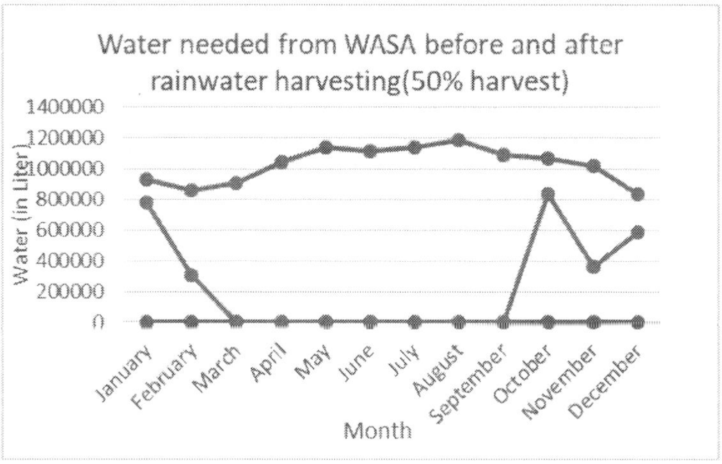

Figure 14.6: Water needed from WASA before and after rainwater harvesting (50 % usage)

Table 14.4: The monthly cost of water at East West University

Month Name	Cost	Water Cost from WASA after rainwater harvesting(50%)	Water Cost from WASA after rainwater harvesting(25%)	Water Cost from WASA after rainwater harvesting(10%)
January	39000	32794.878	35897.439	37758.9756
February	36000	12710.622	24355.311	31342.1244
March	38000	0	11487.185	27394.874
April	44000	0	0	18808.7906
May	48000	0	0	0
June	47000	0	0	0
July	48000	0	0	0
August	50000	0	0	0
September	46000	0	0	0
October	45000	35087.916	40043.958	43017.5832
November	43000	15278.404	29139.202	37455.6808
December	35000	24684.989	29842.4945	32936.9978
Annual Cost	519000	120556.809	170765.5895	228715.0264

- In (October-December) the cost for water will be (15000–35000) tk approx. where it was (35000–40000) tk approx. before harvesting.

From the above Table 14.4, we can see how much annual cost saves in water even in at least 10% rainwater harvesting. Here is a summary presented in Table 14.5.
From Table 14.5, we can see that by using rainwater harvesting techniques our cost of water reduces on a large scale. Again, there is also a cost of setting up the whole

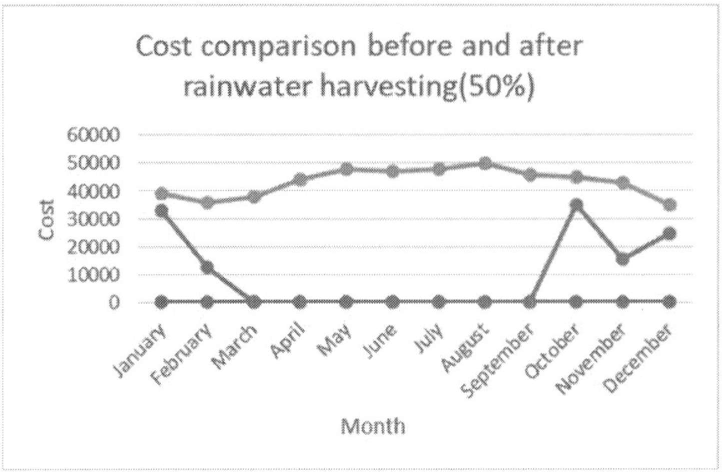

Figure 14.7: Cost comparison before and after rainwater harvesting (50%)

Table 14.5: Total saving comparison of rainwater harvesting with 50%, 25%, and 10% harvesting rates.

Annual Cost(In Taka)	Annual savings after 50% rainwater harvesting (In Taka)	Annual savings after 25% rainwater harvesting (In Taka)	Annual savings after 10% rainwater harvesting (In Taka)
519000	398,443(approximate)	348,234(approximate)	290,284 (approximate)

system but setting costs is a one-time investment and saves a large amount of expense for a lifetime. The cost of PH sensors, chemicals, filters, and setting the environment can be a maximum of 2 lakhs. Even after that, we can save up to 3 lakhs even if we use only 10% of the rainwater. If we can use 50% or more than that it will give more savings. But the amount of savings depends on the sizes of the chambers. So, it's a great way to save money and the earth from natural disasters as well. In Table 14.1 we can see the paper [11] model has saved 2.56 TK every day. While the three components of water utilization in our suggested model may save us 1,092 TK if we utilize 50% of the water (approx). Using 25% will save around 955 TK. Finally, using 10% will save 795.301 TK (approx). The author of the article [12] presented a model that ensures the consumption of rainwater underwater quality. Whereas the PH sensor in our suggested model is designed to purify the water before allowing it to be used. Other papers lack the cost computation that is a key component of our methodology. This proves that our suggested model is superior to other models that have been discussed. Future improvements will also be made on other sides.

14.5 CONCLUSION

After all the analysis and calculation, we have come to this point, there is nothing to lose in this prototype. We have conceded it in 3 versions and they are 10%, 25%, and 50% where we have seen the needed water analysis & the cost. If we want to come to the best outcome that will be the 50% harvesting method, then 25% and 10%. It will not just reduce the cost but also help us to make our university green & more sustainable which will be a great initiative for the upcoming days of the university campus. Increasing the efficiency of the filtering machine and also the model of sending the extra unused water to the ground level can be the extension of this work. In the future, the paper can be further expanded on how we can collect more water in storage and also supply this water and create a business process model.

Bibliography

[1] World Health Organization: WHO. (2019, June 18). 1 in 3 people globally do not have access to safe drinking water – UNICEF, WHO. https://www.who.int/. https://www.who.int/news/item/18-06-2019-1-in-3-people-globally-do-not-have-access-to-safe-drinking-water-unicef-who

[2] Bangladesh's Water Crisis - Bangladesh's Water | Water.org. (n.d.). Bangladesh's Water Crisis - Bangladesh's Water | Water.org. https://water.org/our-impact/where-we-work/bangladesh/

[3] Rainwater harvesting system: The process of maintenance poses a challenge. (2019, October 23). Dhaka Tribune.

[4] A. Khan, Dhaka Water Supply and Sewerage Authority: Performance and Challenges. Dhaka WASA.

[5] Hussain and Ziauddin, 1992. Rainwater harvesting and storage techniques from Bangladesh. Waterlines, 10(3), pp.10–12.

[6] The Financial Express. (n.d.). Rainwater — a prospective source of water security.

[7] Rainwater harvesting in Bangladesh: Building resilience, saving lives. (n.d.). UN Capital Development Fund (UNCDF).

[8] Hashim, Haslenda & Hudzori, Afzan & Yusop, Zulkifli & Ho, Wai Shin. (2013). Simulation-based programming for optimization of large-scale rainwater harvesting system: Malaysia case study. Resources, Conservation, and Recycling. 80. 1–9. 10.1016/j.resconrec.2013.05.001.

[9] Oviedo-Ocaña, E.R., Dominguez, I., Ward, S. et al. Financial feasibility of end-user-designed rainwater harvesting and greywater reuse systems for high water use households. Environ Sci Pollut Res 25, 19200–19216 (2018).

[10] Ashraful, A. M., & Islam, M. M. (2015). A Study on Rain Water Harvesting and Comparative Cost Analysis with Withdrawal of Underground Water around Gazipur City. Journal of Environmental Science and Natural Resources, 8(1), 91–94.

[11] Rahman, S., Khan, M. T. R., Akib, S., Din, N. B. C., Biswas, S. K., & Shirazi, S. M. (2014). Sustainability of rainwater harvesting system in terms of water quality. The Scientific World Journal, 2014.

[12] Rahman, M. M., Afreen, S., & Hussain, M. A. (2011). The potential of rainwater harvesting in Dhaka city. In Proc. Third International Conf. on Water& Flood Management.

[13] Van Meter, K. J., Steiff, M., McLaughlin, D. L., & Basu, N. B. (2016). The socio-ecohydrology of rainwater harvesting in India: understanding water storage and release dynamics across spatial scales. Hydrology and Earth System Sciences, 20(7), 2629–2647.

[14] Paul, J. G. Rainwater utilization: A key measure to improve water resource management.

[15] East West University, Wasa Bill of East West University (June 2021- June 2022), August 2022.

[16] Bangladesh meteorological department. Weather Forecast | Bangladesh Meteorological Department. (n.d.). Retrieved September 13, 2022

[17] Tarek, Y., Debnath, S., Uddin, M., Ahmed, S., Moniruzzaman, M., Jahan, T., Sultana, S., Hasan, M. and Jamal, A., 2022. Harvested Rainwater Quality of Dhaka and Its Importance as the Alternative to Surface and Groundwater. ECS Transactions, 107(1), pp.18813–18821.

[18] Harvested Rainwater quality of Dhaka and its importance as the alternative to surface and groundwater | SPAST abstracts. (n.d.). SPAST Abstracts.

CHAPTER 15

Yoga Pose Classification Using Transfer Learning

M. M. Akash*

*visie.tech, Dhaka, Bangladesh, Email: akash18rayhan@gmail.com, *Corresponding Author*

Rahul Deb Mohalder

Khulna University, Khulna, Bangladesh, Email: rahul@ku.ac.bd

Md. Al Mamun Khan

visie.tech, Dhaka, Bangladesh, Email: almamunkhan04@gmail.com

Laboni Paul

Khulna University, Khulna, Bangladesh, Email: laboni1124@cseku.ac.bd

Ferdous Bin Ali

visie.tech, Dhaka, Bangladesh, Email: hridoyferdous@yahoo.com

CONTENTS

Yoga has recently become an essential aspect of human existence for maintaining a healthy body and mind. People find it tough to devote time to the gym for workouts as their lives get more hectic and they work from home. This kind of human pose estimation is one of the notable problems as it has to deal with locating body key points or joints. Yoga-82, a benchmark dataset for large-scale

DOI: 10.1201/9781003605508-15

yoga pose recognition with 82 classes, has challenging positions that could make precise annotations impossible. We have used VGG-16, ResNet-50, ResNet-101, and DenseNet-121 and finetuned them in different ways to get better results. We also used Neural Architecture Search to add more layers on top of this pre-trained architecture. The experimental result shows the best performance of DenseNet-121 having the top-1 accuracy of 85% and top-5 accuracy of 96% outperforming the current state-of-the-art result.

15.1 INTRODUCTION

Human Activity Recognition (HAR) is a field in computer vision that involves identifying and classifying human activities from sensor data or video streams. This can be done using various techniques, such as machine learning algorithms, pattern recognition, and computer vision.

Yoga poses are a specific type of human activity that can be recognized using HAR techniques. By analyzing sensor data or video streams of a person performing yoga, a HAR system can identify and classify the specific yoga poses being performed. This can be useful for a variety of applications, such as yoga instruction, and injury prevention. HAR system could be used to track a person's progress as they perform a series of yoga poses, providing feedback and guidance on proper form and technique. It could also be used to identify any incorrect or potentially harmful movements, helping to prevent injuries.

Machine learning is being used to help people exercise more effectively through the use of virtual personal trainers. These are systems that use machine learning algorithms to provide personalized exercise routines and coaching based on the user's goals, fitness level, and other factors. By analyzing data such as heart rate, body composition, and exercise performance, these systems can adapt and modify the user's routine in real-time to optimize their results. Wearable technology such as fitness trackers and smartwatches are increasingly using machine learning to improve their accuracy and effectiveness. For example, some fitness trackers use machine learning algorithms to analyze data such as steps taken, calories burned, and heart rate variability to provide more accurate and personalized feedback to the user.

During the COVID-19 pandemic, Yoga is a popular practice that has been shown to reduce stress, anxiety, and depression. However, to get the maximum benefits of yoga, it is crucial to perform the poses correctly, which requires access to a yoga trainer. Unfortunately, not everyone has access to a trainer. For these concerned people, a software application can play the role of an instructor. Traditional pose estimation methods for recognizing and correcting poses are at a disadvantage here [15] because of the diversity of poses on human bodies. Therefore, this paper proposes a novel solution, a classification system that captures the similarities and differences between poses. Instead of producing body key points for human subjects, this method recognizes the overall pose and classifies it. The proposed approach can improve access to yoga for individuals who do not have access to a trainer and enhance the benefits of yoga for all practitioners.

15.1.1 Our Contribution

The key contributions of this research are:

- We have experimented with different types of image preprocessing techniques.
- We have used transfer learning in different stages like using fully imagenet weight, finetuning different layers of the model, and finetuning the whole model.
- We have defined the network using Random Search for architecture.

The paper is organized as follows, the very next section (Section 15.2) contains the previous work description. After that, we introduced our methodology in Section 15.3 followed by result analysis in section 15.4 and the conclusion and future work in Section 15.5.

15.2 RELATED WORK

Chen [5] conducted various deep learning (DL) based techniques to deduce the human pose estimation from images. The usage of Deep learning for human pose estimation was established when AlexNet used CNN and achieved state-of-the-art results. The researchers pointed out that there are now a variety of pose estimate techniques, including pixel-level analysis, body joint point mapping, body-based and model-based human body estimation, and heat-map mapping.

Faisal [7] continued on Chen's body-joint estimation research [5] and mentioned the current state-of-the-art, where they repeated that the body-joint points are identified by employing different gyroscope and joint angle methods for determining joint point angle, as well as multiple sensor fusion.

The authors of [2] repeated Faisal and Chen's findings from [5], [7] by estimating human activity based on poses. It should be noted that prominent posture estimation techniques include template-based, generative model-based, and discriminative model-based techniques. These techniques were later verified in research on human pose estimation published in [6]. Yoga has been more well-liked among people because of its believed benefits of good health and fitness. Nagalakshmi examined the effects of yoga and discovered that remote yoga activities have become significantly more popular since the COVID outbreak, in addition to alleviating musculoskeletal pain and also inspiring a healthy lifestyle [13].

For this goal, Agarwal [1] experimented with various Machine Learning (ML) methods. They created a dataset of 5,500 photos for ten distinct yoga postures and then they used the open pose estimation algorithm to create the skeleton images. On their dataset, they acknowledged that random forest classifiers performed better. Liaqat et al. [11] created a hybrid posture recognition approach by combining deep neural networks with conventional machine learning techniques. The weight learned by the deep learning model and the prediction from the conventional model are combined to produce the final class prediction. In a different paper [10], the authors first extracted key points using OpenPose, and then classified the data using a hybrid CNN-LSTM layer. The researchers on this team worked on sitting pose estimation [9] and concluded a Deep Recurrent Hierarchical Network based on MobileNetV2 and gained 91.47% accuracy. Similarly, the use of SVM and boosting for stance detection is addressed in [14], and Nagalakshmi's work [12] specifically focused on yoga.

Keeping pace with deep neural networks pose estimation models are getting more effective day by day. The mostly accurate keypoints and skeleton annotations are the strength of state-of-the-art pose estimation models. Pose estimation models are widely used in human surveillance [4], and human activity recognition [8] like computer vision problems. So far we are working with yoga poses as a result we can say pose estimation models can make way for this problem. However, the authors [15] opposed the idea of using pose estimation models for the yoga-82 dataset. They showcased some yoga poses that seem too complex from a single point of view because the human body can handle heavy diversity. With the change in image resolutions, it gets worse. Considering these factors they stated that pose models extracted keypoints and skeletons tend to have false and wrong annotations. As a result, they proposed the idea of hierarchical pose classification. They proposed hierarchical labeling where categories are created by the variations in body postures like standing, sitting, etc. And then subcategories are created from those categories.

15.3 METHODOLOGY

The very first step is to acquire the hierarchical poses (Figure 15.1) of the yoga-82 dataset. Before feeding the images into the model we applied some preprocessing techniques to enhance image quality such as contrast enhancement, median filter, and image sharpening. The next involves a deep learning model architecture as a backbone here we have selected DenseNet-121 and added some dense blocks on top of that architecture at the very end adding the classification layer with 82 nodes. By using pre-trained weights on the ImageNet dataset we have fine-tuned our model on the specific yoga-82 dataset. While training the model we applied data augmentation techniques like rotation, zooming and shearing which helped the model generate more generalized predictions.

15.3.1 Dataset

Yoga-82 [15] contains 21,009 train images and 7,469 test images with multilevel class hierarchy depending on the variations of pose. It has 5 classes, standing, sitting, balancing, inverted, reclining, and wheel as a base class. They are divided into 20 subclasses. In the third level, they are divided into 82 classes. We have worked with the third level of annotation. In Figure 15.2 we showed the Yoga-82 dataset description.

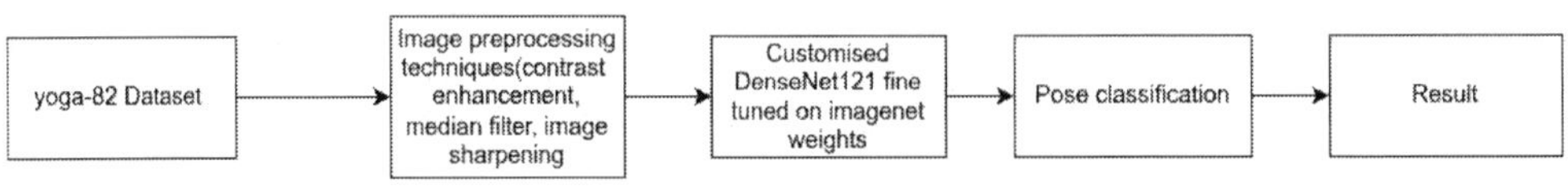

Figure 15.1: Our Proposed Workflow Diagram.

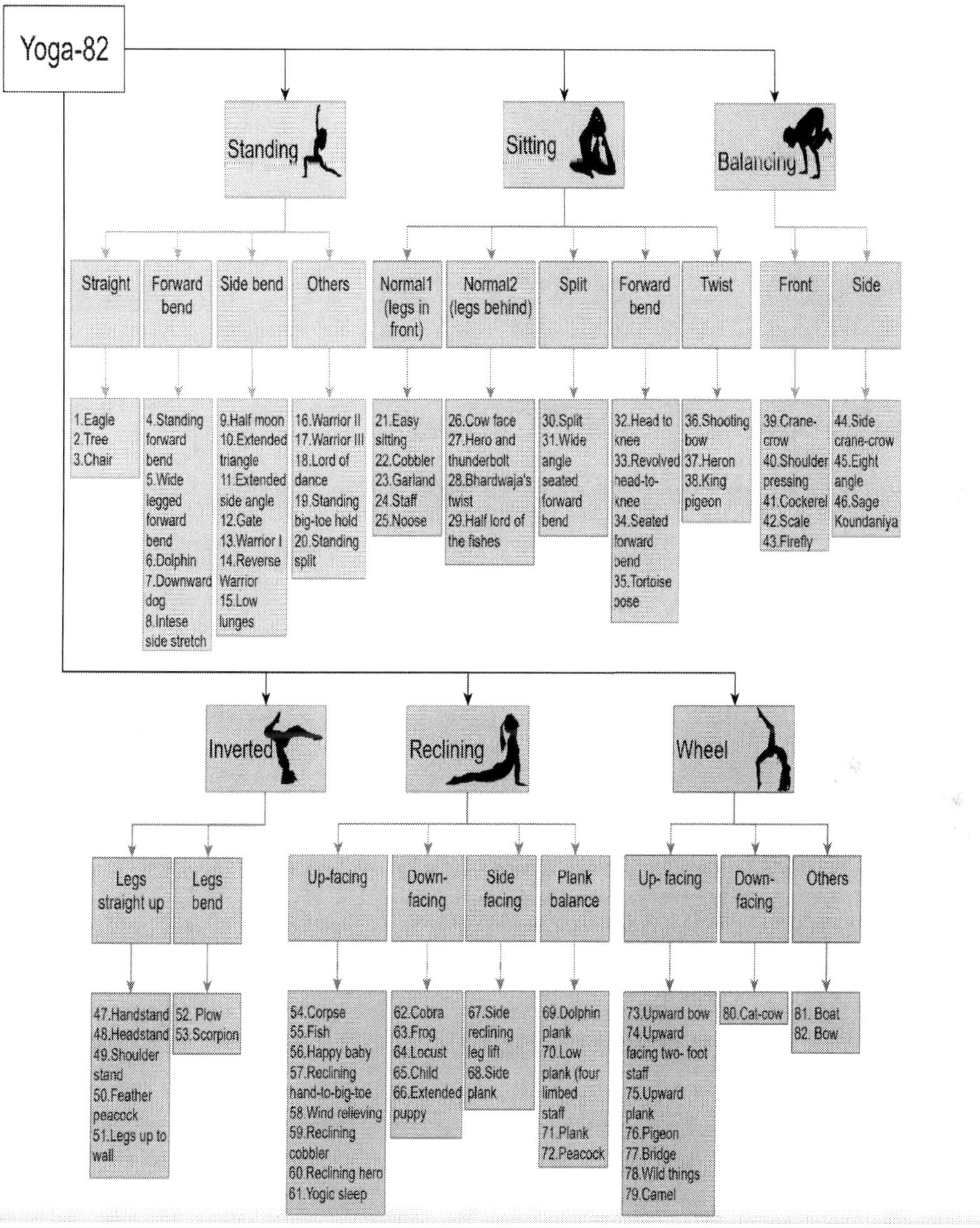

Figure 15.2: Yoga-82 Dataset Description [15].

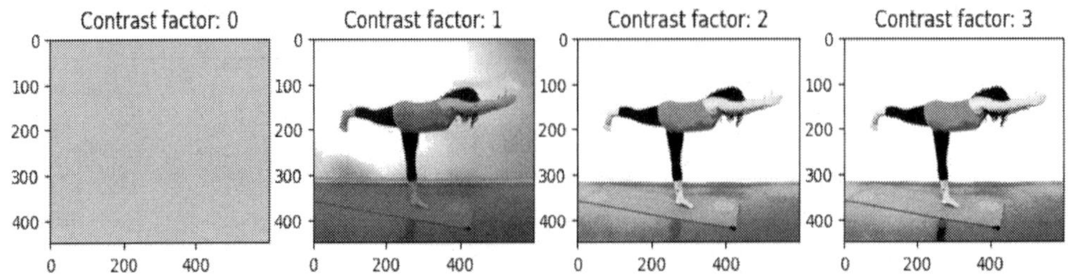

Figure 15.3: Effect of the contrast factor.

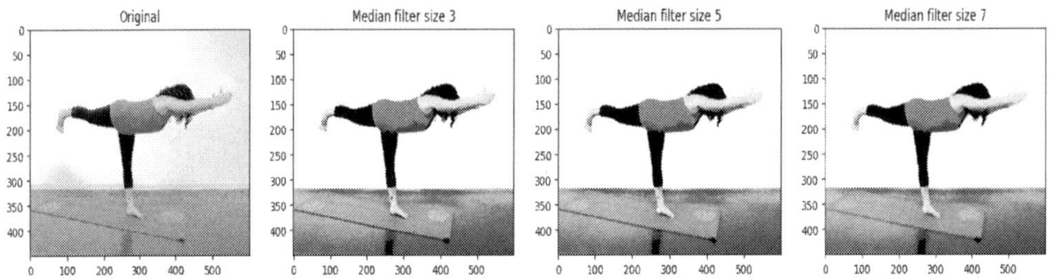

Figure 15.4: Effect of the median filter.

15.3.2 Data Processing

We followed these data processing techniques to process our data.

15.3.2.1 Image Enhancement

From the PIL ImageEnhance class, an object Contrast is instantiated to apply contrast on images. The concept of image contrast is different from brightness. Brightness makes the whole image brighter or darker. On the other hand, the contrast makes the light areas lighter and the dark areas darker. The lighter or darker amount depends on the factor.

Increasing contrast makes different parts of the body more visible. When contrast is increased in an image certain areas like body joints highlight some patterns, and muscle appears more defined. Eventually, it makes it easier for the model to extract better features (Figure 15.3).

15.3.2.2 Median Filter

When we increase the contrast in an image, we are increasing the difference between the light and dark areas of the image. This also leads to an increase in noise. To remove the noise on images we used a median filter (Figure 15.4). The median filter calculates the median of the pixel intensities surrounding the center pixel in a $n \times n$ kernel. The kernel we chose is 3×3.

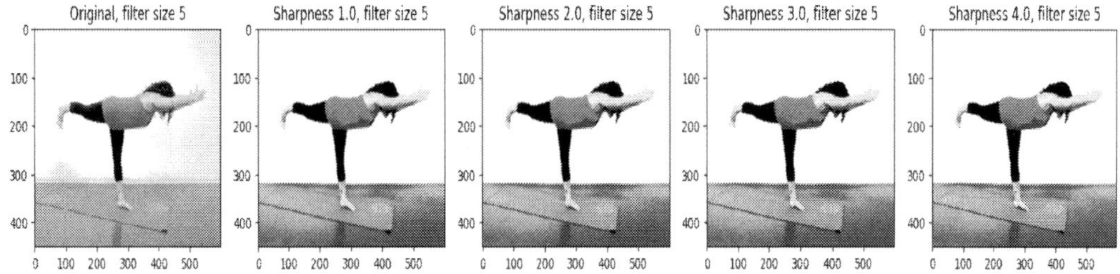

Figure 15.5: Effect of the image sharpening.

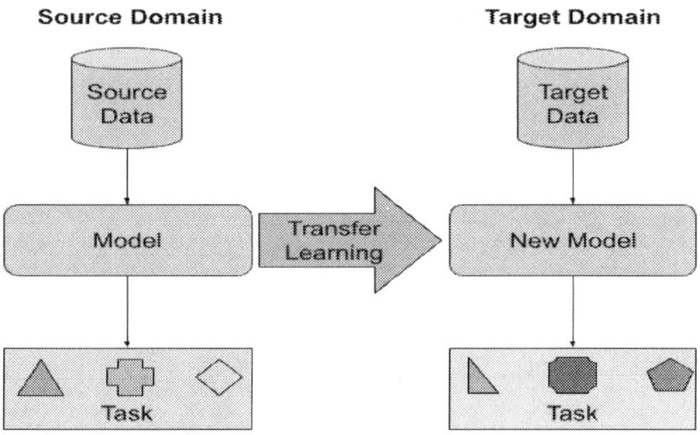

Figure 15.6: Transfer Learning Working Process.

15.3.2.3 Image Sharpening

The median filter also results in a slightly blurred image and some loss of detail and edge sharpness. Applying a sharpening filter to an image increases the contrast between adjacent pixels and enhances the edges in the image, which helps to reduce the appearance of blurriness in the image (Figure 15.5).

15.3.3 Transfer Learning Models

Transfer learning is a machine learning technique that has been successfully applied to many classification tasks. It involves leveraging a pre-trained model to improve the accuracy and efficiency of a new classification task (Figure 15.6).

Consider X be the set of input data, Y be the set of the output labels, and f be a function that represents the pre-trained model. So f is a function that maps input X to output Y, i.e., Y = f(X). The model f has learned some underlying patterns and features that can be reused for the new classification task.

Now, let X' be the input data for the new classification task, and y' be the output labels. The goal is to train a new model g that can accurately classify the input data

X' into the output labels y'. The new model g can be represented as a function that maps input X' to output Y', i.e., Y' = g(X').

To perform transfer learning, the weights of the pre-trained model f are used to initialize the weights of the new model g. This reduces the training time and enables the model to learn the specific features relevant to the new classification task. The fine-tuning process involves updating the weights of the last few layers of the pre-trained model f while keeping the earlier layers frozen. This enables the new model g to learn the specific features relevant to the new classification task while retaining the general features learned by the pre-trained model f.

Mathematically, the fine-tuning process involves minimizing the loss function between the predicted output of g and the actual output of the new classification task dataset. The loss function can be represented as follows:

$$L(Y', Y) = - y' \log(g(X')) + (1 - y') \log(1 - g(X'))$$

Where L is the loss function, y' is the predicted output of g, and y is the actual output of the new classification task dataset. The goal is to minimize the value of L to improve the accuracy of the model.

15.3.4 VGG-16

The Visual Geometry Group at the University of Oxford created VGG-16, a convolutional neural network architecture. It is a deep neural network that in 2014 attained world-leading performance on the ImageNet dataset. The VGG-16 architecture consists of convolutional layers followed by fully connected layers.

The simplicity and clarity of the VGG-16 architecture's design are its assets. This makes it a popular choice among newcomers to deep learning, including researchers and practitioners. In addition, the VGG-16 (Figure 15.7) architecture has been utilized extensively as a pre-trained model for transfer learning, where the weights of the pre-trained network are fine-tuned for a specific task.

Overall, VGG-16 is a potent convolutional neural network architecture that has contributed significantly to the field of computer vision. Its success has prompted additional research into deep learning and created new opportunities for image classification, object recognition, and other computer vision tasks.

15.3.5 ResNet-50

In 2015, ResNet-50 joined the rest of the ResNet family of neural networks. To fix the problem of disappearing gradients, the Microsoft Research team created a convolutional neural network. The vanishing gradient problem was addressed by the use of skip connections.

ResNet-50 consists of 50 layers, some of which are convolutional while others are batch normalization and others are completely linked. The first few layers are responsible for extracting low-level features from the input picture, while the latter layers are in charge of extracting and classifying higher-level characteristics (Figure 15.8). ResNet-50's capacity to accurately train incredibly deep neural networks is one of its greatest capabilities. Image categorization, object recognition, and semantic

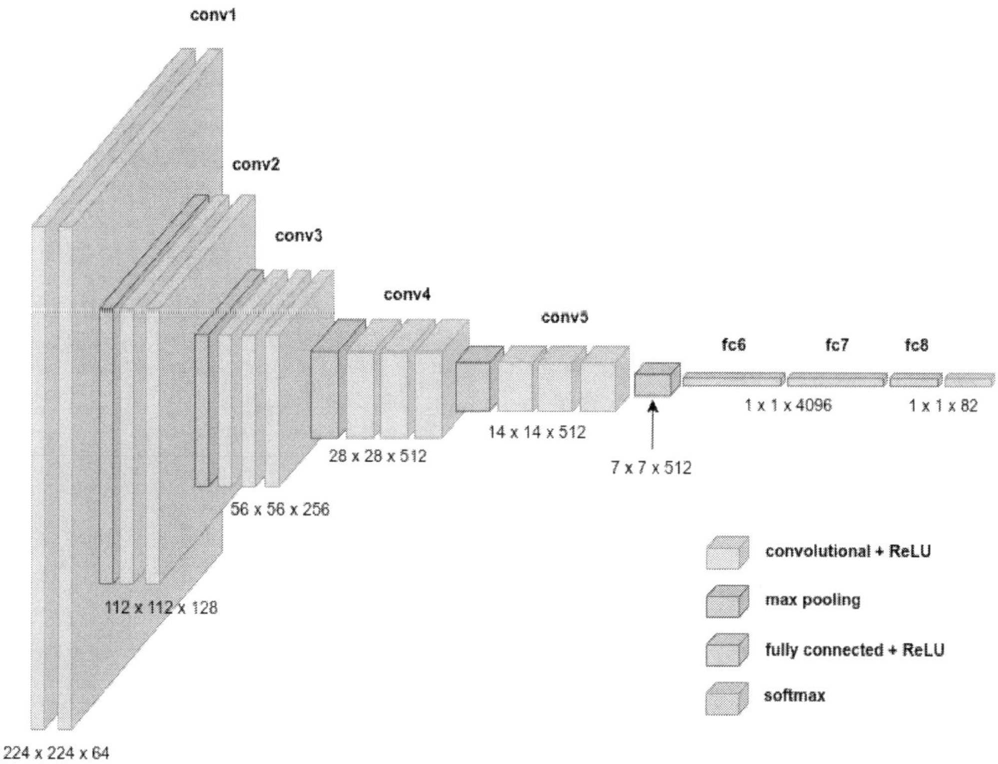

Figure 15.7: Modified VGG-16.

segmentation are just a few of the many computer vision tasks that have benefited from their use.

15.3.6 DenseNet-121

DenseNet-121 is a convolutional neural network architecture developed by University of Toronto researchers in 2017. It is a member of the DenseNet network family, which is notable for its thick connection pattern between layers.

DenseNet-121 consists of 121 layers, which include convolutional layers, batch normalization layers, and pooling layers. The architecture is intended to capture both low-level and high-level information in the input picture, and its dense connection network aids in the prevention of overfitting and the improvement of training efficiency (Figure 15.9).

DenseNet-121 has shown cutting-edge performance on a variety of computer vision applications, including image classification, object identification, and semantic segmentation. It has also been demonstrated to generalize effectively to new tasks when used as a starting point for transfer learning on other datasets.

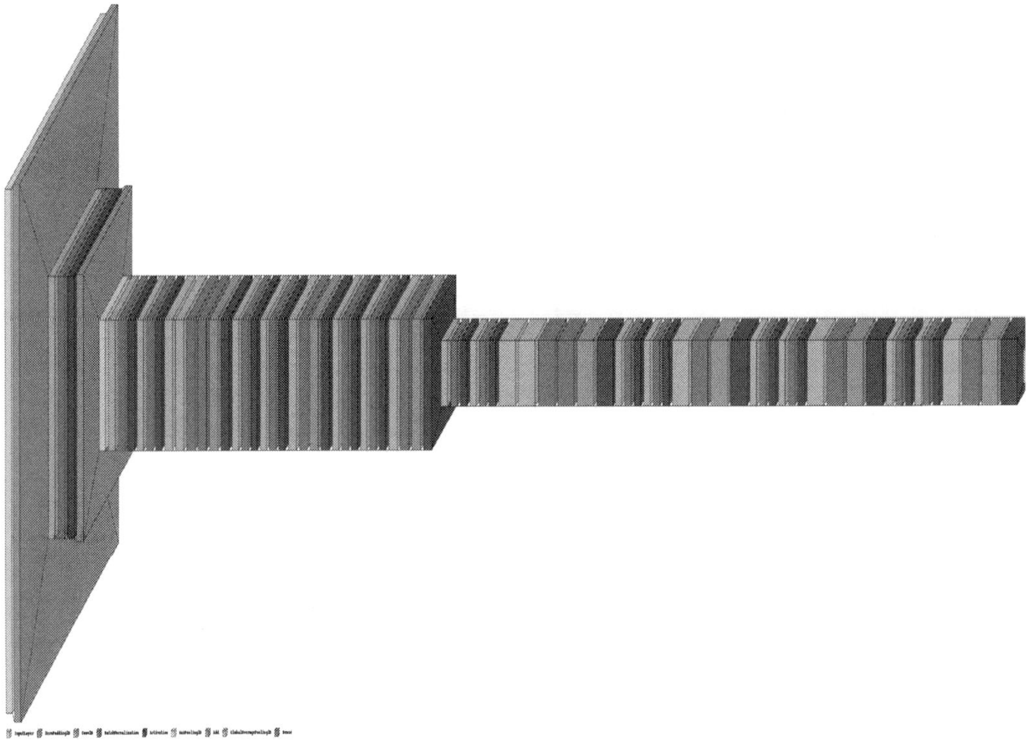

Figure 15.8: Modified ResNet-50.

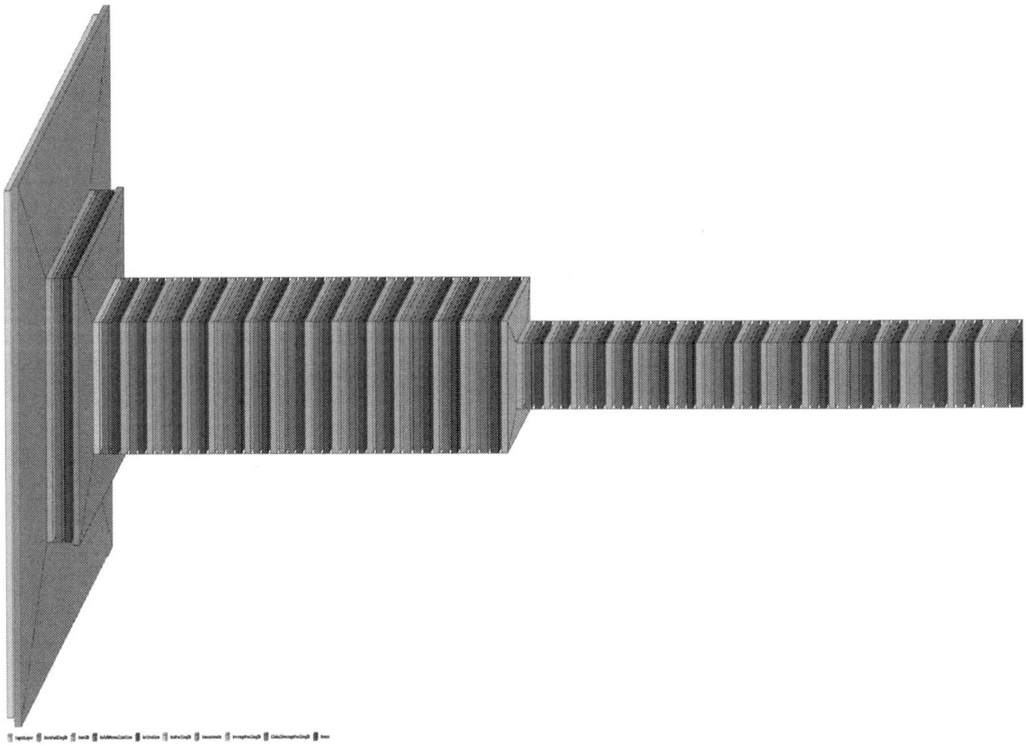

Figure 15.9: Modified DenseNet-121.

15.4 RESULT ANALYSIS AND DISCUSSIONS

The VGG16 network from Keras and its weights after they are pre-trained on the ImageNet dataset. Also, we removed the classifier part of this network. On ConvNet the early layers contain very generic feature maps that will be useful for our dataset even if it is very different from ImageNet, so we started with fine-tuning only the last five layers. As a result, we achieved better accuracy for VGG16 architecture. However, this idea did not perform well when we were working with ResNet-50 architecture. There are 5 stages in the model. Each stage consists of multiple convolution layers and batch normalization. In this architecture, we tried to freeze all the stages except for the last stage layers. But the result turned out to be bad compared to fine-tuning all the layers.

It is possible that the skip connections in ResNet-50 architecture could have played a role in this difference. Skip connections in ResNet-50 allow the gradients to flow directly to the earlier layers, which can help the earlier layers learn some feature maps related to our target dataset. It also means that each layer's output may depend on the earlier layers' output. Therefore, when we fine-tune only the last stage layers, we may not be capturing the full complexity of the images.

But we got our best performance in DenseNet-121 (Figure 15.10 and Figure 15.11). The result in the table shows all our experiment outcomes. We have outperformed the current SOTA result. We have used four metrics to assess our classification models: accuracy, f1-score, precision, and recall. In Table 15.1, we have compared the results among our models. In Table 15.2, we have compared the best performance against the SOTA results.

Python tool packages like Keras, numpy, sci-kit learn, pandas, and Matplotlib were used to set up the project. NVIDIA 3060 GPU was used to run and train the project. The models were trained for 500 epochs with 256-person batches and a 0.01-percent learning rate with exponential decay. There was the use of Adam's planner. Hyperparameter search was done with Kears Autotuner, which had a return method

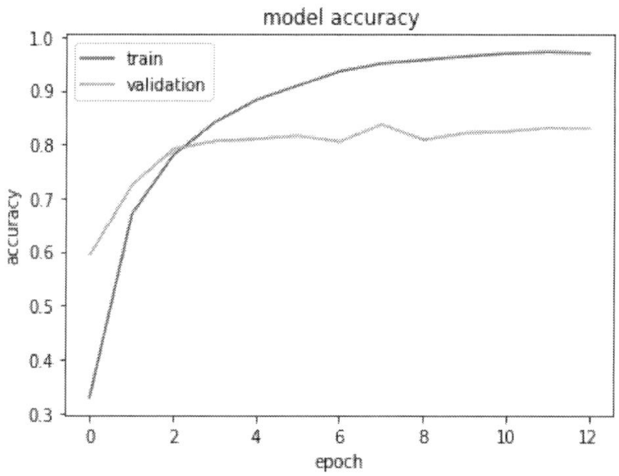

Figure 15.10: Classification accuracy for our modified DensNet-121.

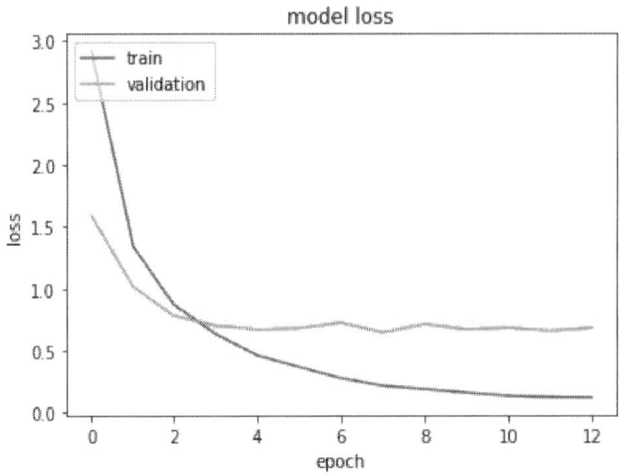

Figure 15.11: Classification loss for our modified DensNet-121.

that kept track of validation loss and accuracy. Models were stopped early by using early stopping with a 15-epoch tolerance number.

Table 15.1: Comparison of Different Experiment Setup.

Used Model	Top-1 Accuracy (%)	Top-5 (%)	Precision	Recall	F1-Score
ResNet-50	77	94	0.81	0.76	0.76
ResNet-101	75	92	0.77	0.73	0.73
VGG-16	74	92	0.73	0.70	0.69
DenseNet-121	85	96	0.87	0.83	0.83

Table 15.2: Comparison with other Methods

Method	Used Model	Top-1 Accuracy (%)	Top-5 (%)
[3]	DCPose(Yoga-82) dataset	77	-
[15]	DenseNet-201	74.91	91.30
[15]	ResNet-50	63.44	82.55

15.5 CONCLUSION AND FUTURE WORK

The main challenges to classifying yoga poses are that they have very repetitive poses and so they are very difficult to annotate as a body key point. That's why the researchers have moved away from key point-based classification to an image classification problem. Due to advancements in deep learning and deep neural networks, these types of multiclass classification problems have become easier. In this work, we have explored the power of transfer learning, used pre-trained deep models, and modified them to serve our purposes. And it has shown promising results in terms of top-1 and top-5 accuracy.

We would like to explore more like ensembling deep learning architectures and different image processing techniques to move forward with the current research. We also want to investigate the explainability of the model using GradCam and similar types of algorithms. We believe this will help us to analyze errors in a more convenient way alongside handling any kind of biases.

Bibliography

[1] Yash Agrawal, Yash Shah, and Abhishek Sharma. Implementation of machine learning technique for identification of yoga poses. In *2020 IEEE 9th international conference on communication systems and network technologies (CSNT)*, pages 40–43. IEEE, 2020.

[2] Sameh Neili Boualia and Najoua Essoukri Ben Amara. Pose-based human activity recognition: a review. In *2019 15th international wireless communications & mobile computing conference (IWCMC)*, pages 1468–1475. IEEE, 2019.

[3] Mustafa Chasmai, Nirjhar Das, Aman Bhardwaj, and Rahul Garg. A view independent classification framework for yoga postures. *SN Computer Science*, 3(6):476, 2022.

[4] Qiuhui Chen, Chongyang Zhang, Weiwei Liu, and Dan Wang. Shpd: Surveillance human pose dataset and performance evaluation for coarse-grained pose estimation. In *2018 25th IEEE International Conference on Image Processing (ICIP)*, pages 4088–4092. IEEE, 2018.

[5] Yucheng Chen, Yingli Tian, and Mingyi He. Monocular human pose estimation: A survey of deep learning-based methods. *Computer Vision and Image Understanding*, 192:102897, 2020.

[6] Atiqul Islam Chowdhury, Mohsena Ashraf, Ashraful Islam, Eshtiak Ahmed, Md Saroar Jaman, and Mohammad Masudur Rahman. hactnet: an improved neural network based method in recognizing human activities. In *2020 4th international symposium on multidisciplinary studies and innovative technologies (ISMSIT)*, pages 1–6. IEEE, 2020.

[7] Abu Ilius Faisal, Sumit Majumder, Tapas Mondal, David Cowan, Sasan Naseh, and M Jamal Deen. Monitoring methods of human body joints: State-of-the-art and research challenges. *Sensors*, 19(11):2629, 2019.

[8] Michael B Holte, Cuong Tran, Mohan M Trivedi, and Thomas B Moeslund. Human pose estimation and activity recognition from multi-view videos: Comparative explorations of recent developments. *IEEE Journal of selected topics in signal processing*, 6(5):538–552, 2012.

[9] Audrius Kulikajevas, Rytis Maskeliunas, and Robertas Damaševičius. Detection of sitting posture using hierarchical image composition and deep learning. *PeerJ computer science*, 7:e442, 2021.

[10] Deepak Kumar and Anurag Sinha. *Yoga pose detection and classification using deep learning.* LAP LAMBERT Academic Publishing, 2020.

[11] Sidrah Liaqat, Kia Dashtipour, Kamran Arshad, Khaled Assaleh, and Naeem Ramzan. A hybrid posture detection framework: Integrating machine learning and deep neural networks. *IEEE Sensors Journal*, 21(7):9515–9522, 2021.

[12] Chirumamilla Nagalakshmi and Snehasis Mukherjee. Classification of yoga asanas from a single image by learning the 3d view of human poses. *Digital Techniques for Heritage Presentation and Preservation*, pages 37–49, 2021.

[13] Dr P Prabhavathy Nagalakshmi Vallabhaneni. The analysis of the impact of yoga on healthcare and conventional strategies for human pose recognition. *Turkish Journal of Computer and Mathematics Education (TURCOMAT)*, 12(6):1772–1783, 2021.

[14] D Panigrahy, PK Sahu, and Felix Albu. Detection of ventricular fibrillation rhythm by using boosted support vector machine with an optimal variable combination. *Computers & Electrical Engineering*, 91:107035, 2021.

[15] Manisha Verma, Sudhakar Kumawat, Yuta Nakashima, and Shanmuganathan Raman. Yoga-82: a new dataset for fine-grained classification of human poses. In *Proceedings of the IEEE/CVF conference on computer vision and pattern recognition workshops*, pages 1038–1039, 2020.

CHAPTER 16

Bornomala: A CNN Ensemble Approach for Bangla Sign Language Detection

Arifur Rahman

Khulna University of Engineering & Technology, Department of Computer Science and Engineering, Khulna, Bangladesh.

Anik Mahmud

Khulna University of Engineering & Technology, Department of Computer Science and Engineering, Khulna, Bangladesh.

Abu Bakar Siddique

Chittagong University of Engineering & Technology, Department of Mechanical Engineering, Chittagong, Bangladesh.

Avizit Nandi

Khulna University of Engineering & Technology, Department of Electronics and Communication Engineering, Khulna, Bangladesh.

Pintu Chandra Shill

Khulna University of Engineering & Technology, Department of Computer Science and Engineering, Khulna, Bangladesh.

CONTENTS

DOI: 10.1201/9781003605508-16

Bangla Sign Language (BdSL) is non-verbal language used by deaf and dumb people to communicate with the rest of the world. Sign language detection is a crucial area of study to facilitate communication with the hard-of-hearing population. Despite Bangla world's sixth most spoken language, much significant work has not been done yet might be due to the unavailability of the proper dataset. Most of the existing works on Bangla Sign Language have done on small datasets and didn't give satisfactory outcomes. In our research, we built a large dataset of hand signs named as 'Bornomala' for both digit and character recognition. Initially, we have constructed a Convolutional Neural Networks (CNNs) model, a customized VGG16 model, and a customized Inception v3 model. Then we utilized Voting Ensemble on these there models (Our proposed CNN model, customized VGG16 model, and customized Inception v3 model). We have also constructed an average CNN model to acquire better performance than individual model. Afterward, we employed multiple classifiers including Random Forest (RF), K-Nearest Neighbor (KNN), Decision Tree (DT), and Support Vector Machine (SVM) in the output layer of the Average CNN model to investigate the performance of each classifier against our dataset. Finally, we have conducted a comparison of the outcomes. We found that the highest accuracy for alphabet detection was 93.80% and for digit detection it was 95.20%. Our proposed Voting Ensemble model's accuracy outperforms all the previous accuracy of existing models.

Keywords: Bangla Sign Language, Alphabets and Digits Recognition, Bangla 'Bornomala' Dataset, Sign Digits and Sign Alphabets, CNN Ensemble, Voting Ensemble.

16.1 INTRODUCTION

Among the top 100 languages spoken worldwide, Bangla is spoken the sixth most frequently. Bengali language is spoken by 265 million native and non-native people worldwide among them 5% of whom have speech and hearing disabilities. Its very difficult for them to interact with the rest of the world. Deaf individuals can socially interact, express their thoughts and feelings, and engage in various activities, including education [38], employment, social interactions, and even military training [32], through the use of sign language. Bangla Sign Language is essential for the deaf community in Bangladesh and is essential in fostering their inclusion and accessibility because of this. Sign language acts as a bridge for communication between people with hearing and speech disabilities. By leveraging a huge dataset, machine learning, deep learning [36], and ensemble algorithms [34], [35], [39] can make it simple to learn sign language. The lack of a perfect dataset in Bangla sign language detection has a significant impact on research in this field. Several key implications arise from the absence of an ideal dataset. Without a comprehensive dataset, the performance of sign language detection models may be limited. Inadequate data can lead to models with limited generalization capability for different sign gestures and variations. A lack of diverse and representative data can lead to overfitting or poor generalization of models to real-world scenarios [33], [37]. Models trained on limited data may fail to handle unseen variations effectively.

Islam et al. [11] created a dataset named as 'Ishara-Lipi'. The dataset consisted of a total of 1800 character images of Bangla Sign Language, representing 36 Bangla basic sign characters. Out of the 36 basic sign characters, 6 were vowels and 30 were consonants characters. In 2019, Islam et al. [12] built another dataset termed 'Ishara-Bachon'. Ishara-Bachon dataset comprises 100 groups, each containing 10 Bangla basic sign digits (0, 1, 2, ..., 9), gathered from various deaf and general volunteers. Talukder et al. [13] developed an image dataset named OkkhorNama, consisting of fingerspelled Bangladeshi Sign Language signs, encompassing 46 signs. Among those 46 basic sign characters, 10 classes were for digits (0, 1, 2, ..., 9), and 36 classes were for characters. Hasan et al. [14] developed a comprehensive dataset of Bangla sign language, covering 36 Bangla letter classes and 10 digit classes. The dataset consists of a total of 5820 images, capturing 150 images for each of digit classes and 120 images for each of the letter classes. By analyzing all of these researches, we have found that the recent researches conducted on less varied dataset. Those datasets have less amount of classes. Moreover, each class is associated with a few amount of images. For the advancement in Bangla Sign Language research, we built a large dataset and gave it the name as 'Bornomala'.

In Table 16.1 we have shown a comparison of our dataset with previously created datasets. For further more study by other researchers, we planned to make our dataset accessible. There are a limited amount of dataset for Bangla Sign Language. However, there aren't many images or character classes in the existing datasets. We have tried to increase the amount of images for each class. To make our numeric dataset, for 10 digits we used a total of 1700 images. Here, for each digit class we used 170 images. Similarly to make our character dataset, for 44 character we used a total of 6160 images. Here, for each digit class we used 140 images. We employed three different CNN models and then we applied Voting Ensemble.

16.2 RELATED WORK

In recent times, sign language recognition has emerged as a prominent area of research, generating significant interest from the research community. A multitude of sign language detection studies is currently available, focusing on diverse sign languages such as Indian Sign Language, British Sign Language, German Sign Language,

Table 16.1: Comparison of Previous Dataset with Our Self-made Dataset.

Dataset Name & Creator Team	**Category**	**Number of Digit Class**	**Number of Character Class**	**Total Digit Images**	**Total Character Images**	**Year**
M.Sanzidul Islam et al. ('Ishara-Lipi')	Characters		36		1800	2018
M.Sanzidul Islam et al. ('Ishara-Bachon')	Digits	10		1000		2019
D.Talukder et al. ('OkkhorNama')	Characters & Digits	10	36			2021
S. N. Hasan et al. ('Shongket')	Characters & Digits	10	36	1500	4320	2021
Our Dataset ('Bornomala')	Characters & Digits	10	44	1700	6160	2023

American Sign Language, Chinese Sign Language, Turkish Sign Language, French Sign Language, and Japanese Sign Language. While many sign languages have been extensively studied and well-documented, there appears to be a noticeable scarcity of research focused on Bangla Sign Language (BSL). This observation suggests that the specific challenges and nuances associated with BSL recognition have not received the same level of attention as other widely recognized sign languages.

Kozyra K et al. [1] conducted a study that aims to design and implement an application proficient in American Sign Language (ASL) sign recognition, utilizing deep learning algorithms built upon convolutional neural network (CNN) architectures. Their project's execution involved constructing a training dataset, preparing a module for transforming images into a format compatible with the artificial neural network, choosing the optimal neural network architecture, and establishing the model. Their proposed network's efficacy for the training set reached 99%. M. A. Hossen et al. [2] introduced a novel method for Bengali Sign Language Recognition, employing Deep Convolutional Neural Networks (DCNN). Initially, they developed a pre-trained network to learn features. They employed a deep convolutional neural network and made use of the acquired features from a pre-trained network, along with fine-tuning the top layers of this network, which resulted in an impressive overall recognition rate of 96.33% on the training dataset and 84.68% on the validation dataset. Sadik et al. [3] employed a novel approach involving hand gesture segmentation, binary masking, and SVM classification to recognize Bangla Sign Language (BdSL). This system implemented the YCbCr color space and Fuzzy C-means clustering techniques to generate binary images, which were then subjected to customized binary masks for precise skin segmentations. Hasan et al. [4] introduced a hand sign language recognition system employing an Artificial Neural Network (ANN) to identify Bangla alphabets. The system extracted and processed the hand sign region from input images through normalization, skin color extraction, and binary conversion. Following this, features were extracted using the Freeman Chain Code (FCC), and the ANN was employed for training and classification. Talukder et al. [5] introduced a real-time BdSL recognition system that interpreted signs from video streams or images and produced textual sentences and speech. The study employed YOLOv4 as the object detection model, introducing three new signs for sentence generation. They used a dataset comprising 12.5k images of 49 classes, encompassing digits, Bangla alphabets, and the proposed signs. Islalm et al. [6] suggested a convolutional neural network model to identify digits and alphabet separately, along with their combined usage. The method was evaluated on their self-made dataset using 10-fold cross-validation. The outcomes demonstrated impressive average recognition accuracies of 99.83%, 100%, and 99.80% respectively, for both primary alphabets and digits, and their joint application. Ahmed et al. [7] investigated a BSL recognition method focused on the positioning of fingertips, employing relative tip positions of five fingers in a two-dimensional space and position vectors to train an artificial neural network (ANN) for recognize signs. The method's efficacy was assessed on a dataset consisting of 518 images depicting 37 signs, and it achieved an impressive recognition rate of 99%. Podder et al. [8] presented a deep transfer learning approach integrating ResNet18 and a color-coded fingertip pattern, enabling the accurate interpretation of the 37 static signs representing the one-handed Bangla sign letters. Training and validation of the model were executed with 5-fold cross-validation for 37-class classification, revealing its exceptional performance in recognizing BdSL signs. Shurid et al. [9] constructed a multi-modal system to identify bangla signs. They basically employed a Convolutional Neural Network model to recognize Bangla Sign Language(BdSL). Wangchuk et al. [10] proposed a Convolutional Neural Network (CNN) based model to recognize Bhutanese Sign Language (BSL) digits to minimize the communication gap between the deaf and the public in Bhutan. The system attained 97.62% training accuracy on a first-ever BSL dataset encompassing 20,000 sign images of 10 static digits collected from volunteers. Various SL models were contrasted with the proposed CNN model and the system's precision, recall, and F1-score were also evaluated.

Taking into account the weaknesses we observed in earlier studies, our objective was to propose a model that can overcomes these weaknesses and provide enhanced solutions.

16.3 METHODOLOGY

In our study, we have built our own dataset named 'Bornomala'. In Figure 16.1 we have shown the entire methodology of our research.

16.3.1 Dataset creation

To do any task the initial step is to collect the necessary data. As we followed, due to lack of sufficient data the previous research on Bangla Sign Language wasn't enough enriched. That's why we built a large dataset for 44 Bangla letters and 10 Bangla digit

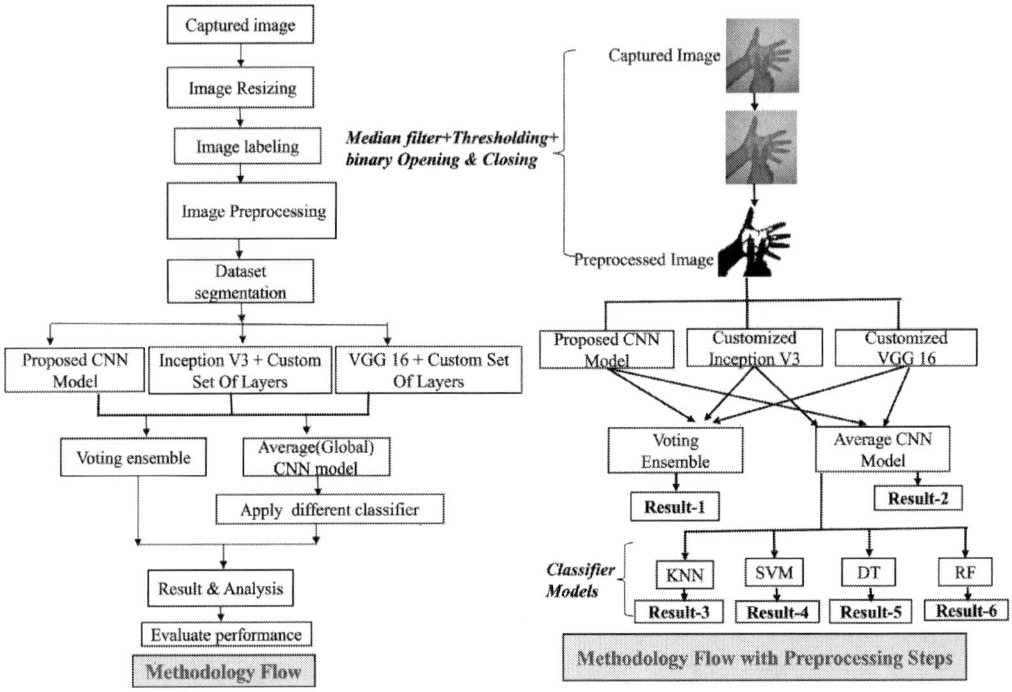

Figure 16.1: Overview of the Proposed Sign Language Recognition Scheme.

classes. Each symbol of a letter and digit has its specific gesture. These gestures act as a communication medium. For better performance to make a variety, we collected those gesture images from various people with various skin colors who helped us in this research. We captured those images in different angles with different light intensities then we applied zoom out, zoom in, right-handed, and left-handed combination to make enough variety. To capture images we used our self mobile device Huawei Y9 Prime Camera (16MP). Figure 16.2 displays the hand gestures corresponding to bangla digits. We have exhibited 44 Bangla characters' corresponding images in Figure 16.3.

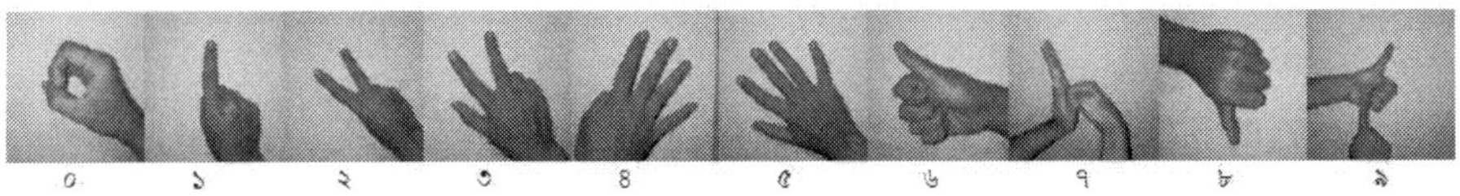

Figure 16.2: Hand Gestures Illustrating Bangla Numerals.

We have precisely executed the steps below in the creation of the model. Figure 16.4 showcases an illustration of the overall preprocessing steps.

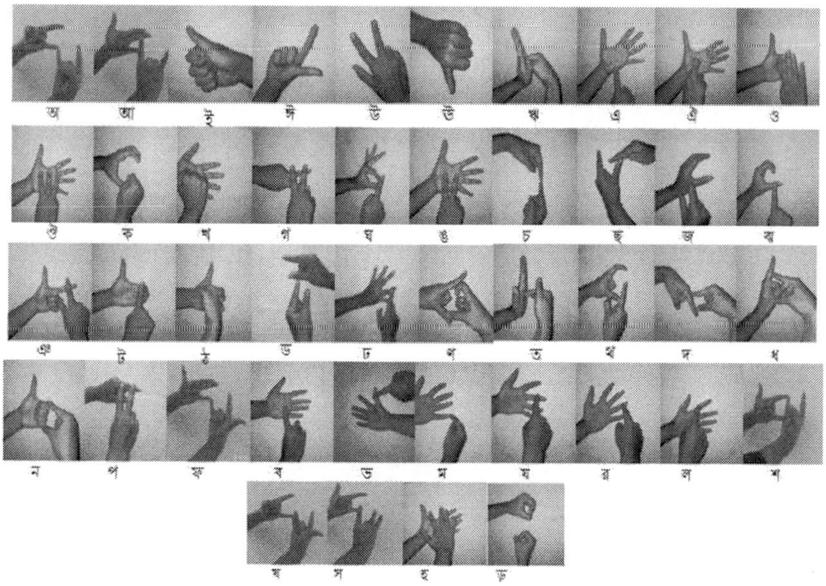

Figure 16.3: Hand Gestures for Bangla Alphabets.

16.3.2 Image Resizing

The images we have taken have had a size of 1600px by 1600px. To reduce time and space complexity we have transformed those images into 128px by 128px. This conversion assisted us in diminishing the size of the file as well as run the proposed model and took less amount of time for training.

16.3.3 Dataset Preprocessing

As raw images increase both time and space complexity, consequently affect the overall performance that's why before using CNN models we preprocessed the images. We applied five steps in pre-processing, those are (1) Image Laveling (2) Gray scaling (3) Thresholding (4) Median filter (5) Opening operation and (6) Closing operation.

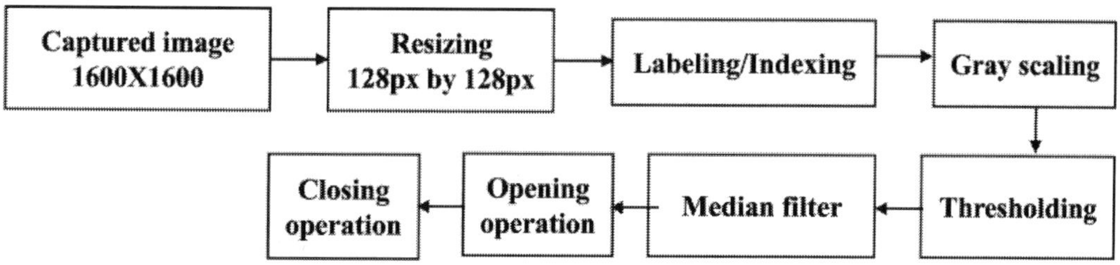

Figure 16.4: Workflow of Preprocessing Steps.

16.3.4 Image Laveling

In classification problems, data laveling/indexing is an important task. Data laveling is the approach for identification the raw data by adding suitable tags or levels to that data to specify it uniquely. Each of the images indicates a specific hand sign gesture. There are two segments of our classification problem one for alphabets and another one for digits. We tagged the 10 classes of digits from 0 to 9 (0, 1, 2, 3.....8, 9). For the characters section, we laveled the 44 classes with their associated index 0 to 43 (0, 1, 2, 3, 4...42, 43). The corresponding indices have been depicted in Figure 16.5.

Characters	0 – অ	1 – আ	2 – ই	3 - ঈ	4 – উ	5 – ঊ
	6 – ঋ	7 – এ	8 – ঐ	9 – ও	10 – ঔ	11 – ক
	12 – খ	13 – গ	14 – ঘ	15 – ঙ	16 – চ	17 – ছ
	18 – জ	19 – ঝ	20 – ঞ	21 – ট	22 – ঠ	23 – ড
	24 – ঢ	25 – ণ	26 – ত	27 – থ	28 – দ	29 – ধ
	30 – ন	31 – প	32 – ফ	33 – ব	34 – ভ	35 – ম
	36 – য	37 – র	38 – ল	39 – শ	40 – ষ	41 – স
	42 – হ	43 – য়				
Digits	0 – ১	1 – ২	2 – ৩	3 – ৪	4 – ৫	5 - ৬
	6 - ৭	7 - ৮	8 – ৯	9 - ১০		

Figure 16.5: Indexed Depiction of Alphabets and Numerals.

16.3.5 Gray Scaling

Each pixel in the RGB color image contains three color channels: red, green, and blue. To convert the image to grayscale, the intensity value of each pixel is calculated by combining the RGB values using a specific formula. We have used the luminosity method, which assigns different weights to the RGB channels according to human perception. The formula is typically as follows:

$$\text{Intensity} = (0.299 \times \text{R}) + (0.587 \times \text{G}) + (0.114 \times \text{B})$$

Where R, G, and B represent the red, green, and blue channel values of the pixel, respectively. As colors have no impact on hand sign gesture classification, grayscale image and colored has to carry the same amount of information. So grayscale image is more effective as it reduces the space complexity.

16.3.6 Thresholding

In our study, we have applied binary thresholding on grayscale images to convert grayscale images into binary images with only two distinct pixel values. It involves setting a threshold value and assigning pixel values above the threshold to one class (e.g., foreground or object) and pixel values below the threshold to another class (e.g., background or non-object). This simplification approach helped us in reducing the complexity of subsequent image processing and analysis tasks.

16.3.7 Median Filter

Applying a median filter in the image classification process can serve several purposes and provide benefits for improving the quality of the image data. This filter is generally effective for noise reduction and edge preservation. Images often contain various types of noise, such as random variations or unwanted artifacts. The median filter is effective in reducing different types of noise, including salt-and-pepper noise, where individual pixels are randomly set to extreme values. By replacing each pixel with the median value of its neighboring pixels, the median filter helps to smooth out noise and improve the overall quality and clarity of the image. It achieves noise reduction while retaining the sharpness and details of edges, making it particularly useful in image classification tasks where edge information is crucial for accurate classification.

16.3.8 Opening and Closing Operation

Opening and closing are essential morphological operations in image processing, used to improve grayscale images. The opening involves erosion to remove noise and dilation to preserve main features while closing performs dilation to fill gaps and erosion to retain main structures. These operations have been applied to grayscale images, facilitating noise removal, segmentation, and feature extraction tasks. By utilizing a structuring element, these operations have manipulated image pixels, playing a crucial role in enhancing images.

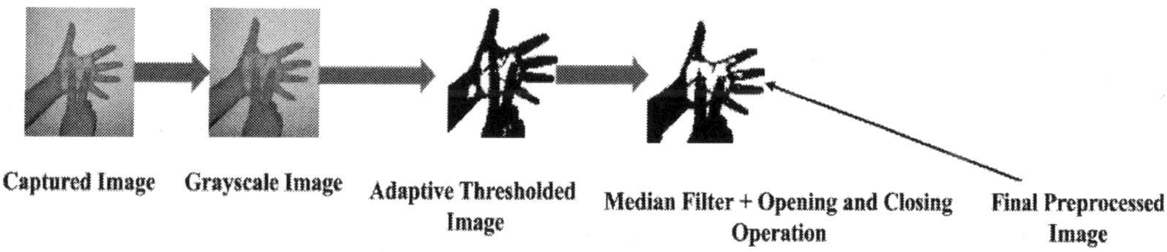

Figure 16.6: Image Preprocessing Workflow for Hand Sign Gestures.

16.3.9 Dataset Segmentation

We have split our dataset into two segments by keeping 80% data to train the model and 20% data to validate the model. For alphabet recognition, we have used a total of 4928 images for training the model and 1232 images for testing the model. For digit recognition, we have utilized a total of 1360 images for the training set and 340 images for the validation set.

Table 16.2: Dataset Overview.

Class	Total Class	Image per Class	Train Image	Test Image	Total Image
Digit	10	170	1360	340	1700
Letter	44	140	4928	1232	6160

16.4 MODEL CONSTRUCTION

In our study, we have employed three CNN models (1) Our Constructed CNN model (2) VGG16 Model and (3) Inception v3 model. Then we built a Global CNN model by averaging all of the three models. To build the Global CNN model we added an extra max-pooling layer, flatten layer, dropout layer and fully connected layer. We also applied a Voting Ensemble approach to the three CNN models.

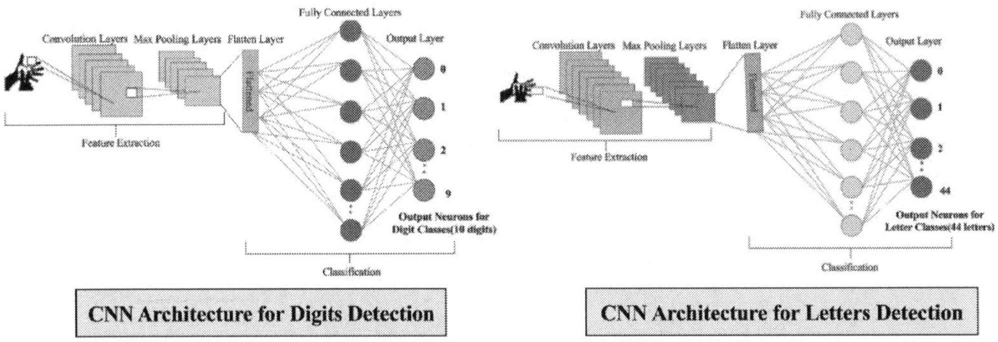

Figure 16.7: Our Proposed CNN Architecture for Sign Language Recognition.

16.4.1 Our Constructed CNN Model

To create our constructed CNN model, we used five 'convolutional layer' and stacked the layers. The key role of convolutional layers is the extraction of features. In the first convolutional layer, we used a total of 16 filters with a kernel size 3×3 and a stride size of 2. To activate the neurons unit we used ReLU activation function (Rectified Linear Unit). The key advantage of using ReLU function is that it is computationally inexpensive than sigmoid and tanh functions as it doesn't activate all the neuron units at the same time. At a time only a fixed amount of neuron units get triggered. The result of ReLU unit is 0 when the input values are less than 0, otherwise the output is 1. After that we used activation layer it discards all images which has ReLU units results less than zero and allowed only those images which has ReLU units value greater than zero. By this filtering way, it converted that to non-linear. Then we employed a max-pooling layer with a pool size 2×2 and stride size of 2. After that, we employed the second convolutional layer which has 32 filters with 3×3 kernel size and with stride size 2×2. Then we applied another max-pooling layer with size 2×2. By this way we applied the third, fourth, and fifth convolutional layers with 32,

64, and 128 filters respectively with the same size of kernel and stride. To decrease overfitting then we used a 50% dropout layer and applied three fully-connected layer with 128, 256, and 512 units of neurons respectively. After that we used another 25% dropout layer. Then we employed a flatten layer to convert the 2-dimensional arrays to 1-dimensional vector. At the end, in the output layer we used 10 units for digit recognition and 44 units for character recognition.

16.4.2 Customized VGG16

We have incorporated VGG16 using TensorFlow and Keras libraries. After that by adding an extra custom set of layers we have developed a customized VGG16 model. Initially, the deep features extracted from VGG16 have passed through a 'Max-pooling' layer with a pool size of 2 × 2 and 'ReLu' activation function. Following that, we have applied a dropout layer with a dropout rate of 10%. Then, we utilized three fully connected layers with 128, 256, and 512 unit neurons respectively, followed by another dropout layer with a 10% dropout rate. Next, a flattened layer was incorporated. Finally, an output layer was used with 'softmax' activation function. For digits and alphabets detection, there are a total of 10 and 44 neuron units in the output layer respectively.

16.4.3 Customized Inception v3

We have included Inception v3 using TensorFlow and Keras libraries. Then we have added an extra set of custom layers to develop a customized Inception v3 model. Firstly, the deep features extracted by Inception v3 have passed through a 'Max-pooling' layer of a pool size of 2 × 2 with the 'ReLu' activation function. Then we have utilized a dropout layer with a dropout rate of 25%. Then, we have utilized three fully connected layers with 128, 256, and 512 unit neurons respectively, followed by another dropout layer with a 10% dropout rate. Next, a flattened layer was incorporated. Lastly, an output layer has been employed with the 'softmax' activation function. For digits and alphabets detection, there are a total of 10 and 44 neuron units in the output layer of customized Inception V3 model respectively.

We have designed customized VGG16 and customized Inception v3 models for 20 epochs. As in our study, we have classified multiple classes, that's why we have employed 'softmax' activation function for the CNN models. To compile our model we have utilized the 'adam' optimizer associated with 'categorical cross-entropy' loss function. In Figure 16.8 we have shown the architecture of customized VGG16 and customized Inception V3 models.

16.4.4 Average CNN Model

Global CNN Model has been used to avoid overfitting and biased decision problems since we know a single model can take a biased decision. In this global model, we used the same optimizer, the same number of epochs that we also used in the previous models. To build this model we used extra max-pooling layer, flatten layer, dropout layer and fully connected layer. Then lastly we applied different machine

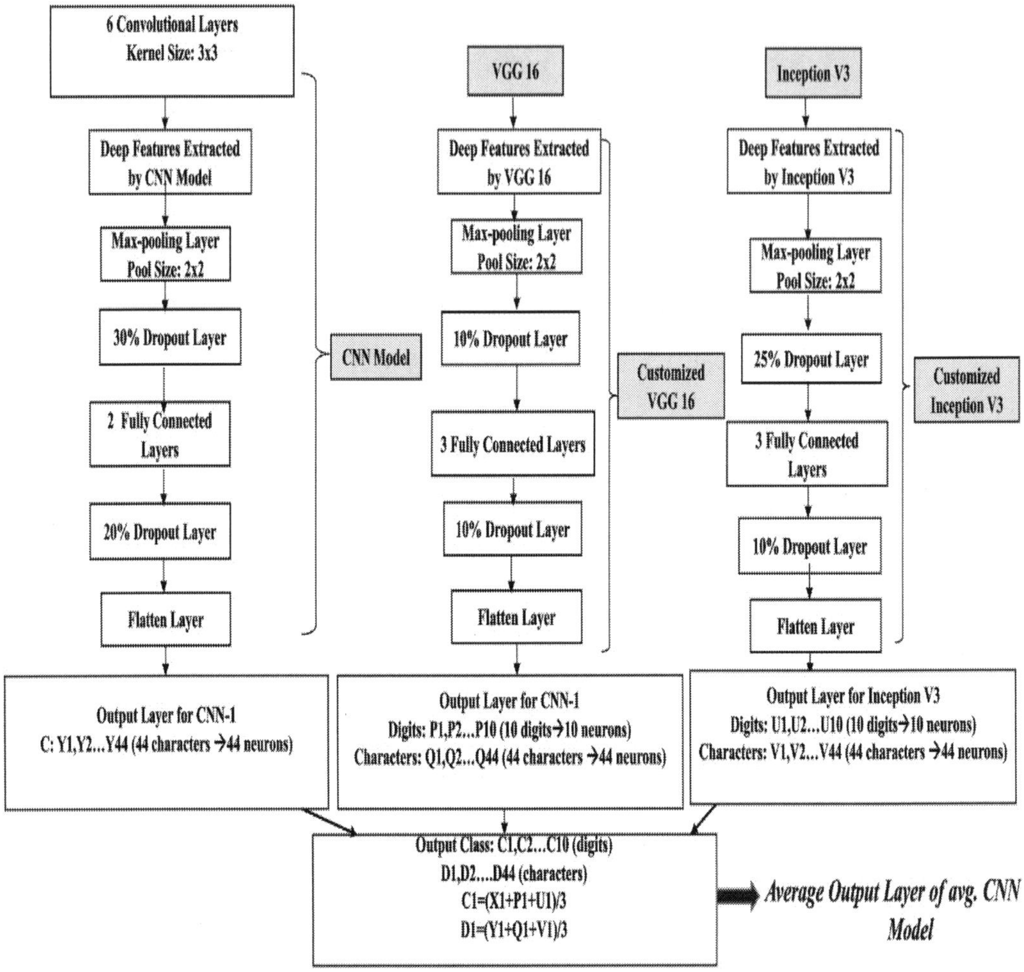

Figure 16.8: Average CNN Model Architecture.

learning classifiers for performance analysis. These classifiers are Decision tree (DT), K-Nearest Neighbour (KNN), Support Vector Machine (SVM), and Random Forest (RF).

Decision Tree is a graphical representation algorithm that constructs a tree of all feasible solutions of a decision. In the tree from the root node, the algorithm divides a dataset into segments and sub-segments. To determine the output value of a feature set, this algorithm uses the subset information. The convolutional layer functions as an input for this classifier model. We have deployed our decision tree model by using the Python scikit-learn toolkit.

K-Nearest Neighbour (KNN) is a lazy learner, non-parametric algorithm. This algorithm does not make any assumptions on the underlying data. It saves all available categorizes and cases. In KNN each data point that is neighbors to another belongs to the same class. Here number K represents how many nearest neighbor data points we will consider to predict the class of a test data point. The convolutional layer acts as an input for this model. Using the Python scikit-learn toolkit we implemented the KNN classifier. In our model for the k value 7 we got the best accuracy.

Support Vector Machine (SVM) classifier model is used to create decision boundary or hyperplane. This hyperplane defines the class of the data points. In this classifier, we used a linear activation function with '12' regularization. Implementation of the model is done by using the model using the 'Adam' optimizer and the 'squared hinge' loss function.

$$k(v) = \sum_{p=1}^{Q} \alpha_p b_p K(V, V_p) + t$$

α_p means individual weight

t means bias term

b_p means target

Q sum of number of vectors

V_p means support vectors

In this formula ,$\alpha_p b_p K(V, V_p)$ means the whole weight of each node

We got best hyperplane for RBF kernel hyperparameters C=1.4 and γ=0.2.C represents the penalty for a data points misclassification. Larger C value accepts smaller margin. Larger margin gives a better solution. Decision region determines by γ value. Radial Basis Function (RBF) creates a "bump" around each data point. Where, $K(Z, Z_j) = e^{-\gamma||Z-Z_j||^2}, \gamma > 0$

Random forest(RF) combines the output of multiple decision trees to produce a single result which is generally used to generate accurate predictions for bigger dataset to avoid overfitting. Here in our model 60 trees have been used.

16.4.5 Voting Ensemble

In the Voting Ensemble model we used VGG16 CNN models, our proposed CNN model, and Inception v3 model. After applying the three models, we got three different predicted classes. Then we applied the 'Majority Rule' to get a single prediction result. If two or all three models predict the same class then that class will be considered as the voting ensemble's predicted class. If all three models predict three different classes then the voting ensemble's predicted class will be the most accurate model's predicted class.

16.5 MODEL EVALUATION

In this study, firstly we divided our dataset into two segments one segment is for training and another segment is for validating the model. We used a total of 7860 images to build up our dataset. Then we took 80% of our whole data as the training set and 20% of our total data as the testing set. For characters, we used a total of 6160 images, and for digits, we used a total of 1700 images. According to data segmentation, for characters 4928 images were used for the training set, and 1232

Table 16.3: Performance of Various Models over the Alphabet and Numeral Dataset.

Model	Characters Accuracy	Digit Accuracy
Our Proposed CNN Model	86.10%	87.60%
Customized VGG16	83.90%	84.80%
Customized Inception v3	82.30%	83.50%
Average CNN Model	92.30%	93.40%
Average CNN Model + KNN Classifier	91.40%	91.80%
Average CNN Model + SVM Classifier	86.50%	88.40%
Average CNN Model + RF Classifier	91.10%	91.60%
Average CNN Model + Decision Tree	88.70%	89.40%
Voting Ensemble	93.80%	95.20%

images were used for the testing set. On the other hand, for digits 1360 images were used for the training set and 340 images were used for the testing set. We achieved a maximum of 97.5% validation accuracy for our digit dataset and 95.1% validation accuracy for our character dataset. Table 16.3 depicts all models and models with classifier accuracy for both the digit dataset and letter dataset.

After applying all models, we also applied various machine learning classifiers for performance evaluation of our dataset against those classifiers. We run total 30 epochs. We followed after a tiny amount of epochs, the accuracy was enhanced and training loss decreased this phenomenon indicated that our models trained perfectly against the dataset without occurring under-fitting, over-fitting problems. We calculated the accuracy by divides the total number of predictions by the number of right predictions. According to the Table 16.3 we followed among the all classifier KNN classifier did best as the average CNN and KNN classifier's accuracy is 94.20% and 95.60% for character recognition and Digit recognition respectively. We also followed among the classifier SVM performed worst. Among the models Average (Global) CNN model performed best and the second highest accuracy goes to our proposed model. Our proposed model gave 93.20% and 95.6% for character and digit accuracy respectively.

As shown in Table 16.3 our digits recognition accuracy always outperforms the characters recognition accuracy. Behind this occurrence there are some reasons. One of the reasons is that digit recognition hand gesture signs differ from each other on the other hand character recognition hand gesture signs has many similarity among them. Another reason is that the digit recognition system has less amount of class than character recognition system. After all our character's gesture images are double-handed so hand overlapping could also occur in the preprocessing stage.

Based on our accuracy Table 16.3, we can say that the Average CNN Model performed best in our dataset. The histogram depiction of the accuracy of various models for digit and character recognition have shown in Figure 16.9.

In our study, we have achieved satisfactory outcomes with different classifiers. In our digit dataset, the KNN model delivered an accuracy of 91.80%, followed by the Random Forest model at 91.60%, the Decision Tree model at 89.40%, and the SVM

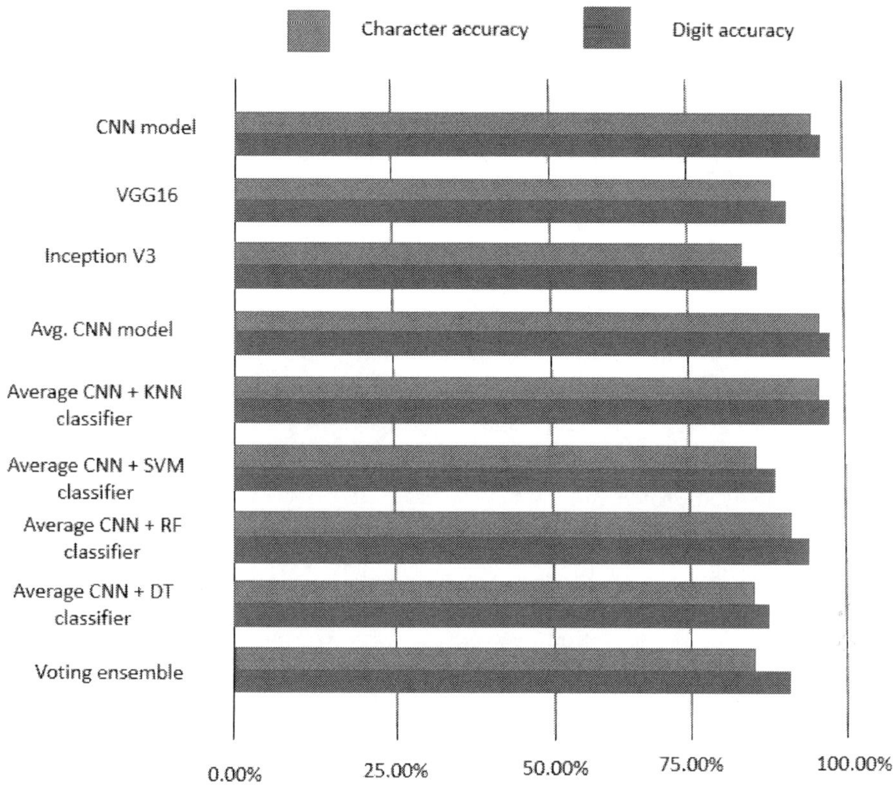

Figure 16.9: Comparison of Accuracy among Various Models.

model at 88.40%. On the other hand, the KNN model yielded an accuracy of 91.40% on our letter dataset, while the Random Forest model achieved 91.10%, the Decision Tree model attained 88.70%, and the SVM model obtained 86.50%.

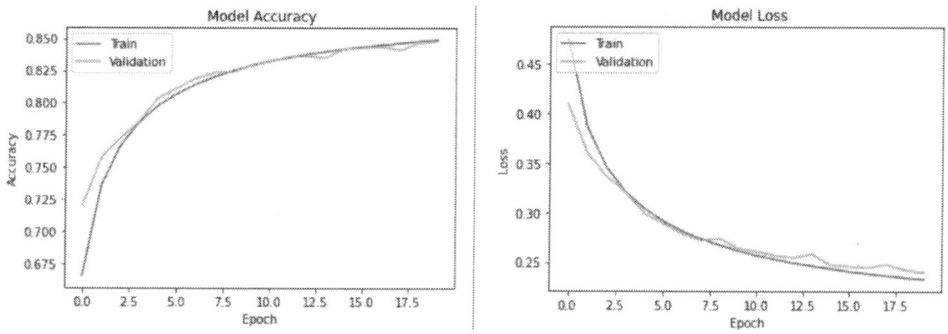

Figure 16.10: Voting Ensemble Models Training Accuracy and Training Loss against Our Letter Dataset over 20 Epochs.

Based on the aforementioned accuracy, it is evident that the digit dataset consistently outperforms the letter dataset in terms of accuracy. The difference in class quantity between sign digit recognition and sign letter recognition influences the superior accuracy achieved in the recognition process. Our model is effectively trained due to the generation of a dataset with sufficient diversity. Data preprocessing brings about substantial benefits that significantly affect the models training. We anticipate that our inclusive and large dataset in Bangla Sign Language Detection will be beneficial to researchers in the field.

16.6 CONCLUSION AND FUTURE WORKS

In our study, we generated our own Bangla sign language dataset. We created two datasets one for digit detection and another one for character recognition. For better accuracy we took those images in different light intensity, from different people with different skin colors to make a variety. We created a large dataset with sufficient amount images. We plan to make public our dataset so that in future researchers can enhance the research in the field of sign language detection. Our dataset accuracy outperforms all the existing model's accuracy. To enhance the performance in the future, we will try future to make large amount of variety in the images by taking images from more different people with different skin colors. We will also include more images in the dataset. Our model can't able to perform in real time. So in the future our aim is to establish a model with real-time detection capabilities. Another thing is that we created our dataset only for 44 different characters, but there is a total of 52 different characters in Bangla language, so in the future our aim is to extend our dataset with all 52 classes. We also didn't apply cross validation as we tried a lot to make our dataset balanced. So in the future, our aim is to apply cross validation techniques. Our aim is also to apply an object detection algorithm.

Bibliography

[1] Kozyra, K., Trzyniec, K., Popardowski, E., et al. Application for Recognizing Sign Language Gestures Based on an Artificial Neural Network. Sensors [Internet]. 2022;22:9864. Available from: http://dx.doi.org/10.3390/s22249864.

[2] M. A. Hossen, A. Govindaiah, S. Sultana and A. Bhuiyan, "Bengali Sign Language Recognition Using Deep Convolutional Neural Network," 2018 Joint 7th International Conference on Informatics, Electronics & Vision (ICIEV) and 2018 2nd International Conference on Imaging, Vision & Pattern Recognition (icIVPR), Kitakyushu, Japan, 2018, pp. 369–373, doi: 10.1109/ICIEV.2018.8640962.

[3] F. Sadik, M. R. Subah, A. G. Dastider, S. A. Moon, S. S. Ahbab and S. A. Fattah, "Bangla Sign Language Recognition with Skin Segmentation and Binary Masking," 2019 IEEE International WIE Conference on Electrical and Computer Engineering (WIECON-ECE), Bangalore, India, 2019, pp. 1–5, doi: 10.1109/WIECON-ECE48653.2019.9019931.

[4] M. M. Hasan, M. Khaliluzzaman, S. A. Himel and R. T. Chowdhury, "Hand sign language recognition for Bangla alphabet based on Freeman Chain

Code and ANN," 2017 4th International Conference on Advances in Electrical Engineering (ICAEE), Dhaka, Bangladesh, 2017, pp. 749–753, doi: 10.1109/ICAEE.2017.8255454.

[5] D. Talukder and F. Jahara, "Real-Time Bangla Sign Language Detection with Sentence and Speech Generation," 2020 23rd International Conference on Computer and Information Technology (ICCIT), DHAKA, Bangladesh, 2020, pp. 1–6, doi: 10.1109/ICCIT51783.2020.9392693.

[6] M. S. Islalm, M. M. Rahman, M. H. Rahman, M. Arifuzzaman, R. Sassi and M. Aktaruzzaman, "Recognition Bangla Sign Language using Convolutional Neural Network," 2019 International Conference on Innovation and Intelligence for Informatics, Computing, and Technologies (3ICT), Sakhier, Bahrain, 2019, pp. 1–6, doi: 10.1109/3ICT.2019.8910301.

[7] S. T. Ahmed and M. A. H. Akhand, "Bangladeshi Sign Language Recognition using fingertip position," 2016 International Conference on Medical Engineering, Health Informatics and Technology (MediTec), Dhaka, Bangladesh, 2016, pp. 1–5, doi: 10.1109/MEDITEC.2016.7835364.

[8] Podder, Kanchon Kanti & Chowdhury, Muhammad & Mahbub, Zaid & Kadir, Muhammad. (2020). Bangla Sign Language Alphabet Recognition Using Transfer Learning Based Convolutional Neural Network. The Bangladesh journal of scientific research. 31–33. 20–26.

[9] S. A. Shurid et al., "Bangla Sign Language Recognition and Sentence Building Using Deep Learning," 2020 IEEE Asia-Pacific Conference on Computer Science and Data Engineering (CSDE), Gold Coast, Australia, 2020, pp. 1–9, doi: 10.1109/CSDE50874.2020.9411523.

[10] Karma Wangchuk, Panomkhawn Riyamongkol, Rattapoom Waranusast, Real-time Bhutanese Sign Language digits recognition system using Convolutional Neural Network, ICT Express, Volume 7, Issue 2, 2021, Pages 215–220, ISSN 2405–9595, https://doi.org/10.1016/j.icte.2020.08.002 (https://www.sciencedirect.com/science/article/pii/S2405959520301685)

[11] M. Sanzidul Islam, S. Sultana Sharmin Mousumi, N. A. Jessan, A. Shahariar Azad Rabby and S. Akhter Hossain, "Ishara-Lipi: The First Complete MultipurposeOpen Access Dataset of Isolated Characters for Bangla Sign Language," 2018 International Conference on Bangla Speech and Language Processing (ICBSLP), Sylhet, Bangladesh, 2018, pp. 1–4, doi: 10.1109/ICBSLP.2018.8554466.

[12] M. Sanzidul Islam, M. S., Mousumi, S. S. S., Jessan, N. A., Rabby, A. S. A., Abujar, S., Hossain, S. A. (2019). Ishara-Bochon: The First Multipurpose Open Access Dataset for Bangla Sign Language Isolated Digits. In: Santosh, K., Hegadi, R. (eds) Recent Trends in Image Processing and Pattern Recognition. RTIP2R 2018. Communications in Computer and Information Science, vol 1035. Springer, Singapore.

[13] D. Talukder, F. Jahara, S. Barua and M. M. Haque, "OkkhorNama: BdSL Image Dataset For Real Time Object Detection Algorithms," 2021 IEEE Region 10 Symposium (TENSYMP), Jeju, Korea, Republic of, 2021, pp. 1–6, doi: 10.1109/TENSYMP52854.2021.9550907.

[14] S. N. Hasan, M. J. Hasan and K. S. Alam, "Shongket: A Comprehensive and Multipurpose Dataset for Bangla Sign Language Detection," 2021 International Conference on Electronics, Communications and Information Technology (ICECIT), Khulna, Bangladesh, 2021, pp. 1–4, doi: 10.1109/ICECIT54077.2021.9641137.

[15] Bhavana, D., Kumar, K. K., Chandra, M. B., Bhargav, P. S. K., Sanjanaa, D. J. and Gopi, G. M., 2021. Hand sign recognition using CNN. International Journal of Performability Engineering, 17(3), pp.314–321.

[16] Molchanov, P., Gupta, S., Kim, K. and Kautz, J., 2015. Hand gesture recognition with 3D convolutional neural networks. In Proceedings of the IEEE conference on computer vision and pattern recognition workshops (pp. 1–7).

[17] Chen, L., Fu, J., Wu, Y., Li, H. and Zheng, B., 2020. Hand gesture recognition using compact CNN via surface electromyography signals. Sensors, 20(3), p.672.

[18] Nagi, J., Ducatelle, F., Di Caro, G. A., Cireşan, D., Meier, U., Giusti, A., Nagi, F., Schmidhuber, J. and Gambardella, L. M., 2011, November. Max-pooling convolutional neural networks for vision-based hand gesture recognition. In 2011 IEEE international conference on signal and image processing applications (ICSIPA) (pp. 342–347). IEEE.

[19] Breland, D. S., Dayal, A., Jha, A., Yalavarthy, P. K., Pandey, O. J. and Cenkeramaddi, L. R., 2021. Robust hand gestures recognition using a deep CNN and thermal images. IEEE Sensors Journal, 21(23), pp.26602–26614.

[20] D. Perera, B. Kanchana, R. Peiris, K. Madushan and D. Kasthurirathna, "Review On Hand Gesture Recognition for Bengali Sign Language," 2022 2nd International Conference on Advanced Research in Computing (ICARC), Belihuloya, Sri Lanka, 2022, pp. 194–199, doi: 10.1109/ICARC54489.2022.9753791.

[21] Zhang, Z., Tian, Z. and Zhou, M., 2018. Latern: Dynamic continuous hand gesture recognition using FMCW radar sensor. IEEE Sensors Journal, 18(8), pp.3278–3289.

[22] Tan, Y.S., Lim, K.M., Tee, C., Lee, C.P. and Low, C.Y., 2021. Convolutional neural network with spatial pyramid pooling for hand gesture recognition. Neural Computing and Applications, 33, pp.5339–5351.

[23] Mujahid, Abdullah, et al. "Real-time hand gesture recognition based on deep learning YOLOv3 model." Applied Sciences 11.9 (2021): 4164.

[24] Ozcan, Tayyip, and Alper Basturk. "Transfer learning-based convolutional neural networks with heuristic optimization for hand gesture recognition." Neural Computing and Applications 31 (2019): 8955–8970.

[25] Hu, Z., Hu, Y., Liu, J., Wu, B., Han, D. and Kurfess, T., 2018. 3D separable convolutional neural network for dynamic hand gesture recognition. Neurocomputing, 318, pp.151–161.

[26] Kim, Seo Yul, et al. "A hand gesture recognition sensor using reflected impulses." IEEE Sensors Journal 17.10 (2017): 2975–2976.

[27] Devineau, Guillaume, et al. "Deep learning for hand gesture recognition on skeletal data." 2018 13th IEEE International Conference on Automatic Face & Gesture Recognition (FG 2018). IEEE, 2018.

[28] Nunez, Juan C., et al. "Convolutional neural networks and long short-term memory for skeleton-based human activity and hand gesture recognition." Pattern Recognition 76 (2018): 80–94.

[29] C. Wang, H. Mark Liao, Y. Wu, P. Chen, J. Hsieh and I. Yeh, "CSPNet: A New Backbone that can Enhance Learning Capability of CNN", 2020 IEEE/CVF Conference on Computer Vision and Pattern Recognition Workshops (CVPRW), 2020. Available: 10.1109/cvprw50498.2020.00203 .

[30] Napier, J. Sign Language Interpreter Training, Testing, and Accreditation: An International Comparison. Am. Ann. Deaf. 2004, 149, 350–359.

[31] Tasmere2020BanglaSD, Bangla Sign Digits: A Dataset For Real Time Hand Gesture Recognition, Dardina Tasmere and Boshir Ahmed and Md. Mehedi Hasan, 2020 11th International Conference on Electrical and Computer Engineering (ICECE), 2020, 186–189.

[32] A. Rahman, A. Mahmud and P. C. Shill, "A CNN Ensemble Approach to Provide Non-combat and Support Roles Training Deaf and Dumb Individuals using Hand Signs," 2023 14th International Conference on Computing Communication and Networking Technologies (ICCCNT), Delhi, India, 2023, pp. 1–7, doi: 10.1109/ICCCNT56998.2023.10307387.

[33] A. Rahman, S. Zaman, S. Parvej and H. M. Abdul Fattah, "Heart Disease Prediction Using Ensemble Solutions and Exploring the Effect of Target Engineering with Pearson Correlation Based Feature Selection," 2024 6th International Conference on Electrical Engineering and Information & Communication Technology (ICEEICT), Dhaka, Bangladesh, 2024, pp. 1407–1412, doi: 10.1109/ICEEICT62016.2024.10534517.

[34] A. Rahman, S. Zaman, S. Parvej and H. M. Abdul Fattah, "Advancements in Breast Cancer Diagnosis: Integrating Classifier Algorithms, Neural Network and Ensemble Learning with PCA, VIF for Feature Selection and Dimensionality Reduction," 2024 6th International Conference on Electrical Engineering and Information & Communication Technology (ICEEICT), Dhaka, Bangladesh, 2024, pp. 893–898, doi: 10.1109/ICEEICT62016.2024.10534427.

[35] A. Rahman, S. Zaman and D. Das, "Cracking the Genetic Codes: Exploring DNA Sequence Classification with Machine Learning Algorithms and Voting Ensemble Strategies," 2024 International Conference on Advances in Computing, Communication, Electrical, and Smart Systems (iCACCESS), Dhaka, Bangladesh, 2024, pp. 1–6, doi: 10.1109/iCACCESS61735.2024.10499483.

[36] A. Rahman, A. Mahmud and P. Chandra Shill, "Neural Network-based Approach to Predict Protein Secondary Structure," 2023 2nd International Conference on Applied Artificial Intelligence and Computing (ICAAIC), Salem, India, 2023, pp. 135–140, doi: 10.1109/ICAAIC56838.2023.10140404.

[37] D. Das, M. J. Hasan, S. Nahid Hasan, A. Rahman and K. M. Rokibul Alam, "Secure Land Purchasing using Different Multi-Party Skyline Queries with Anonymous Information," 2023 26th International Conference on Computer and Information Technology (ICCIT), Cox's Bazar, Bangladesh, 2023, pp. 1–6, doi: 10.1109/ICCIT60459.2023.10441132.

[38] N. Kamal, F. Sarker, A. Rahman, S. Hossain and K. A. Mamun, "Recommender System in Academic Choices of Higher Education: A Systematic Review," in IEEE Access, vol. 12, pp. 35475–35501, 2024, doi: 10.1109/ACCESS.2024.3368058.

[39] A. Rahman, S. Parvej, K. S. Alam and H. M. Abdul Fattah, "Optimizing SMS Spam Detection: Comparative Analysis of Hybrid Voting Ensembles and Bi-LSTM Networks with Stratified Cross-Validation," 2024 5th International Conference on Data Intelligence and Cognitive Informatics (ICDICI), Tirunelveli, India, 2024, pp. 1030–1035, doi: 10.1109/ICDICI62993.2024.10810777.

CHAPTER 17

Fuzzy-based Model for Perceived Value and Customer Satisfaction in Fine Dining

Dipanjali Kundu

Dept. of Computer Science and Engineering, National Institute of Textile Engineering and Research (NITER), Constituent Institute of the University of Dhaka, Savar, Dhaka, 1350, Bangladesh.

Diganta Das

Jahangirnagar University, Dept. of Statistics, Dhaka, Bangladesh.

Tanoy Debnath

Green University, Dept. of CSE, Dhaka, Bangladesh.

Anichur Rahman*

Dept. of Computer Science and Engineering, National Institute of Textile Engineering and Research (NITER), Constituent Institute of the University of Dhaka, Savar, Dhaka, 1350, Bangladesh.

CONTENTS

DOI: 10.1201/9781003605508-17

Fuzzy Logic is a leading method for the analysis of sentiment from linguistic terms. The development of Fuzzy logic tools in the field of study of Fuzzy has been widespread. In the field of recommendation systems for the food industry, restaurant, university subject ranking or university ranking, it can be used to reach the delicate balance between human reasoning and predicting the output. In this proposal, we are concerned about the growing industry of restaurants. As with the proliferation of different formats of online expressions, such as reviews and ratings, the recommendation process in the competitive world of business is becoming more complicated. It is difficult to determine the truc intent of the customer because of the large number of false reviews available on the internet that hinder the way of actual ratings of a service. To solve this issue and analyze the actual feedback of the customers to provide a good review, soft computing can be best suited. In this regard, the aim of this study is to analyze the factors that may have an impact on the perceived value and customer satisfaction in the fine dining system. A fuzzy-based model has been proposed in this study to build a recommendation system. To see a comparative stand and to choose the best one, different membership functions and defuzzification methods are employed, and a comparative analysis has been provided based on the results obtained from the study. A comprehensive example is provided to illustrate the decision procedure.

17.1 INTRODUCTION

The process of collecting, examining, and making decisions based on consumer data is known as data-driven innovation (DDI) [1]. Emerging technologies like blockchain, cloud computing, artificial intelligence, data analytics, and the Internet of Things provide a wealth of data that companies may use to develop new products or enhance old ones [2]. Besides, we cannot imagine those online services without the use of recommendation systems. In a nutshell, recommendation systems are algorithms designed to suggest relevant items to users [3]. The recommendation system is beneficial to both users and service providers because it reduces the transaction costs associated with selecting and finding items from any web service. It aids in increasing sales, making sound decisions in online transactions, and rearranging the user's experience with web services [4], [5].

Data and analytics are regarded as the primary components of innovation and the development of new services, products, or business models from the perspective of academic research [6]. The literature has also emphasized how crucial innovation is to an organization's ability to maintain a competitive advantage [7]. Therefore, a consumer-centric perspective is essential to ensure the commercial success of innovations, emphasizing the idea that an innovation's success is ultimately determined by the end user [8]. Again, the ability of a person, group, corporation, or even an entire economy to innovate, adapt, produce a new product or service, or apply a new idea is referred to as innovativeness [9]. Although innovation is valued across all industries, studies on innovation have rarely focused on the service sector [10]. Finally, there are few studies that look at how DDI is used in fine dining establishments to increase value for patrons and strengthen competitive advantages.

There are various techniques for making recommendations, such as content-based filtering, collaborative filtering hybrid filtering [11], [12], and so on. Collaborative filtering is the most commonly used technique because it recommends items by identifying other users who have similar tastes. In content-based filtering [13], a user profile is created based on his or her browsing history or data provided by the user, and then likable contents are suggested to the user. Content-based filtering, collaborative filtering, and other approaches are combined in hybrid filtering [14]. However, most recommendation systems work with a binary value of 0 or 1 [15]. That is, after filtering the contents and performing calculations, If the value becomes 1, the item will be suggested; otherwise, it will not. However, there may be values ranging from 0 to 1 that define how close the content is to being suggested. Then came fuzzy logic to define the values ranging from 0 to 1. It has been widely used in the design of a recommender system to more accurately suggest items and to deal with uncertainty, impreciseness in item features, and user behaviour.

The design of the complex system has been simplified by leveraging the explicability of the FL system, and additionally, the exploration of different rule-based methods has also been simplified and applied in different research fields [16], [17]. In the sector of prediction or recommendation, there are many ways like, ANN [18], FIS [19], ANFIS [20], and so on. Among them, FIS is the simplest way to do human reasoning that resembles human thinking [21]. For the evaluation of the customer's performance expectancy, a fuzzy-based system has been proposed. This section presents the details of the proposed fuzzy logic-based model. Figure 17.1 describes the framework of a fuzzy logic-based model. Here the input data enters a block that denotes fuzzification, resulting in a fuzzy set that can be used for subsequent processing. We designed the rule base using the Mamdani inference system. The output of this inference system is the fuzzy output which needs to be defuzzfied using a defuzzification process that converts the fuzzy set to a crisp set. In this study, we have applied several defuzzification methods; COA, BOA, MOM, LOM, and SOM. To summarize, the contributions of this paper are as follows:

- We provide an FLS based model for the analysis of the impact of customer satisfaction and the perceived value on the intention of customers to recommend a restaurant.

- We analyze the impact of the factors such as; food quality, facility comfort, cleanliness, and timelines on perceived value and customer satisfaction.

- We formulate 27 rules for the three cases to compute the sentiment of a user in the review system and finally make a comparative analysis with the aid of different defuzzification methods.

Organization: The rest of the paper has been organized as follows: We discuss the related work in Section 1.2. Section 1.3 presents the system framework for fine dining recommendations with fuzzy-based case analysis. Then, the result analysis and Discussion have been covered in Section 1.4. Finally, this paper concludes with Section 1.5.

17.2 RELATED WORK

In recent decades, there has been a growing interest in using Fuzzy techniques for different recommendation systems, particularly in the area of fine dining applications. In this section, we review some of the literature work on Fuzzy-based systems that use the dining system of customers.

According to Leppäniemi [22], customer-perceived value has a major influence on customer satisfaction. According to Palominos et al. [23], it is important to include the consumer perspective when managing performance metrics. The problem of quantifying perceived value has emerged from recent attempts to do so. In this regard, Trigos et al. [24] show that by employing approaches to enhance the performance of operational indicators that the customer may directly affect, it is also possible to enhance the customer's perception of the value that was supplied. The fuzzy inference system (FIS), first proposed by Zadeh [26], has found widespread use in multicriteria decision-making due to its capacity to model uncertainty [27] and simulate human thinking using fuzzy if-then rules [28]. Restaurants are an instance of the customer service industry. As a result, we must analyze its characteristics, such as intangibility, perishability, inseparability, and heterogeneity.

On the other hand, the authors of this [29] worked to improve ubiquitous hotel recommendations for online applications. They created the FUTCHR system, which is a fuzzy widespread traveler clustering and recommendation system for hotels. This system grouped travellers based on their decisions and past data. They then solved it using a fuzzy mixed binary-nonlinear programming model. In this [30], the authors worked to create a Fuzzy-Based Recommendation System using a fuzzy inference engine, which is a decision support system that assists students in choosing their major subject at university. They used a clustering-based technique to determine the students' preferred majors. Moreover, the authors of this [31] worked to recommend a framework that helps users monitor their calories based on their BMI and also provides food suggestions based on their history and preferences. This recommender framework is built using personal preferences, Collaborative Filtering, and Fuzzy logic. Users can also use the Android application or Pedometer to track their steps or workouts.

In summary, we can see there are many related works about fuzzy systems used in different sectors. But there is very little evidence of fuzzy-based methods for recommending fine dining. In this regard, we have proposed a fuzzy-based system to analyze the recommendation procedure of a restaurant depending on customer satisfaction, perceived value, and behavioural intention.

17.3 PROPOSED METHODOLOGY FOR FINE DINING RECOMMENDATION THROUGH FUZZY SYSTEM

To recommend the perceived value and customer satisfaction more effectively, the authors present a fuzzy logic-based architecture as depicted in Figure 17.1. The first task required for the framework is converting crisp input into a fuzzy input set. The authors chose two scenarios for this work., the first step is to determine the perceived value based on the quality of food served, the comfort facilities available, and the time it takes to serve food, which is measured from the time the order is placed to the time the food is served on the table. The second step is to determine customer satisfaction based on the following factors: food quality, facility comfort, and cleanliness. The inputs are linguistic forms that are represented as a set of fuzzy with membership degrees. The triangular membership function was used in this work to compute MF. Then the logic needs to be set up that resembles human reasoning and comparatively, the IF-THEN form of rules is considered to be the simplest way of expressing human thinking. In the Fuzzy logic-based system, the antecedents, as well as the consequents, are the linguistic variables and this characteristic makes the FIS system most useful for building a complex system. For simplicity, this work considered the Mamdani FIS method, which uses the IF-THEN format to build the rules, among the three methods of inference system.

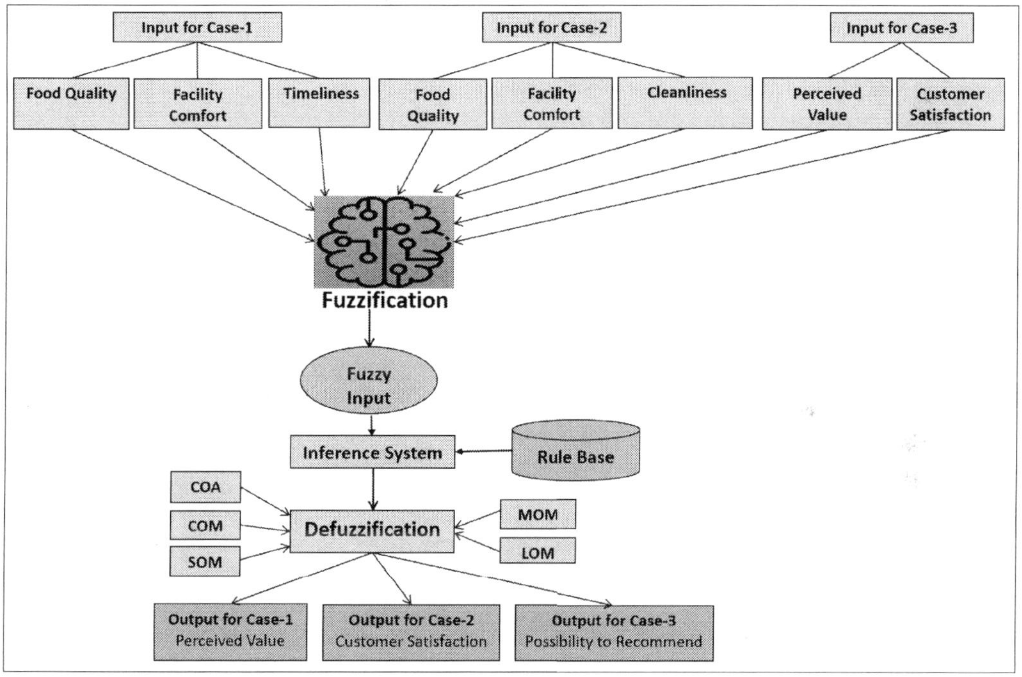

Figure 17.1: Proposed Model for recommendation system of Fine Dinning

17.3.1 Hardware and Software used

We have used MATLAB Fuzzy Logic toolbox for the implementation of the code. The system specification is depicted in Table 17.1.

17.3.2 Development Process of Fuzzy System

For the formulation of the Fuzzy Rule Base system, we have used the popular fuzzy inference method, which is known as Mamdani Model. In order to develop a fuzzy

Table 17.1: Environment parameters Setup based on technology

	Parameter	Value
Server	Platform	MATLAB Fuzzy logic Toolbox
	Operating System	Windows 10
	Hardware	Intel Core i5 (10th Gen)
		8GB DDR4 3200 Mhz RAM

Table 17.2: Input and output variable for Case-1

Input/Output	Linguistic Variable	Linguistic Values
Input	Food Quality	High, Low or Medium
	Facility Comfort	Poor, Acceptable or Amazing
	Timeliness	Poor, Medium, Fast
Output	Perceived Value	Very low, Low, Medium, High, Very high

system, the first step is to recognize the input and output linguistic terms, as well as their associated values (as depicted in Table 17.2 and Table 17.3). After that, identification of fuzzy sets and membership functions. After designing the rule base, the defuzzification method is applied to obtain the output crisp value. This process is simply the opposite of the fuzzification procedure.

Table 17.3: Input and output variable for Case-2

Input/Output	Linguistic Variable	Linguistic Values
Input	Food Quality	High, Low or Medium
	Facility Comfort	Poor, Acceptable or Amazing
	Cleanliness	Poor, Acceptable or Amazing
Output	Customer Satisfaction	Satisfied, Neutral, Dissatisfied

For evaluating the performance expectancy, we have subdivided our task into three parts as follows:

- Case:1 The Impact of Service Quality on Consumer Perception
- Case:2 The effect of service quality and perceived value on customer satisfaction
- Case:3 The impact of perceived value and customer satisfaction on behavioral intention.

Data set for Case-1: According to the analysis of [32] the factors in the selected factors to analyze the perceived values are: **Food Quality** that has three linguistic variables High with a range of 6-10, Medium 2-8, and Low.**Facility Comfort** with linguistic variable Amazing (6-10), Acceptable (2-8), and Poor (0-4). **Timeliness** with linguistic variables Fast (6-10), Medium (2-8), and Poor (0-4). Another variable **Perceived Value** is used for the measurement of customers' behavioral intention based on the input variables that have the range of 0-10 with five linguistic terms; Very low, Low, Medium, High, and Very High.

Data set for Case-2: Further, the factors selected to analyze the Customer satisfaction values are: **Food Quality** that has three linguistic variables High with a range of 6–10, Medium 2–8 and Low. **Facility Comfort** with linguistic variable Amazing (6–10), Acceptable (2–8) and Poor (0–4). **Cleanliness** with linguistic variables Amazing (6–10), Acceptable (2–8), and Poor (0–4). Another variable **Customer Satisfaction** is used for the measurement of customers' behavioral intention based on the input variables that have the range of 0–10 with three linguistic terms; Satisfied, Neutral, and Dissatisfied.

Data set for Case-3: According to the study in [32] customer satisfaction and perceived value, both have a positive impact on the intention of a customer to recommend a place or not recommend it. Sometimes there might be some cases where the food quality and price of the food do not match rather may unreasonable. In that case the perceived value of a customer changes that impact the recommendation procedure. They might be satisfied with the food or the place but the perceived value may restrict them to recommend the place with a good score. In this scenario, this section presents the input and output variables as shown in Table 17.4 and values for the analysis of the behavioral intention of a person based on two factors perceived value and customer satisfaction.

Table 17.4: Input and Output variable for Behavioural Intention

Input/Output	Linguistic Variable	Linguistic Values
Input	Perceived Value	Very low, Low, Medium, High, Very high
	Customer Satisfaction	Satisfied, Neutral, Dissatisfied
Output	Possibility to Recommend	Will not recommend, less possibility of Recommendation and Will Recommend

17.3.3 Representation of Considering Fuzzy Membership Function

Facility Comfort, Food Quality, Cleanliness has impact on the perceived value. Customer's perceived value changes when the place is not clean or the arrival of food is not on time. On the other customer satisfaction rely **Facility Comfort, Food Quality, Cleanliness** in this sector timeliness does not have that high impact because in a fine dining system, the service and the quality matter the most [32].

17.3.4 Fuzzy Rule Base Formulation

In this work, we have formulated fuzzy rules separately for the three cases.

Case-1: As there are three membership variables, each having three values, we have formed a rule base with 3 * 3 * 3 forming twenty-seven rules. For example: If Food Quality is High, Facility Comfort is Amazing and Timeliness is Fast then Perceived Value is Very High. The total rule base is shown in Table 17.5.

Case-2: As there are three membership variables each having three values, we have formed a rule base with 3 * 3 * 3 forming twenty-seven rules. For example: If

Food Quality is High, Facility Comfort is Amazing and Cleanliness is Amazing then the Customer is Satisfied. The total rule base is shown in Table 17.6.

Case-3: As discussed in the previous section, in this case, there are two input variables having five and three input values, so the total rules will be 3* 5, which means fifteen rules will be formed in this scenario as depicted in Table 17.7. For example: **If Customer Satisfaction is Neutral and Perceived Value is Medium then Customer intention is Less possibility to recommend**

Table 17.5: Fuzzy Rules for Case-1

Food Quality	**Facility Comfort**		
	Poor	Acceptable	Amazing
	1. Timeliness: Poor		
Low	Very Low	Very Low	Low
Medium	Low	Low	Medium
High	Medium	Medium	High
	2. Timeliness: Medium		
Low	Low	Medium	Medium
Medium	Low	High	High
High	Medium	High	Very High
	3. Timeliness: Fast		
Low	Low	Medium	Medium
Medium	Medium	High	Very High
High	Medium	Very High	Very High

Table 17.6: Fuzzy Rules for Case-2

Food Quality	**Facility Comfort**		
	Poor	Acceptable	Amazing
	1. Cleanliness: Poor		
Low	Dissatisfied	Dissatisfied	Dissatisfied
Medium	Dissatisfied	Dissatisfied	Neutral
High	Neutral	Neutral	Satisfied
	2. Cleanliness: Acceptable		
Low	Dissatisfied	Neutral	Neutral
Medium	Dissatisfied	Satisfied	Satisfied
High	Neutral	Satisfied	Satisfied
	3. Cleanliness: Amazing		
Low	Dissatisfied	Neutral	Neutral
Medium	Neutral	Satisfied	Satisfied
High	Neutral	Satisfied	Satisfied

17.3.5 Rule Aggregation and Defuzzification

Rule aggregation is the step of obtaining the output membership function. After the aggregation procedure then comes the defuzzification step. It is basically the opposite of the fuzzification process and produces crisp output that a user can understand. In this process of producing crisp output fuzzy sets along with corresponding membership degrees are provided and based on these the crisp output is produced. For the

Table 17.7: Fuzzy Rules for Case-3

Customer Satisfaction	**Perceived Value**				
	Very Low	Low	Medium	High	Very High
Dissatisfied	Will not Recommend	Will not Recommend	Will not Recommend	Less possibility to Recommend	Less possibility to Recommend
Neutral	Will not Recommend	Will not Recommend	Less possibility to Recommend	Will Recommend	Will Recommend
Satisfied	Less possibility to Recommend	Less possibility to Recommend	Will Recommend	Will Recommend	Will Recommend

sake of achieving the goal in this work, the authors proposed five types of defuzzification methods (COA, BOA, MOM, LOM, SOM).

COA Method This is the most widely used defuzzification method, also known as the Centroid method. The COA is calculated using the following equation 1 after taking into account the range of output linguistic variables under the membership function.

$$COA(x) = \frac{\int_{min}^{max} \mu_A(x)dx}{\int_{min}^{max} \mu_A(x)dx} (1)$$

BOA Method In this method, the actual task is done by dividing the total fuzzy sets into 2 equal regions as shown in Equation 2.

$$\int_{min}^{x} \mu_A(x)dx = \int_{x}^{max} \mu_A(x)dx (2)$$

Mean of Maximum Method In this method, the defuzzified output is the element with the largest MF value but in case of more than one having maximum, it calculates the mean value for finding the output.

$$MOM(x) = \frac{\int_M x_i dx}{|Mean|} (3)$$

17.4 RESULTS ANALYSIS AND DISCUSSION

To obtain the final result this model is divided into three parts. First, the perceived value is calculated then customer satisfaction is evaluated based on the factors of facility comfort, timeliness, cleanliness, and food quality. After the defuzzification of both cases, the defuzzified output enters the final stage. Then further processing is continued where based on the perceived value and customer satisfaction level the intention of a person to recommend a place is observed. For the analysis of customer satisfaction first, this method generated the membership values and presented them graphically for better visualization.

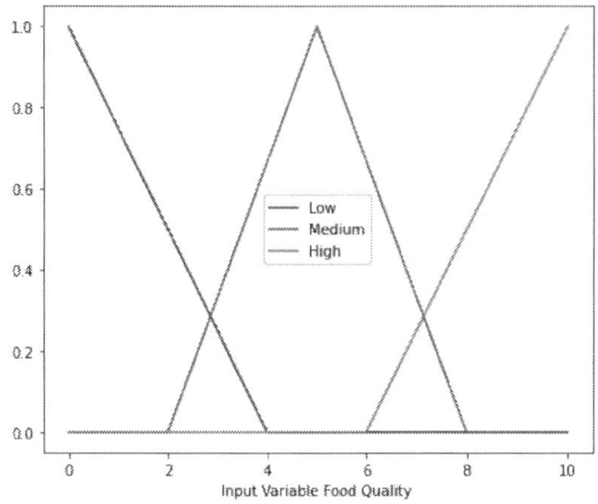

Figure 17.2: Input Membership Function: Food Quality for Case-1

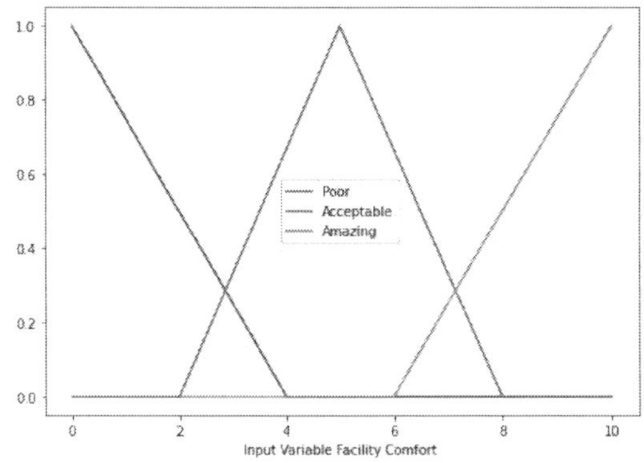

Figure 17.3: Input Membership Function: Facility Comfort for Case-1

Case-1: In this scenario, the input and output membership functions are generated to analyze the impact of the factors on Perceived Value. As depicted in Figure 17.2, Input Membership Function: Facility Comfort, the membership function of the input variable facility comfort is shown in terms of Poor, Acceptable, and Amazing. In the same way, Figure 17.3 and Figure 17.4, are the graph showing the input membership function for food quality and Timeliness that is used for the evaluation of Perceived Value. Again, Figure 17.5 represents the output membership function having the linguistic value Very Low, Low, Medium, High, and Very High. Finally, the output membership activity of the customer satisfaction based on the Facility Comfort = 5, Food quality = 6, and Timeliness = 7 input terms with the five defuzzification methods COA, BOA, MOM, LOM, and SOM applied and graphically represented in Figure 17.6. The impact of different defuzzification methods on output with the change of input is shown in Figure 17.6. The first 5 input and their corresponding defuzzified output are presented in Table 17.8.

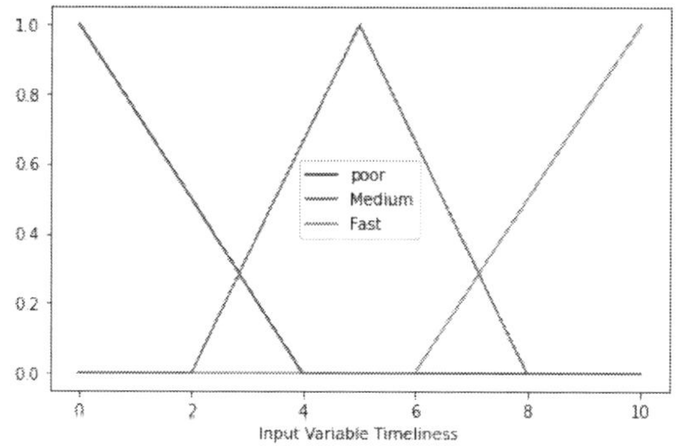

Figure 17.4: Input Membership Function: Timeliness for Case-1

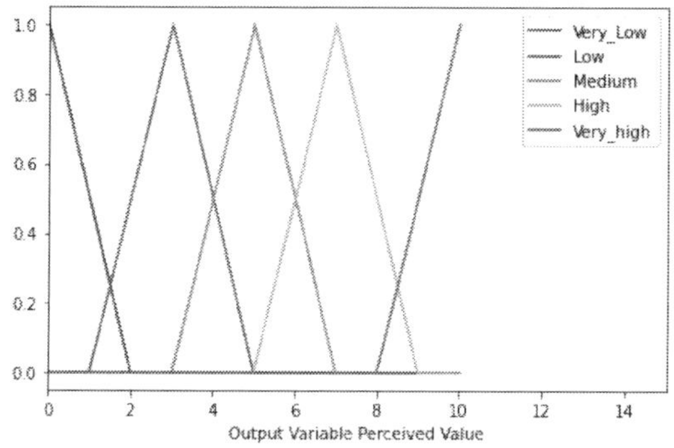

Figure 17.5: Output Membership Function: Perceived Value for Case-1

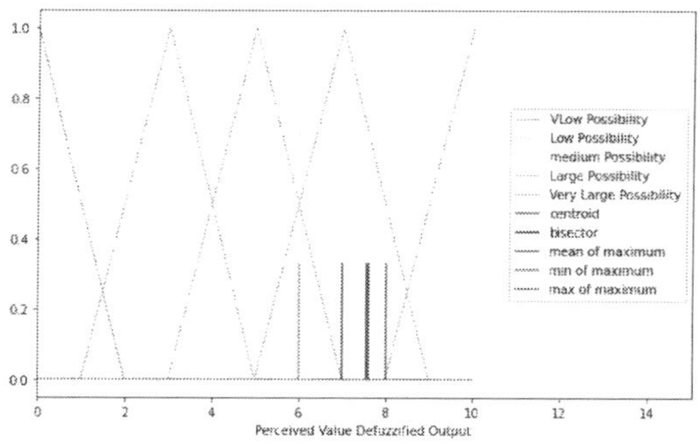

Figure 17.6: Output Membership Activity for Case-1

Case-2: Similarly, for Case-2 the input and output membership function was generated to analyze the impact of the factors on Customer Satisfaction. As depicted in Figure 17.7, Input Membership Function: Facility Comfort, the membership function of the input variable facility comfort is shown in terms of "Poor", Acceptable and, Amazing. In the same way, Figure 17.8 and Figure 17.9, are the graphs showing the

Table 17.8: Defuzzified Values Derived Through Different Defuzzification Methods (Case-1)

Food Quality	Facility Comfort	Timeliness	Op. Centroid	Op. Bisector	Op. LOM	Op. MOM	Op. SOM
1	2	1	.77	.75	1	0	0
2	2	4	.77	.75	1	0	0
4	4	4	7	7	7	7	7
2.5	3	1	2.39	2.437	4	3	2
6	.5	.5	2.99	3	3	3	3

input membership function for food quality and cleanliness that is used for the evaluation of customer satisfaction. Again, Figure 17.10 represents the output membership function having the linguistic values Dissatisfied, Satisfied and Neutral. Finally, the output membership activity of the customer satisfaction based on the Facility Comfort = 5, Food quality = 6, and "Cleanliness" = 7 input terms with the five

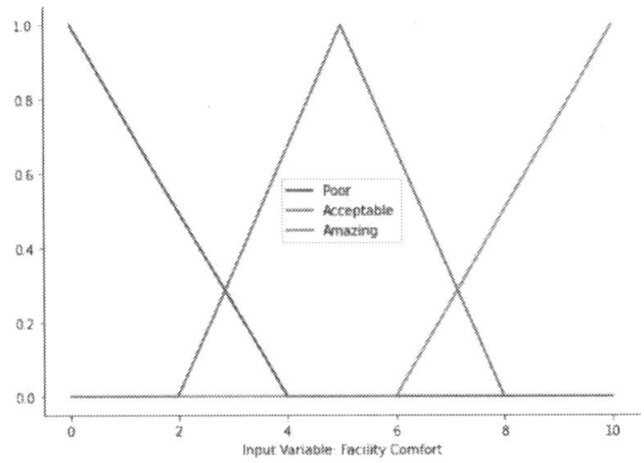

Figure 17.7: Input Membership Function: Facility Comfort for Case-2

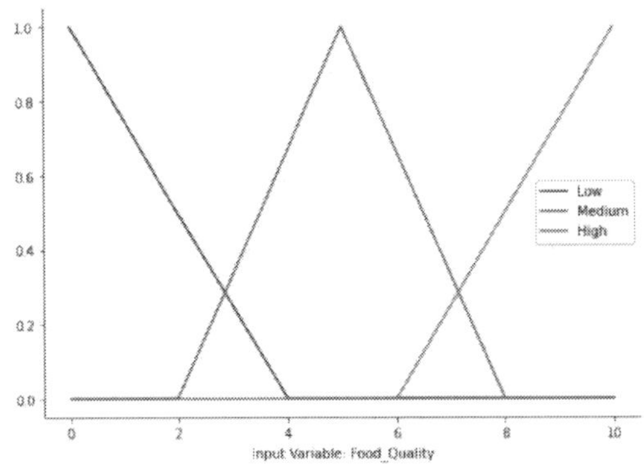

Figure 17.8: Input Membership Function: Food Quality for Case-2

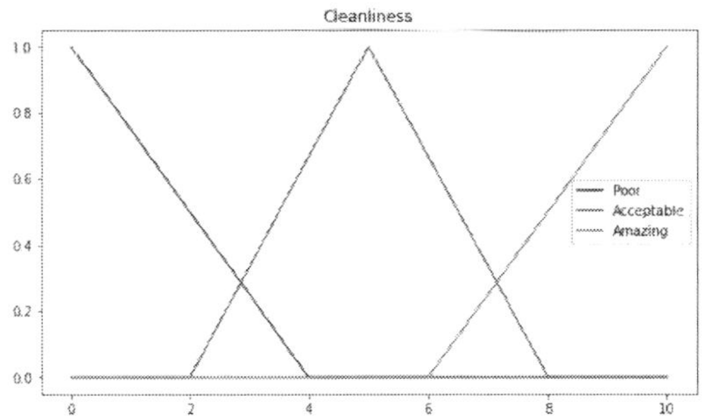

Figure 17.9: Input Membership Function: Cleanliness for Case-2

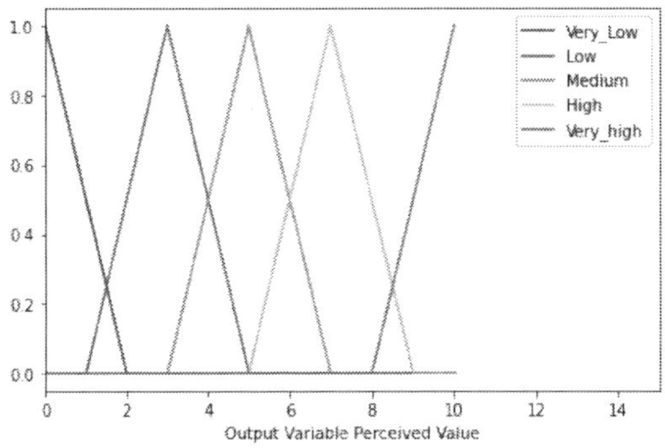

Figure 17.10: Output Membership Function: Perceived Value for Case-2

defuzzification methods COA, BOA, MOM, LOM, and SOM applied as depicted in, Figure 17.11 Output Membership Activity. The impact of different defuzzifaction methods on output with the change of input is shown in Table 17.2. The first 5 input is presented in Table 17.9.

Case-3: Again, for the final stage, the input and output membership functions are defined the same as in the previous cases. In this case, as input, the output from

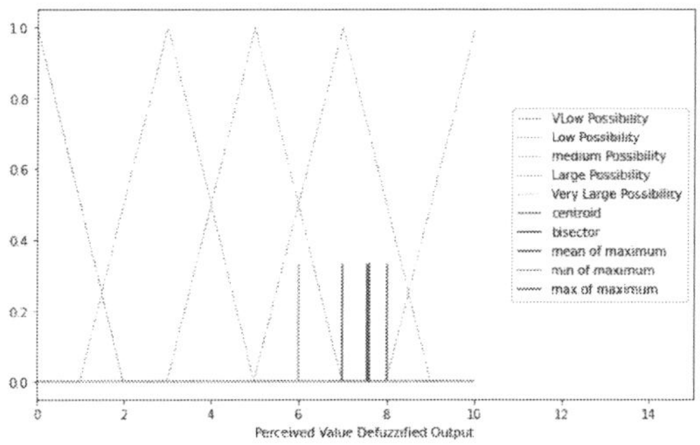

Figure 17.11: Output Membership Activity for Case-2

Table 17.9: Defuzzified Values Derived Through Different Defuzzification Methods (Case-2)

Food Quality	Facility Comfort	Timeliness	Op. Centroid	Op. Bisector	Op. LOM	Op. MOM	Op. SOM
1	2	1	1.5	1.5	2	1	0
2	2	4	1.5	1.5	2	1	0
4	4	4	8.55	8.67	10	9.5	7
2.5	3	1	1.67	1.62	2	1	0
6	.5	.5	1.44	1.325	1	.5	0

the above two cases is considered: customer satisfaction and perceived value, and the output is the possibility of recommendation based on the input terms. Figure 17.12 is the input membership function that defines customer satisfaction; Figure 17.13 depicts the input membership function of perceived value, and Figure 17.14 shows the output membership function. In Figure 17.15 the defuzzified output membership activity is shown for the input values Perceived Value = 6.66 and Customer Satisfaction = 8.37, which means that with a high perceived value and being satisfied, the possibility of recommending a place becomes high. Finally, the 3D plot of the input

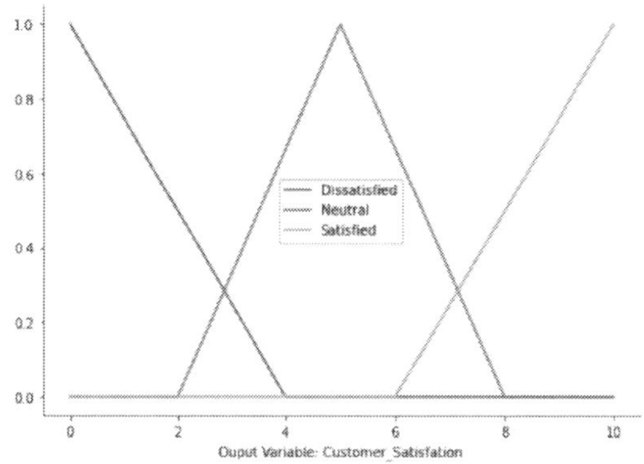

Figure 17.12: Input Membership Function: Customer Satisfaction for Case-3

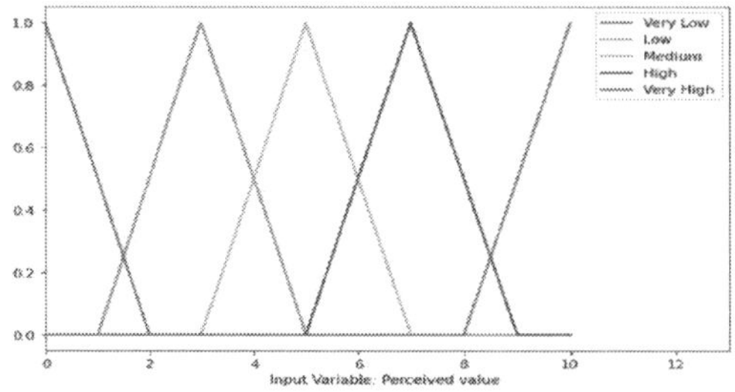

Figure 17.13: Input Membership Function: Perceived Value for Case-3

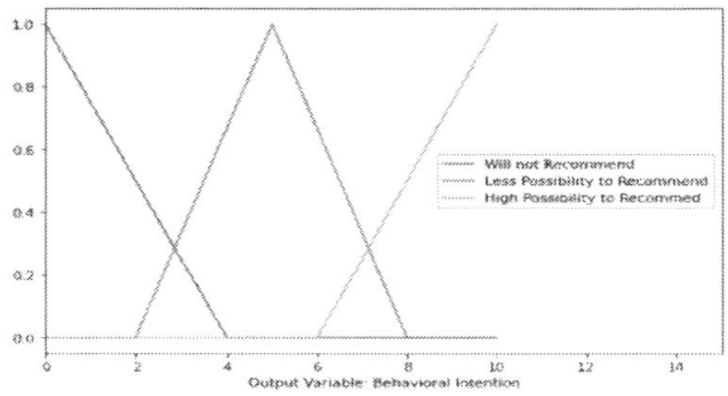

Figure 17.14: Output Membership Activity Behavioral Intention for Case-3

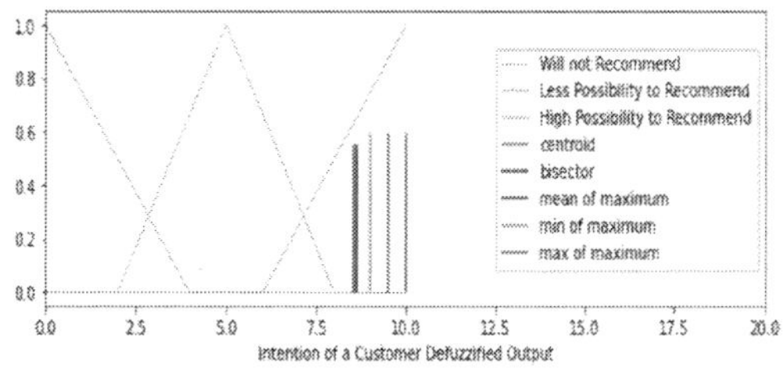

Figure 17.15: Output Membership Activity for Case-3

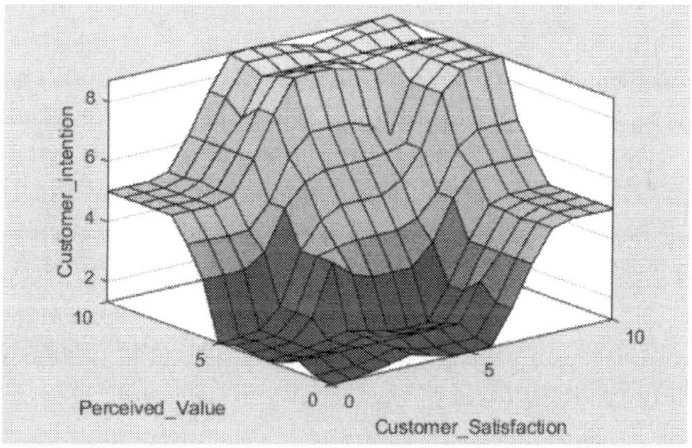

Figure 17.16: 3D Surface View of the Fuzzy-based Customer Satisfaction System

perceived value, customer satisfaction, and output customer intention are shown in Figure 17.16.

Further is Table 17.10 the comparative analysis based on parameters analyzed in existing works with respect to our proposed system has been shown.

Table 17.10: Existing systems vs. Proposed system

Assessment Parameters	[27]	[29]	[33]	[34]	[23]	[22]	Proposed System
Food Quality	X	X	X	✓	X	X	✓
Facility Comfort	X	X	X	✓	✓	X	✓
Timeliness	✓	X	X	X	X	X	✓
Cleanliness	X	X	X	X	X	X	✓
Perceived Value	X	X	X	✓	X	✓	✓
Customer Satisfaction	X	X	✓	X	X	X	✓

17.5 CONCLUSION

Finding a better place to eat has been mentioned as a source of entertainment. Many consumers consider evaluations left by other people who have been to an area and provided their comments. It can be difficult to evaluate a place based on the enormous number of reviews. This paper offered a fuzzy-based strategy for consumer sentiment analysis and restaurant recommendation. User won't have to read as many entries regarding a location's ratings if they use this recommendation method. Once more, this study looked at the recommendation system from two perspectives: the worth of a place with reliable pricing and quality, as well as the satisfaction of the customer who visits that place with excellent service. The end result makes it evident that customer happiness and perceived value both have a favorable influence on a customer's total desire to provide a positive review. Even if a customer is delighted with the service they received, the likelihood that they will recommend a place may decrease if there is any mismatch between the service provided and the perceived value. However, despite making some assumptions about potential values that might have an impact on the variables, this work suggested one possible recommendation system strategy. We only considered the factors that might have the most impact on the fine dining system in order to simplify computation. Future research will examine how the outcomes differ by including additional factors like cost and attractiveness. In order to evaluate how effectively the system functions, it will also research various defuzzification approaches and compare them.

Bibliography

[1] Bhatti, S., Hussain, W., Khan, J., Sultan, S. & Ferraris, A. Exploring data-driven innovation: What's missing in the relationship between big data analytics capabilities and supply chain innovation?. *Annals Of Operations Research.* pp. 1–26 (2022)

[2] Akter, S., Michael, K., Uddin, M., McCarthy, G. & Rahman, M. Transforming business using digital innovations: The application of AI, blockchain, cloud and data analytics. *Annals Of Operations Research.* pp. 1–33 (2022)

[3] Reza, M., Debnath, T. & Sultan, S. Outliers Elimination: A Modified Clustering Technique of K-means Algorithm.

[4] Al-Dossari, H. & Alyahya, S. Using Fuzzy-Logic in Decision Support System based on Personal Ratings. *International Journal Of Advanced Computer Science And Applications.* **11** (2020)

[5] Rahman, A., Hossain, M., Muhammad, G., Kundu, D., Debnath, T., Rahman, M., Khan, M., Tiwari, P. & Band, S. Federated learning-based AI approaches in smart healthcare: concepts, taxonomies, challenges and open issues. *Cluster Computing.* pp. 1–41 (2022)

[6] Debnath, T., Reza, M., Rahman, A., Beheshti, A., Band, S. & Alinejad-Rokny, H. Four-layer ConvNet to facial emotion recognition with minimal epochs and the significance of data diversity. *Scientific Reports.* **12**, 6991 (2022)

[7] Khan, S., Shahrior, A., Karim, R., Hasan, M. & Rahman, A. MultiNet: A deep neural network approach for detecting breast cancer through multi-scale feature fusion. *Journal Of King Saud University-Computer And Information Sciences.* **34**, 6217–6228 (2022)

[8] Tan, T. & Staats, B. Behavioral drivers of routing decisions: Evidence from restaurant table assignment. *Production And Operations Management.* **29**, 1050–1070 (2020)

[9] Acar, O., Tarakci, M. & Van Knippenberg, D. Creativity and innovation under constraints: A cross-disciplinary integrative review. *Journal Of Management.* **45**, 96–121 (2019)

[10] Khan, M., Rahman, A., Debnath, T., Karim, M., Nasir, M., Band, S., Mosavi, A. & Dehzangi, I. Accurate brain tumor detection using deep convolutional neural network. *Computational And Structural Biotechnology Journal.* **20** pp. 4733–4745 (2022)

[11] Rezaimehr, F. & Dadkhah, C. A survey of attack detection approaches in collaborative filtering recommender systems. *Artificial Intelligence Review.* **54** pp. 2011–2066 (2021)

[12] Zhang, Z., Dong, M., Ota, K., Zhang, Y. & Ren, Y. LBCF: A link-based collaborative filtering for overfitting problem in recommender system. *IEEE Transactions On Computational Social Systems.* **8**, 1450–1464 (2021)

[13] Niyazov, A., Mikhailova, E. & Egorova, O. Content-based music recommendation system. *2021 29th Conference Of Open Innovations Association (FRUCT).* pp. 274–279 (2021)

[14] Rajendran, D. & Sundarraj, R. Using topic models with browsing history in hybrid collaborative filtering recommender system: Experiments with user ratings. *International Journal Of Information Management Data Insights.* **1**, 100027 (2021)

[15] Nilashi, M., Yadegaridehkordi, E., Ibrahim, O., Samad, S., Ahani, A. & Sanzogni, L. Analysis of travellers' online reviews in social networking sites using fuzzy logic approach. *International Journal Of Fuzzy Systems.* **21** pp. 1367–1378 (2019)

[16] Islam, M., Rahman, A., Kabir, S., Karim, M., Acharjee, U., Nasir, M., Band, S., Sookhak, M. & Wu, S. Blockchain-SDN-Based Energy-Aware and Distributed Secure Architecture for IoT in Smart Cities. *IEEE Internet Of Things Journal.* **9**, 3850–3864 (2022)

[17] Shayshab Azad, K., Hossain, N., Islam, M., Rahman, A. & Kabir, S. Preventive Determination and Avoidance of DDoS Attack with SDN over the IoT Networks. *2021 International Conference On Automation, Control And Mechatronics For Industry 4.0 (ACMI).* pp. 1–6 (2021)

[18] Sood, S. & Saini, M. Hybridization of cluster-based LDA and ANN for student performance prediction and comments evaluation. *Education And Information Technologies.* **26** pp. 2863–2878 (2021)

[19] Forouzandeh, S., Berahmand, K., Nasiri, E. & Rostami, M. A hotel recommender system for tourists using the Artificial Bee Colony Algorithm and Fuzzy TOPSIS Model: a case study of tripadvisor. *International Journal Of Information Technology & Decision Making.* **20**, 399–429 (2021)

[20] Sukmawan, D., Yudono, M., Purwanto, D., Permadi, D., Suryana, A., Saputri, U. & Artiyasa, M. Adaptive Neuro-Fuzzy Inference System (ANFIS) Method for Developing a Decision Support System for Determining Landslide Susceptibility. *Fidelity: Jurnal Teknik Elektro.* **5**, 1–8 (2023)

[21] Rojas, J., Espitia, H. & Bejarano, L. Design and optimization of a fuzzy logic system for academic performance prediction. *Symmetry.* **13**, 133 (2021)

[22] Leppäniemi, M., Karjaluoto, H. & Saarijärvi, H. Customer perceived value, satisfaction, and loyalty: the role of willingness to share information. *The International Review of Retail, Distribution and Consumer Research.* **27**, 164–188 (2017)

[23] Palominos, P., Quezada, L. & Gonzalez, M. Incorporating the voice of the client in establishing the flexibility requirement in a production system. *International Journal of Production Economics.* **211** pp. 34–43 (2019)

[24] Trigos, F., Vazquez, A. & Cárdenas-Barrón, L. A simulation-based heuristic that promotes business profit while increasing the perceived quality of service industries. *International Journal of Production Economics.* **211** pp. 60–70 (2019)

[25] Tseng, M., Wu, K., Hu, J. & Wang, C. Decision-making model for sustainable supply chain finance under uncertainties. *International Journal of Production Economics.* **205** pp. 30–36 (2018)

[26] Zadeh, L. Fuzzy sets. *Inform Control.* **8** pp. 338–353 (1965)

[27] Farajpour, F., Taghavifard, M., Yousefli, A. & Taghva, M. Information sharing assessment in supply chain: hierarchical fuzzy rule-based system. *Journal of Information & Knowledge Management.* **17**, 1850002 (2018)

[28] Khan, S., Kusi-Sarpong, S., Arhin, F. & Kusi-Sarpong, H. Supplier sustainability performance evaluation and selection: A framework and methodology. *Journal of Cleaner Production.* **205** pp. 964–979 (2018)

[29] Chen, T. A fuzzy ubiquitous traveler clustering and hotel recommendation system by differentiating travelers' decision making behaviors. *Applied Soft Computing.* **96** pp. 106585 (2020)

[30] Alghamdi, S., Alzhrani, N. & Algethami, H. Fuzzy-Based Recommendation System for University Major Selection.. *IJCCI.* pp. 317–324 (2019)

[31] Shrimal, M., Khavnekar, M., Thorat, S. & Deone, J. Nutriflow: A Diet Recommendation System. *Proceedings of the 4th International Conference on Advances In Science & Technology (ICAST2021).* (2021)

[32] Tuncer, İ., Unusan, C. & Cobanoglu, C. Service quality, perceived value and customer satisfaction on behavioral intention in restaurants: An integrated structural model. *Journal of Quality Assurance in Hospitality & Tourism.* **22**, 447–475 (2021)

[33] Asghar, M., Subhan, F., Ahmad, H., Khan, W., Hakak, S., Gadekallu, T. & Alazab, M. Senti-eSystem: a sentiment-based eSystem-using hybridized fuzzy and deep neural network for measuring customer satisfaction. *Software: Practice and Experience.* **51**, 571–594 (2021)

[34] Vo-Thanh, T., Zaman, M., Thai, T., Hasan, R. & Senbeto, D. Perceived customer journey innovativeness and customer satisfaction: a mixed-method approach. *Annals of Operations Research.* pp. 1–26 (2022)

CHAPTER 18

An Improved Ensemble Model for Intent Classification of Bangla Chatbot

Solaiman Ahmed

Dhaka University of Engineering & Technology, Dept. of CSE, Gazipur, Bangladesh. Email: hira@duet.ac.bd

Md. Mostafizer Rahman

Dhaka University of Engineering & Technology, Gazipur, Bangladesh. Email: mostafiz26@gmail.com, Corresponding author

Fazlul Hasan Siddiqui

Dhaka University of Engineering & Technology, Dept. of CSE, Gazipur, Bangladesh. Email: siddiqui@duet.ac.bd

CONTENTS

With the rise of Artificial Intelligence and Natural Language Processing, chatbots have become an integral part of the customer service industry. However, developing chatbots in non-English languages poses unique challenges such as a lack of dedicated pipelines, NLP resources, tokenizers, etc. The focus of the study is on developing a chatbot that can interact with users in low-resource languages like Bangla with high confidence and accuracy. In this study, we proposed an ensemble machine

DOI: 10.1201/9781003605508-18

learning pipeline for intent classification of Bangla language chatbots. The proposed pipeline is developed using spaCy, BERT, and fastText vector on the Rasa framework. spaCy is a powerful NLP library used to classify user input for low-resource languages, and BERT is a transformer-based language model. To improve the performance, the fastText vector is combined with spaCy and BERT. We performed all experiments with different pipelines on the student support services dataset in the Bangla language. The proposed ensemble pipeline achieved high performances in accuracy, precision, recall, and F1 score of 93.75%, 93.05%, 93.75%, and 92.36%, respectively. The results indicate that the proposed pipeline can significantly improve the performance of intent classification for Bangla chatbots. The results suggest that the model can be applied to the development of chatbots in other languages.

18.1 INTRODUCTION

With the increasing demand for automated customer support services, chatbots have become a popular choice for various industries. Chatbots use Natural Language Processing (NLP) techniques to understand and respond to customer queries, making them an excellent choice for customer support services. In recent years, the use of chatbots has increased significantly in education sectors, and many universities and educational institutions have started using them to provide support services to their students [1, 2]. However, developing a chatbot for a specific language can be challenging, especially for languages that do not have adequate NLP resources. One such language is Bangla, which is the seventh most spoken language in the world, with approximately 230 million speakers [3]. Despite the significant number of speakers, the development of chatbots in Bangla has been relatively limited due to the lack of NLP resources and research in this language [4].

Recently, ChatGPT [5] has emerged as a revolutionary AI tool that allows users to interact in a human-like conversational style. It has a wide range of applications, from text generation to computer problem solving. However, selecting large language models (LLMs) such as ChatGPT, GPT-4, and BERT for a particular task remains challenging. Numerous factors such as data size, training time, flexibility, scalability, and contextual understanding are considered when selecting an appropriate LLM for a task. For a small-scale task such as intent classification of a Bangla language chatbot, BERT [6] can be an efficient choice. BERT is effective on small datasets and requires less time for model training. Also, BERT has pre-trained models and is useful for less sophisticated tasks compared to GPT-based models.

In this study, we propose an ensemble pipeline for Bangla language chatbots using the spaCy library, fastText vector, and BERT for student support service queries. Our proposed pipeline aims to overcome the challenges associated with developing a chatbot in Bangla language and provide an efficient solution for student support services. To use fastText word vectors in spaCy, we define a custom component in the pipeline that loads fastText word vectors. Once the custom component is added to the pipeline, we use it in the same way as any other pre-trained spaCy models. Next, we tokenize the text, and the BERT model is used to extract additional features from the tokens. Moreover, various NLP tasks such as part-of-speech tagging, named entity recognition, and text classification are applied. Finally, a combination of predefined templates and user information is used to generate an appropriate response.

We prepared a useful NLP solution for languages that have no pre-trained language models in spaCy by combining spaCy's flexible pipeline architecture with fastText's word embedding capabilities and BERT language model. To evaluate the performance of our proposed pipeline, we conducted experiments on a custom dataset of the Bangla language based on student support service queries. The results of our experiments show that the proposed pipeline outperforms existing Bangla language chatbots [4] in terms of accuracy, precision, recall, and F1 score.

In summary, the proposed ensemble pipeline for Bangla language chatbots using the spaCy library, fastText vector, and BERT language model has become an effective model for providing student support services in Bangla. Our research can help bridge the gap in NLP resources for the Bangla language and contribute to the development of chatbots in other languages with limited NLP resources.

18.2 BACKGROUND AND RELATED WORK

Several studies have been conducted in the field of intent classification for chatbots using different techniques and approaches. In this section, we review some of the relevant literature on the topic. The history of AI chatbots dates back to the mid-1960s when Joseph Weizenbaum developed ELIZA, a rule-based chatbot that mimicked a Rogerian psychotherapist. ELIZA [7] was one of the first chatbots to simulate human conversation and generated responses based on a set of pre-programmed rules. Over time, chatbots became more sophisticated and began to incorporate machine learning (ML) techniques. In the 1970s, PARRY was developed to simulate a person with paranoid schizophrenia [8], while in the 1990s, ALICE was created to simulate a natural language conversation [9].

Moreover, SmarterChild was introduced in the early 2000s as an instant messaging chatbot and gained popularity for its ability to provide information, play games, and answer questions [10]. Apple's Siri and Google's Assistant are two examples of chatbots that became widely popular as personal assistants on mobile devices [11]. Today, chatbots are used in a wide range of application domains, from customer service to healthcare, and continue to evolve with advancements in ML technologies. Meanwhile, there are many chatbot frameworks available, both open source and proprietary. These frameworks provide developers with the ability to build intelligent chatbots that can understand user input and respond appropriately. Some of the most popular chatbot frameworks include Rasa, Dialogflow, Microsoft Bot Framework, etc. [12]. Rasa is an open-source framework for building chatbots and virtual assistants [13]. It provides tools for natural language understanding, dialogue management, and integration with messaging channels and APIs. It allows developers to train models on their own data and provides more control over the behavior of the chatbot. Moreover, Rasa offers more privacy than cloud-based services like Dialogflow, as it can be run on-premises or in a private cloud [14]. In addition, Rasa's open-source model allows developers to contribute to the framework's improvements, and a large community of users can support and share best practices.

In a recent study [4], the authors proposed a model for developing a Bangla chatbot that can communicate in low-resource languages like Bangla and also Bangla Transliteration in English with reasonable performance. The authors used Rasa Open Source Framework, fastText embeddings, Polyglot embeddings, Flask, and other systems as building blocks for the chatbot. They performed an extensive comparative analysis of various pipelines and their components. In their study, two components CustomTokenizer and fastTextFeaturizer were introduced for a low-resource language like Bangla to boost performance. But their proposed best fit pipeline does not use those components. LanguageModelFeaturizer uses the BERT model to support 100 languages including Bangla with the largest Wikipedia [15]. Also, the LanguageModelFeaturizer leverages Language-agnostic BERT Sentence Embedding as model_weights, which is an effective tool for encoding texts into a single shared embedding space. Another component LanguageModelTokenizer is already deprecated [16]. Therefore, any suitable tokenizer can be substituted.

Meshram et al. developed a chatbot "College Enquiry Chatbot" using the Rasa Framework for academic purposes [17]. In particular, they envisioned automating conversations with students regarding admissions-related inquiries. The study evaluated the performance of the chatbot by measuring precision, accuracy, and F1 score and found that the performance was moderate. However, the proposed chatbot was limited to English only.

Overall, the existing literature highlights the importance of developing accurate and efficient intent classification pipelines for chatbots in low-resource languages. Our proposed pipeline builds on previous studies and utilizes spaCy and fastText libraries to enhance the accuracy of intent classification in Bangla chatbots.

18.3 PROPOSED METHODOLOGY

NLP is a rapidly growing field of Artificial Intelligence that can process, analyze, and understand human language. NLP is used in various applications such as chatbots, virtual assistants, sentiment analysis, language translation, and so on. Despite the vast activity of NLP in various languages, it falls short in Bangla due to the insufficient resources of the Bangla language for model training. This study proposed an ensemble pipeline for processing student queries asked in Bangla related to student support services. The development steps of the proposed chatbot are described in the following sections.

18.3.1 Dataset Preparation

The first step in building a chatbot is to prepare a dataset that contains the user queries and corresponding responses. The lack of resources in Bangla makes it difficult to create a dataset. For this study, data was collected from user interactions on the web and manual forms related to student inquiries about student support services. Rasa uses YAML (YAML Ain't Markup Language) files such as *nlu.yml*, *stories.yml*, *domain.yml*, *config.yml*, etc. to define various aspects of a conversational AI project, including training data, configuration settings, and domain knowledge. To create the dataset, first, the *nlu.yml* file is used to define the training data for the NLU module. It contains examples of user inputs, along with their corresponding intents and entities. We have added more than 40 intents with 220+ examples and 10 entities. Some examples of intents and stories in our dataset are shown in Figure 18.1. By providing accurate training data, it can improve the chatbot's ability to understand user input and provide appropriate responses.

```
version: "3.1"

nlu:

- intent: ask_casr_meeting
  examples: |
  - সিএএসআর সভার তারিখ
  - কোন আসন্ন সিএএসআর মিটিং আছে?
  - পরবর্তী সিএসআর সংখ্যা কত?
  - সিএসআর শেষ তারিখ কত?
  - সিএসআর তারিখ চেক করতে আমাকে সাহায্য করতে পারেন?

- intent: ask_appointment_of_supervisor
  examples: |
  - সুপারভাইজার নিয়োগ
  - সুপারভাইজার আবেদন
  - আমি কিভাবে সুপারভাইজার জন্য আবেদন করতে পারি?
  - আমি কি সুপারভাইজার নিয়োগের আবেদন করার যোগ্য?
  - আমার সিজিপিএ কি সুপারভাইজার নিয়োগের জন্য যথেষ্ট?
  - সুপারভাইজার আবেদন করার জন্য কত ক্রেডিট সম্পূর্ণ করতে হবে?
  - সুপারভাইজার আবেদন করার মানদণ্ড কি?
  - কখন আমার থিসিস সুপারভাইজার পাওয়ার চেষ্টা করা উচিত?

- intent: ask_change_of_supervisor
  examples: |
  - সুপারভাইজার পরিবর্তন
  - সুপারভাইজার পরিবর্তনের আবেদন
  - আমার সুপারভাইজার পরিবর্তনের জন্য কিভাবে আবেদন করতে পারি?
  - আমার থিসিসের বিষয় আমার সুপারভাইজারের আগ্রহের ক্ষেত্রগুলির সাথে সম্পর্কিত নয়
  - আমার থিসিস সুপারভাইজার ছুটিতে আছেন
  - আমার সুপারভাইজার পৌঁছানো যায় না
```

```
version: "3.1"

stories:

- story: seeking help path
  steps:
  - intent: greet
  - action: utter_greet

- story: acpg meeting
  steps:
  - intent: greet
  - action: utter_greet
  - intent: ask_acpg_meeting
  - action: utter_ask_department
  - intent: ask_department
  - action: utter_acpg_meeting
  - action: utter_did_that_help

- story: appointment of supervisor
  steps:
  - intent: greet
  - action: utter_greet
  - intent: ask_appointment_of_supervisor
  - action: utter_ask_student_id
  - intent: ask_student_id
  - action: utter_appointment_of_supervisor
```

Figure 18.1: Some examples of intents (nlu.yml) and stories (stories.yml) from the student support services dataset

Next, the *stories.yml* file is used to add more than 40 stories that represent different scenarios related to student queries. The stories contain a sequence of user intents, entities, and corresponding responses (see Figure 18.1). These stories were used to train the Rasa model to identify the user's intent and generate appropriate responses. Finally, the *domain.yml* file is used to define the chatbot's domain, including the intents, entities, actions, and responses. It is considered a central hub of all the information for the chatbot to function properly. The file includes information such as the chatbot's name, default response templates, and custom actions. Defining the domain in the *domain.yml* file ensures consistency and appropriate response of the chatbot to user input.

18.3.2 Modeling in Machine Learning

Figure 18.2 illustrates the workflow of our proposed model. In the second step, the ML model is trained using the dataset to create a chatbot. Rasa provides a flexible and customizable architecture for building chatbot models. Rasa has two main modules, Rasa NLU (Natural Language Understanding) and Rasa Core. The NLU module of

Rasa is an essential component of the Rasa chatbot framework. It is also responsible for understanding user input and extracting relevant information to determine the user's intent and the proper response.

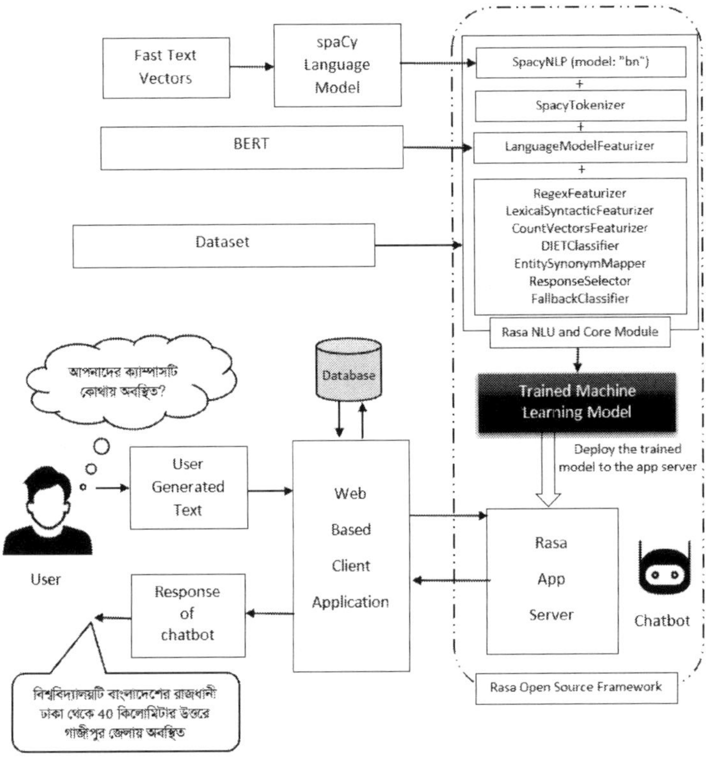

Figure 18.2: Workflow of the proposed model

In order to develop a Bangla chatbot, the Rasa NLU module uses various components to preprocess, tokenize, and featurize the user's input before passing it to the intent classifier. Some of the commonly used components are *WhitespaceTokenizer*: breaks down the user's input into individual words based on whitespace characters. *RegexFeaturizer*: uses regular expressions to extract specific patterns or keywords from the user's input, which can then be used as features for the intent classifier. *LexicalSyntacticFeaturizer*: uses a combination of rule-based and statistical techniques to extract syntactic and lexical features from the user's input. *CountVectorFeaturizer*: converts the user's input into a vector representation. *DIETClassifier* is the primary intent classifier used by Rasa. It is a neural network architecture that combines a pre-trained language model (such as BERT or GPT) with a multi-task learning approach to jointly learn intent and entity recognition tasks. *EntitySynonymMapper* maps different entity values to a single canonical value to improve the entity extraction's accuracy. *FallbackClassifier* is used to handle incoming messages with low NLU confidence. When all other intent predictions fall below the specified confidence threshold, this configuration will forecast the intent *nlu_fallback*.

On the other hand, Rasa Core handles the dialogue management aspect of a chatbot. It is responsible for selecting the appropriate response to a user's input based on the current state of the conversation. Rasa Core works by defining a set of rules or stories that represent different paths through a conversation. Each story consists of a sequence of user inputs and the corresponding bot responses. The stories can also include conditions or checkpoints that allow the bot to branch to different parts of the conversation based on the user's input or other factors. Rasa Core is an essential module in Rasa that allows chatbots to engage in dynamic and context-aware conversations with users. It works in conjunction with Rasa NLU to provide a complete conversational AI solution.

In our research, the Rasa NLU module uses fastText vectors released by Facebook, which provides a variety of pre-trained vectors out of the box in many languages [18]. We use Bangla fastText vectors in this research. Although Rasa doesn't natively support fastText as a backend, it is still possible to use fastText vectors by loading them into spaCy, which is a supported backend. This can be accomplished by creating a spaCy language model from the fastText vectors and then enabling the model into spaCy, in our research, we referred to the model as "*bn*". The fastText vectors can be used for tokenization, featurization, and entity recognition within the Rasa NLU module after enabling the model. The responsibility of the Rasa Core is to determine what should happen next after the Rasa NLU has interpreted the messages. Rasa Core can take some actions such as collecting information from the previous history, getting the required data from the database, etc. to generate accurate responses.

18.3.3 Deployment

The third step in building a chatbot is to deploy the model on a Rasa app server. We developed a web-based client application using Apache, PHP, and MySQL whose front end is used to communicate with users, and the back end is connected to the Rasa app server. The Rasa app server is connected with the web-based client application through API. The user can interact with the chatbot by making queries on the web client application and the chatbot model will respond with appropriate answers over the application. Figure 18.3 shows the user interaction with the chatbot.

Figure 18.3: User interface of the client application

Furthermore, the chatbot has the capability to handle custom actions. For example, if the user asks for the contact details of the student support services, the chatbot can provide the contact details by invoking a custom action. This custom action is defined in the domain.yml file and executed by the chatbot when the user's query matches the intent associated with the custom action. To enhance the user experience, we also implemented a feature to provide feedback to the user. The user can rate the chatbot's response as either helpful or not. These feedbacks are stored in a database and can be used to improve the chatbot's performance in the future.

18.3.4 Evaluation Methods

The following evaluation methods such as precision, Recall, F1 score, and Accuracy are used to measure the classification performance of the pipelines. The definition and description of these methods can be found in [19, 20].

$$Precision = \frac{TP}{TP + FP} \tag{18.1}$$

$$Recall = \frac{TP}{TP + FN} \tag{18.2}$$

$$F1\ Score = 2 \times \frac{Precision \times Recall}{Precision + Recall} \tag{18.3}$$

$$Accuracy = \frac{TP + TN}{TP + TN + FP + FN} \tag{18.4}$$

Where TN is the true negative, TP is the true positive, FN is the false negative, and FP is the false positive.

18.4 EXPERIMENTAL RESULTS

In this study, four pipelines were developed for the experiment, as shown in Table 18.1. The two most common techniques such as Held-Out Test Set and Cross-Validation are used to evaluate the Rasa model [21]. The Held-Out Test Set involves splitting the dataset into one training and one test set. The model is trained on the training set, and the test set is used to evaluate the model's performance. In this approach, it is recommended that the data be shuffled and split using the Rasa data split method each time during evaluation, rather than using a static NLU test set.

Cross-Validation is another technique to evaluate the Rasa model that involves splitting the data into multiple train/test splits and averaging the results of evaluations on each split. This technique is particularly useful when significant changes are made to the NLU training data, such as splitting an intent into two intents or adding a lot of training examples. Cross-validation evaluates all the data in the dataset, making it the most comprehensive method to automatically test the Rasa model.

In this study, the Held-Out Test Set technique is used to split the annotated data into training and test sets at a ratio of 80:20. Custom spaCy language models were

Table 18.1: List of the pipelines

Pipelines	Pipelines Components
default_pipeline	WhitespaceTokenizer + RegexFeaturizer + LexicalSyntacticFeaturizer + CountVectorsFeaturizer + DIETClassifier + EntitySynonymMapper + ResponseSelector + FallbackClassifier
spacy_pipeline	SpacyNLP (model: “bn”) + SpacyTokenizer + SpacyFeaturizer + RegexFeaturizer + LexicalSyntacticFeaturizer + CountVectorsFeaturizer + DIETClassifier + EntitySynonymMapper + ResponseSelector + FallbackClassifier
white_bert_pipeline	WhitespaceTokenizer + LanguageModelFeaturizer (model_name: “bert”, model_weights: “rasaLaBSE”) + RegexFeaturizer + LexicalSyntacticFeaturizer + CountVectorsFeaturizer + DIETClassifier + EntitySynonymMapper + ResponseSelector + FallbackClassifier
spacy_bert_pipeline (**proposed**)	SpacyNLP (model: “bn”) + SpacyTokenizer + LanguageModelFeaturizer (model_name: “bert”, model_weights: “rasaLaBSE”) + RegexFeaturizer + LexicalSyntacticFeaturizer + CountVectorsFeaturizer + DIETClassifier + EntitySynonymMapper + ResponseSelector + FallbackClassifier

used in some of the pipelines, specifically those designed for the Bangla corpus. Additional parameters and arguments were fixed for some of the components. For the *DIETClassifier*, we set the epochs to 100 and included the “constrain_similarities” parameter as true. *CountVectorsFeaturizer* utilized “char_wb” as the analyzer and set the minimum n-gram [22, 23] to 1 and the maximum n-gram to 4. For *FallbackClassifier*, we set the threshold to 0.3 and added an ambiguity threshold of 0.1 to address the ambiguity. The policies were kept constant for all experiments, and the following policies were employed: *MemoizationPolicy*, *TEDPolicy* (with a maximum history of 5 and 200 epochs), and *RulePolicy*.

To determine which parts of the pipeline are most effective for our tasks, we conducted four experiments. Table 18.1 shows the corresponding pipeline configurations. We evaluate the results obtained to show how and why a particular factor affects the results. As we mentioned in Table 18.1, default_pipeline uses several featurizers and a *DIETClassifier* for intent classification as well as entity recognition. The pipeline also includes an *EntitySynonymMapper* and *ResponseSelector* for handling entities and selecting responses, respectively. Finally, a *FallbackClassifier* is used to handle ambiguous user inputs.

Table 18.2: Comparative result analysis for intent classification based on micro-average scores

NLU Pipelines	Accuracy	Precision	Recall	F1-Score
default_pipeline	75%	68.75%	75%	69.72%
spacy_pipeline	78.12%	73.61%	79.16%	74.3%
white_bert_pipeline	81.25%	72.91%	79.16%	73.47%
spacy_bert_pipeline (**proposed**)	93.75%	93.05%	93.75%	92.36%

Based on our experiments, the default pipeline performed the worst among the tested pipelines. This result is not surprising, as the default pipeline uses relatively simple featurizers and does not include a pre-trained language model like BERT or spaCy. Experimental results show that it achieves only 75% accuracy, as shown in Table 18.2, and has a high percentage of misclassifications, as depicted in Figure 18.4 and Figure 18.5. Basically, this pipeline is the default pipeline in the Rasa framework.

In our second pipeline is spaCy_pipeline that uses the spaCyNLP component to extract linguistic features from user inputs. It also includes several featurizers and a *DIETClassifier* like the previous pipeline. Experimental results show that the spaCy_pipeline performed slightly better than the default_pipeline and achieved an accuracy of 78.12%, as shown in Table 18.2. In this pipeline, *spaCyTokenizer* is used to split the input text into individual tokens, which can be used as inputs for further NLP processing. Finally, the *spaCyFeaturizer* transforms the user message and response into a vector representation. The confusion matrix and intent histogram of spacy_pipeline are shown in Figure 18.6 and Figure 18.7, respectively.

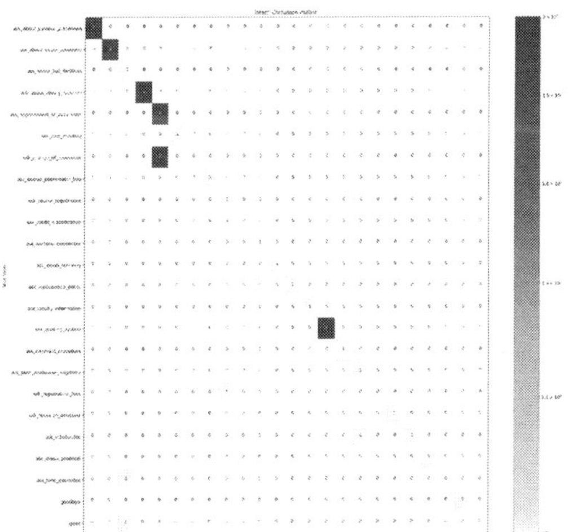

Figure 18.4: Confusion matrix using default_pipeline

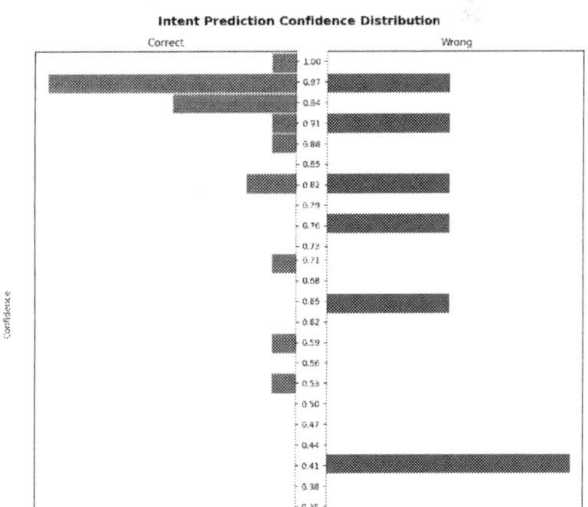

Figure 18.5: Intent Histogram using default_pipeline

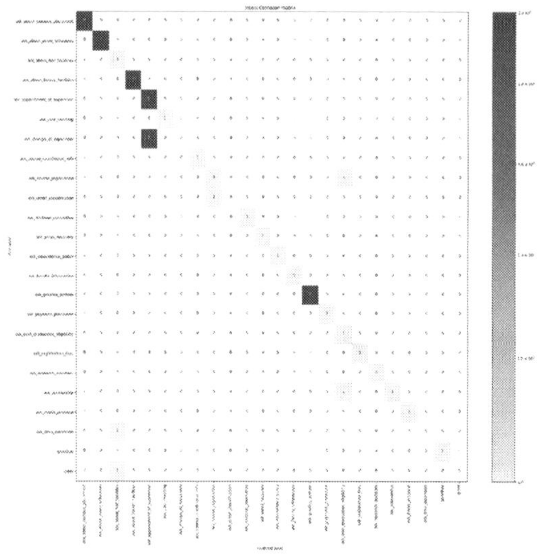

Figure 18.6: Confusion matrix using spacy_pipeline

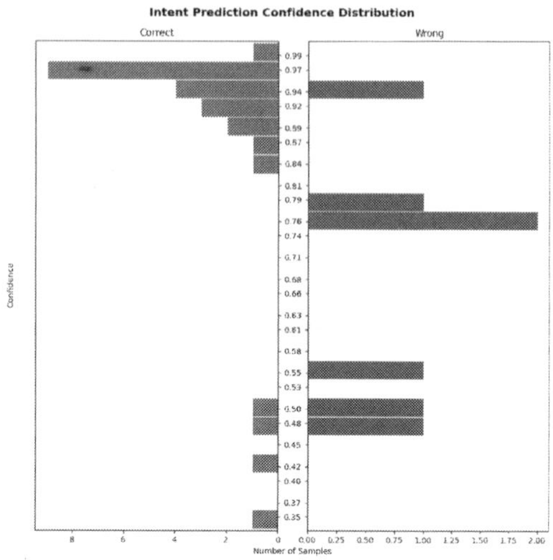

Figure 18.7: Intent Histogram using spacy_pipeline

Next, the white_bert_pipeline includes a *LanguageModelFeaturizer* that uses a pre-trained BERT as a model and LaBSE as model_weights to extract features. This pipeline included several other featurizers and classifiers. The experimental results in Table 18.2 show that the white_bert_pipeline performed better than the default_pipeline and spaCy_pipeline. This result suggests that the use of a pre-trained language model like BERT is important for improving the performance of NLP pipelines. The confusion matrix and intent histogram of white_bert_pipeline are shown in Figure 18.8 and Figure 18.9, respectively. The spaCy_bert_pipeline includes the same components as the default pipeline but replaces the *WhitespaceTokenizer* with *spaCyTokenizer*. In addition, it includes a *LanguageModelFeaturizer* that uses a pre-trained BERT model to extract features.

Our experimental results show that the spaCy_bert_pipeline outperforms other pipelines in terms of precision, recall, F1 score, and accuracy measures (see Table 18.2). In Figure 18.10 and Figure 18.11, the confusion matrix and histogram show that a few responses are incorrect and misclassified when using spaCy_bert_pipeline.

Moreover, Figure 18.12 shows the superiority of the proposed spacy_bert_pipeline in computing the F1 score with training data. The increasing growth of the F1 score of the spacy_bert_pipeline ensures that the performance of the pipeline continues to improve with more training data. On the other hand, the F1 score of the other pipelines fluctuated for different numbers of test data, which means that adding more training data does not improve the performance. The results we obtained demonstrate that the fastText vector, the spaCy library, and BERT can considerably increase the accuracy of intent classification in the Bangla language. This finding indicates that adopting a pre-trained language model, such as BERT, can greatly boost the efficiency of NLP pipelines, and switching from the basic WhitespaceTokenizer to a more sophisticated tokenizer, such as spaCyTokenizer, can boost accuracy even more.

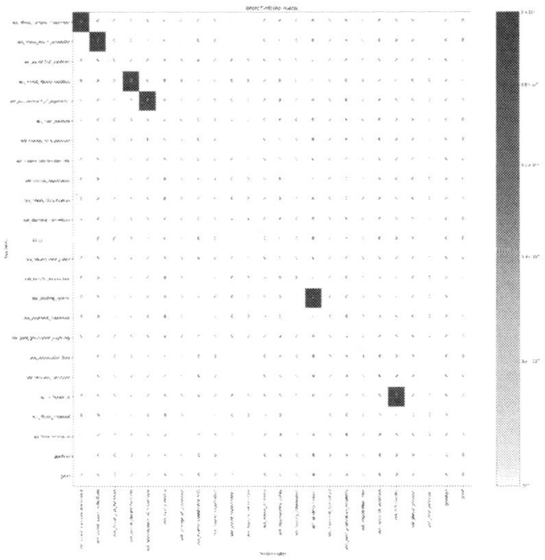

Figure 18.8: Confusion matrix using white_bert_pipeline

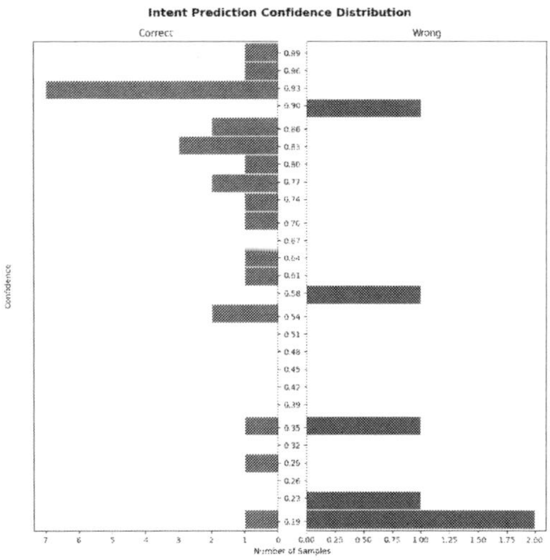

Figure 18.9: Intent Histogram using white_bert_pipeline

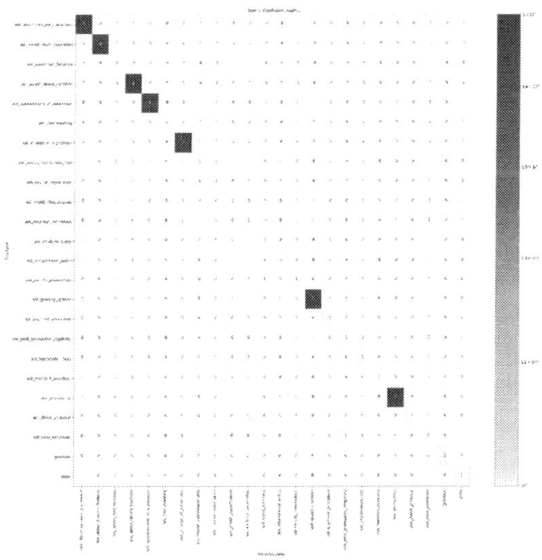

Figure 18.10: Confusion matrix using spacy_bert_pipeline

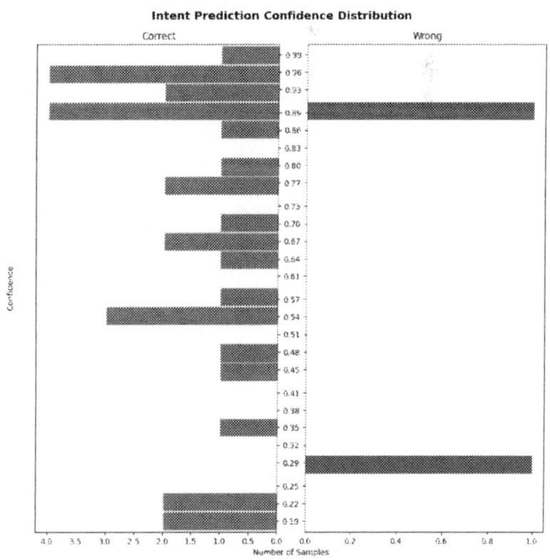

Figure 18.11: Intent Histogram using spacy_bert_pipeline

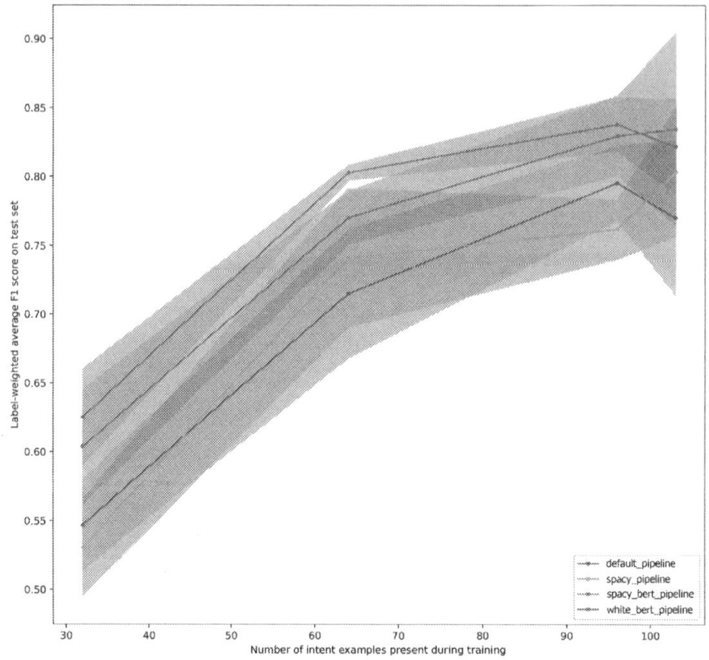

Figure 18.12: Performance of the pipelines based on the F1-score

18.5 CONCLUSION

The Rasa framework is a popular open-source tool used for building chatbots and virtual assistants. However, for low-resource languages like Bangla, the default components of Rasa perform poorly due to a lack of adequate training data and models. This highlights the need for custom pipelines and models specifically designed for Bangla language processing. To overcome this challenge, specialized components need to be built particularly for Bangla to improve chatbot performance. Custom pipelines can be created using resources such as spaCy tokenizer incorporated with BERT and Bengali word vectors to train the chatbot and improve its accuracy and precision in understanding and responding to user queries. The proposed ensemble pipeline for developing a Bangla language chatbot for student support services is based on the Rasa open-source framework. The proposed pipeline includes several components such as *spaCyTokenizer* and *LanguageModelFeaturizer*. Furthermore, we created a dataset for model training and evaluation. The experimental results confirm that the combination of *spaCyTokenizer* and *LanguageModelFeaturizer* yields the best results in terms of accuracy, precision, F1 score, and recall. The proposed pipeline can be used to develop Bangla language and other low-resource language chatbots. In the future, the chatbot will be further developed for low-resource languages using the deep learning model.

Bibliography

[1] Parkar, R., Payare, Y., Mithari, K., Nambiar, J., & Gupta, J. AI and web-based interactive college enquiry chatbot. *In 2021 13th International Conference on Electronics, Computers and Artificial Intelligence (ECAI)*. pp. 1–5 (2021)

[2] Perdana, R. S., Adikara, P. P., & Kurnianingtyas, D. Knowledge-Enriched Domain Specific Chatbot on Low-resource Language. *In 2022 11th Electrical Power, Electronics, Communications, Controls and Informatics Seminar (EECCIS)*. pp. 310–315 (2022)

[3] Siddique, M. R., & Mahmood, M. A. An absolute Optical Character Recognition system for Bangla script Utilizing a captured image. *In 2021 Emerging Technology in Computing, Communication and Electronics (ETCCE)*. pp. 1–6 (2021)

[4] Khan, F. S., Al Mushabbir, M., Irbaz, M. S., & Al Nasim, M. A. End-to-End Natural Language Understanding Pipeline for Bangla Conversational Agents. *In 2021 20th IEEE International Conference on Machine Learning and Applications (ICMLA)*. (pp. 205–210) (2021)

[5] Rahman, M. M., & Watanobe, Y. ChatGPT for education and research: Opportunities, threats, and strategies. *Applied Sciences*, 13(9), p. 5783 (2023)

[6] Devlin, J., Chang, M. W., Lee, K., & Toutanova, K. Bert: Pre-training of deep bidirectional transformers for language understanding. *arXiv preprint* arXiv:1810.04805 (2018)

[7] Sambhe, G., Awaze, S., Bobade, S., Bhagwatkar, S., & Bhoyar, P. A. Review On College Enquiry Chatbot. *In 2023 IEEE International Students' Conference on Electrical, Electronics and Computer Science (SCEECS)*. pp. 1–3 (2023)

[8] Zemcik, M. T. A brief history of chatbots. *DEStech Transactions on Computer Science and Engineering*. 10 (2019)

[9] Bani, B. S., & Singh, A. P. College enquiry Chatbot using ALICE. *International Journal of New Technology and Research*, 3(1). p. 263368 (2017)

[10] Molnar, G., & Szüts, Z. The role of chatbots in formal education. *In 2018 IEEE 16th International Symposium on Intelligent Systems and Informatics (SISY)*. pp. 000197–000202 (2018)

[11] Rahman, M. M., Watanobe, Y., Matsumoto, T., Kiran, R. U., & Nakamura, K. Educational data mining to support programming learning using problem-solving data. *IEEE Access*, 10, pp. 26186–26202 (2022)

[12] Shah, S., & Shah, S. A comparison of various chatbot frameworks. J. *Multi-Criteria Decis. Anal*, 6, pp. 375–383 (2019)

[13] Bocklisch, T., Faulkner, J., Pawlowski, N., & Nichol, A. Rasa: Open source language understanding and dialogue management. *arXiv preprint* arXiv:1712.05181 (2017)

[14] Perez-Soler, S., Juarez-Puerta, S., Guerra, E., & de Lara, J. Choosing a chatbot development tool. *IEEE Software*, 38(4), pp. 94–103 (2021)

[15] Alammary, A. S. BERT models for Arabic text classification: a systematic review. *Applied Sciences*, 12(11), p. 5720 (2022)

[16] Rasa. Components, Available: `https://rasa.com/docs/rasa/2.x/components/` (2023)

[17] Meshram, S., Naik, N., Megha, V. R., More, T., & Kharche, S. (2021, August). College enquiry chatbot using rasa framework. *In 2021 Asian Conference on Innovation in Technology (ASIANCON).* pp. 1–8 (2021)

[18] fastText. Word vectors for 157 languages, Available: `https://fasttext.cc/docs/en/crawl-vectors.html` (2023)

[19] Watanobe, Y., Rahman, M. M., Amin, M. F. I., & Kabir, R. Identifying algorithm in program code based on structural features using CNN classification model. *Applied Intelligence*, 53(10), pp. 12210–12236 (2023)

[20] Rahman, M. M., Watanobe, Y., Szmeja, P., Sowinski, P., Paprzycki, M., & Ganzha, M. Code Semantics Learning with Deep Neural Networks: An AI-Based Approach for Programming Education. *In International Conference on Computational Science.* pp. 737–750 (2023)

[21] Rasa. Testing Your Assistant, Available: `https://rasa.com/docs/rasa/testing-your-assistant`. (2023)

[22] Rahman, M. M., Watanobe, Y., & Nakamura, K. (2020). Source code assessment and classification based on estimated error probability using attentive LSTM language model and its application in programming education. *Applied Sciences*, 10(8), p. 2973 (2020)

[23] Rahman, M. M., Watanobe, Y., & Nakamura, K. (2021). A bidirectional LSTM language model for code evaluation and repair. *Symmetry*, 13(2), p. 247 (2021)

Index